AQUA INCOGNITA:
WHY ICE FLOATS ON WATER
AND GALILEO 400 YEARS ON

AQUA INCOGNITA:
WHY ICE FLOATS ON WATER
AND GALILEO 400 YEARS ON

Pierandrea Lo Nostro & Barry W Ninham

Editors

Connor Court Publishing

Ballarat

Published in 2014 by Connor Court Publishing Pty Ltd.

Copyright © 2014 Pierandrea Lo Nostro and Barry W Ninham Eds

ALL RIGHTS RESERVED. This book contains material protected under International and Federal Copyright Laws and Treaties. Any unauthorized reprint or use of this material is prohibited. No part of this book may be reproduced or transmitted in any form or by any means, electronic or mechanical, including photocopying, recording, or by any information storage and retrieval system without express written permission from the publisher.

Connor Court Publishing Pty Ltd.
PO Box 224W
Ballarat VIC 3350
sales@connorcourt.com
www.connorcourt.com

ISBN: 9781925138214   (pbk.)

Cover design by Ian James

Cover picture Dawes Glacier carving July 2012 Alaska Inside Passage
Copyright Barry W. Ninham

Picture of Statue of Galileo Galilei, Loggiato degli Uffizi-Firenze. With permission from the Italian "Ministero dei beni e delle attività culturali e del turismo" (N. ord. 164 P, 15 Nov 2013). This picture cannot be reproduced or duplicated by any means.

Printed in Australia

CONTENTS

Introductory remarks
Barry W. Ninham, Pierandrea Lo Nostro — xi

I - From water to the stars: a reinterpretation of Galileo's style
Louis Caruana SJ — 1

II - The conservation of wall paintings: the conflictual relationship with water
Piero Baglioni, David Chelazzi, Rodorico Giorgi — 18

III - The state of water in living systems: from the liquid to the jellyfish
Marc Henry — 34

IV - Water, salt and oil. An exploration of the foundations of molecular forces
Barry W. Ninham, Pierandrea Lo Nostro — 103

V - Hofmeister effects in living organisms
Niccolò Peruzzi, Barry W. Ninham, Pierandrea Lo Nostro — 127

VI - *Aqua reticulata*: topology of liquid water networks
Stephen T. Hyde — 145

VII - Supercooled water: two liquids?
Francesco Sciortino — 176

VIII - Is water chiralic? (Experimental evidence of chiral preference of water)
Yosef Scolnik — 183

IX - The topological and quantum structure of zoemorphic water
Marc Henry — 197

X - *Aqua extrema* and *vita incognita* deep below the waves
Shigeru Deguchi, Keigo Kinoshita, Takaaki Kubota — 240

XI - Water talks to water. Might we listen in?
D. James Morré, Dorothy M. Morré — 256

XII - Aqua incognita: liquor aquae superficies
Richard J. Saykally — 293

XIII - Helen Keller problem: tactile texture of water isn't necessarily favorable
Yoshimune Nonomura, Rina Saito, Takashi Maeno — 306

XIV - Recognition and patterning of molecules on water surface: reconstruction of some fundamental features of biomembranes
Toyoki Kunitake — 320

XV - Interfacial osmotic pressure
Martin Chaplin — 329

XVI - On the surface tension of electrolyte solutions
Vincent S. J. Craig, Jian Cui, Thomas G. Brazier — 341

XVII - The use of air bubbles to desalinate seawater without boiling
Muhammad Shahid, Richard M. Pashley — 350

XVIII - Water as a probe for the fractal structure evolution of cement: the effect of organic additives
Francesca Ridi, Emiliano Fratini, Piero Baglioni — 367

XIX - The emergence of structure from randomness in aqueous aggregation equilibria
Blake M. Rankin, Dor Ben-Amotz   387

XX - Theory and modeling of ion specific hydration
Yu Shi, Travis Pollard, Thomas L. Beck   407

XXI - Comparison of mechanical and thermodynamical evaluations of electrostatic potential differences between electrolyte solutions
Xinli You, Mangesh I. Chaudhari, Lawrence R. Pratt   434

XXII - Modelling water as a continuum solvent to understand ion-specific effects
Tim Duignan, Drew Parsons, Barry Ninham   443

XXIII - The impact of ionic solvation energy and water structure on forces
Drew F. Parsons   456

XXIV - Nanobubbles of dissolved gas in bulk aqueous solutions of electrolyte
Nikolai F. Bunkin, Alexey V. Shkirin, Valeriy A. Kozlov, Artyom L. Sendrovitz   469

*Picture of Statue of Galileo Galilei, Loggiato degli Uffizi-Firenze.
With permission from the Italian "Ministero dei beni e delle attività culturali
e del turismo" (N. ord. 164 P, 15 Nov 2013).
This picture cannot be reproduced or duplicated by any means.*

Dedication

Dedicated to Professor Enzo Ferroni (1921-2007), Dean of the University of Florence (1976-1979), Dean of the Faculty of Sciences, Dean of the Institute of Physical Chemistry, and founder of the current Department of Chemistry.

Teacher and researcher extraordinaire, the pioneer in Italy of Physical Chemistry of Colloids and Interfaces, and after the 1966 flood of Florence the first scientist to apply the science of colloids and interfaces to the restoration and conservation of works of art.

And to Dava Sobel for her marvellous book, *Galileo's Daughter* which gave us the human side.

Acknowledgement

We acknowledge the Enzo Ferroni Foundation (*http://apple.csgi.unifi.it/~fondazione/index.html*) for financial support.

Preamble

We remembered that 400 years ago, on October 2, 1611 AD, Cosimo II de' Medici, Grand Duke of Tuscany and patron of the arts and sciences, sponsored a debate on the topic: Why Ice Floats on Water.

The then young Galileo led this famous debate.

Cardinal Barberini, a great admirer of Galileo the soon to become Pope Urban VIII – who eventually put him under house arrest – in a tragedy too well known to recount, was there.

We could not let the occasion pass.

So a small group of us, research scientists, took upon ourselves to organise a different kind of small meeting, in Florence, July 15-19, 2013 to commemorate the occasion.

The motivation derived too from our observation that at practically all scientific meetings these days, mutual incomprehension is almost mandatory. The focus seems always on specialised technical minutae of our disciplines, physical, colloid surface chemistry, and more. It is the same for all fields of science and technology.

The reader may be surprised to learn that the answer to the ice question is still unresolved.

So are the answers to a myriad questions about water. In the simpler terminology of an older world, our discussions would be about water, salt, air and light. And biology. And geology and, and ... From that perspective we are all at sea.

It is an extraordinary and not widely recognised fact that our present, classical, theories that underpin the entire enabling disciplines of physical, colloid and surface chemistry lack predictability, from biology to earth sciences and chemical engineering. These theories are undergoing a paradigm shift from the beginnings up.

The same is true in many other disciplines like cosmology and particle physics. It is true also for molecular biology.

We wanted to have an overview debate and discussion between scientists at the cutting edge from around the world, on where we are now on WATER. And to record the situation. Hence, on **Aqua Incognita, Galileo 400 years on**.

Some technical topics that we addressed included:

1. water, its properties and structure, and in biology
2. solutions of electrolytes in water and non-aqueous solvents
3. specific ion effects in the bulk and at interfaces, particularly in medicine and biology
4. effects of light and magnetic fields on water and processes in water dispersion
5. dissolved gases, how they affect aqueous media
6. hydration and, in general, solvation and crystallisation
7. challenges to the foundations and the emerging new theories in physical chemistry

But the enquiry expanded to a more eclectic collection of papers, on Fresco restoration, on metaphysics and Galileo's problems, on chirality, on the history of water throughout civilisation, on promising advances in desalination, and deep sea water, and human sweat and touch!

Venue

The conference was held in the ancient convent "Convitto della Calza" in Florence, founded in 1362 as St. John the Baptist hospital.

The furniture for our meeting dated back 400 years, the frescos covering the walls more venerable, our visits to Galileo's house and to the Museum of the History of Science that has much of his original equipment focussed the mind.

Our debates were informal and our participants were limited to 20, with occasional attendance by wives and non specialists as for the original debates 400 years ago.

Florence itself has many attractions, being the center of the Renaissance.

So it was difficult for such a meeting to fail.

This book represents the contributions from participants and their affiliations.

Some contributions were from people unable to attend.

These and their topics follow.

Some are easily understood by laymen and non specialists, some highly technical, some very practical. Some may seem to be crazy - but nothing ventured nothing gained.

We think we achieved our aim. And hope these contributions will provide a useful perspective and introduction for anyone interested in water and its manifold manifestations.

And an insight into the science of water for the third meeting 400 years hence.

At least we are confident that Galileo would have been very pleased.

Any discussion that touches on Galileo and scientists and the Roman Catholic Church has the potential to become explosive. The Church has had a deserved very bad press for its treatment of Galileo. It has apologised for this and admits error, which is remarkable. But the regret and the stain are with us still. Many scientists remain outraged.

At first sight it is difficult to enter the mindset of western european society of 400 years ago, devastated by internal wars and attacks from the east, and dominated by a faith and dogma that to present "rational scientific minds" appear absurd.

But not much reflection is needed to remind us that we are not so rational either. Proofs of human behaviors and habits not driven by rationality cross the centuries and pervade our current history. Wars in the name of religion at the present time are the rule rather than the exception. Philosophies and ideologies carry the heaviest burden of responsibility for the recent crimes against humanity. And sometimes science has been involved: eugenics developed in the second half of the XIX century in France, the UK, Germany, and the US as the attempt to extrapolate and apply Darwinism to humankind. The implications of science in politics, economics, and social views are significant and represent a delicate matter for our societies.

As time goes, memory needs to be preserved as a warning for the present and the future.

Sciences and faiths can honestly cooperate for the disclosure of truth and common, peaceful good for all mankind. Unprejudicially.

Pierandrea Lo Nostro, University of Florence and Enzo Ferroni Foundation
Barry Ninham, Australian National University
Firenze, 30 November 2013

List of Participants

Piero Baglioni: Dept. of Chemistry and CSGI, Università di Firenze via della Lastruccia 3, 50019 Sesto Fiorentino (Florence) Italy, baglioni@csgi.unifi.it

Thomas Beck: Department of Chemistry, University of Cincinnati, Cincinnati, OH 45221-0172, USA, thomas.beck@uc.edu

Dor Ben-Amotz: Purdue University, Dept. of Chemistry, 560 Oval Drive, West Lafayette, IN 47907, USA, bendor@purdue.edu

Nikolai Bunkin: Prokhorov General Physics Institute, 119991, Vavilov Str., 38, Moscow, Russia, nbunkin@kapella.gpi.ru

Louis Caruana: Faculty of Philosophy, Gregorian University, Piazza della Pilotta 4, 00187 Rome, Italy, caruana@unigre.it

Martin Chaplin: London South Bank University, London, SE1 0AA, United Kingdom, martin.chaplin@btinternet.com

Vincent Craig: Dept. of Applied Mathematics, Res. School of Physical Sciences and Eng., Australian National University, Canberra, ACT 0200, Australia, Vince.Craig@anu.edu.au

Shigeru Deguchi: Institute of Biogeosciences, Japan Ag. Marine-Earth Science & Technol., 2-15 Natsushima-cho, Yokosuka, 237-0061, Japan, shigeru.deguchi@jamstec.go.jp

Timothy T. Duignan: Dept. of Applied Mathematics, Res. School of Physical Sciences and Eng., Australian National University, Canberra, ACT 0200, Australia, tim@duignan.net

Sarah Everts: C&EN, Berlin, Germany, saraheverts@gmail.com

Juan Manuel García-Ruiz: Laboratorio de Estudios Cristalográficos, IACT (CSIC-UGR), Av. de las Palmeras 4, 18100 Armilla, Spain, jmgruiz@ugr.es

Marc Henry: Université de Strasbourg, Institut Le Bel, Lab. Chimie Moléculaire de l'État Solide, 4, Rue Blaise Pascal, CS 90032, 67081 Strasbourg Cedex, France, henry@unistra.fr

Stephen T. Hyde: Dept. of Applied Mathematics, Res. School of Physical Sciences and Eng., Australian National University, Canberra, ACT 0200, Australia, stephen.hyde@anu.edu.au

Toyoki Kunitake: Kitakyushu Foundation for the Advancement of Industry, Science and Technology, Hibikino, Kitakyushu, 808-0135, Japan, kunitake@ruby.ocn.ne.jp

Pierandrea Lo Nostro: Dept. of Chemistry and CSGI, Università di Firenze, via della Lastruccia 3, 50019 Sesto Fiorentino (Firenze), Italy, PLN@csgi.unifi.it

Dorothy M. Morré: Mor-NuCo, Inc., Purdue Research Park, West Lafayette, Indiana, USA, dj_morre@yahoo.com

D. James Morré: Mor-NuCo, Inc., Purdue Research Park, West Lafayette, Indiana, USA, dj_morre@yahoo.com

Barry W. Ninham: Res. School of Physical Sciences and Eng., Australian National University, Canberra, ACT 0200, Australia, barry.ninham@anu.edu.au.

Yoshimune Nonomura: 4-3-16 Jonan, Yonezawa 992-8510, Japan, nonoy@ya.yamagata-u.ac.jp

V. Adrian Parsegian: 301 Hasbrouck Lab. Physics, UMass, Amherst, MA 01003, USA, parsegian@physics.umass.edu

Drew F. Parsons: Dept. of Applied Mathematics, Res. School of Physical Sciences and Eng., Australian National University, Canberra, ACT 0200, Australia, Drew.Parsons@anu.edu.au

Richard M. Pashley: School of Physical, Envir. and Math. Sciences, University of New South Wales, Canberra, ACT 0200, Australia, R.Pashley@adfa.edu.au

Niccolò Peruzzi: Dept. of Chemistry and CSGI, Università di Firenze, via della Lastruccia 3, 50019 Sesto Fiorentino (Florence), Italy, peruzzi@csgi.unifi.it

Lawrence R. Pratt: Herman and George R. Brown Chair, Dept. Chemical & Biomolecular Eng., 300 Lindy Boggs Center, Tulane University, New Orleans, LA 70118, USA, lpratt@tulane.edu

Francesca Ridi: Dept. of Chemistry and CSGI, Università di Firenze via della Lastruccia 3, 50019 Sesto Fiorentino (Florence), Italy, ridi@csgi.unifi.it

Richard J. Saykally: Dept. of Chemistry, Univ. of California and Chemical Sciences Division, Lawrence Berkeley National Laboratory, Berkeley, CA 94720-1460, USA, saykally@berkeley.edu

Francesco Sciortino: Dipartimento di Fisica, Sapienza Università di Roma, Piazzale Aldo Moro 2, 00165 Roma, Italy, francesco.sciortino@uniroma1.it

Yosef Scolnik: IYAR, Israel Institute for Advanced Research, Rehovot, Israel, yosefsc@walla.com

I
From water to the stars: a reinterpretation of Galileo's style

Louis Caruana SJ

Faculty of Philosophy, Pontificia Università Gregoriana,
00187 Rome, Italy.
caruana@unigre.it

Galileo Galilei's contribution during the early stages of the scientific revolution and his clash with the Catholic Church have been discussed, studied, and written about for many decades. There are indications however that recent work in this area has tended to underestimate the fact that Galileo had a particular style. By style here I mean a particular combination of behavioural features that are specific to a person or a historical period. Style of course can be related to behaviour in general, but what is relevant in this paper is the combination of dispositions that determine a particular way of engaging in science, as discussed by scholars like A. C. Crombie.[1] Galileo, I will argue, had a scientific style marked by overconfidence. He tended to downplay the importance of obvious contradictory evidence that undermined his claims, and he did this by producing auxiliary hypotheses that sometimes verged on the extravagant. If we focus on this somewhat neglected aspect of his style, some interesting new questions emerge: To what extent did Galileo depend on such auxiliary hypotheses? How insecure did they render his position? And how *ad hoc* were they? In this paper, I explore these questions by comparing two important debates: one about the nature of water and buoyancy, the other about cosmology. Since the main features of the cosmology debate, the one involving Galileo's defence of heliocentrism, are well known, I will dedicate more time to the water debate, before proceeding to highlight the elements of style that are common to both debates, and to evaluate the relevance of these elements for current understanding of scientific practice.

1. The buoyancy debate

First, a word about Galileo's social and cultural situation. The way empirical inquiry used to be motivated and propagated at that time, when what we now call the scientific revolution was at its infancy, differed considerably from the way it is today. In that context, the driving force used to originate mainly not from scientific questioning as such but from what the major patrons of individual scholars regarded as marvels and curiosities, from what these patrons considered worthy of exciting debates and controversy. The question "Why does ice float on water?" was one clear example of an exciting question because we all know that ice is in fact nothing more than water. The overall social, political, and cultural context in the seventeenth century was such that science was dependent to a very large extent on what patrons wanted, and this meant that natural philosophers, or anyone we would now recognize as a scientist, could never be fully in control of their research. Patron-dependence was crucial: through financial support, it made the scientist's work possible. But it produced a number of difficulties as well, mainly because the general habitat for science, where science happened, was not the isolated laboratory but pubic disputation, and this mode of scientific practice usually drew attention not to careful and technical understanding but to quick, publicly accessible answers. Moreover, during the period when Galileo flourished, mathematics was still considered a discipline that was less important than Aristotelian philosophy within the overall hierarchy of knowledge. Galileo had to struggle hard against this mindset. The only way he could gain a hearing was to make himself philosophically versatile enough to engage with the Aristotelians on the same level.[2-6]

With this background in mind, we can now appreciate better the various forces at work during the debate that concerns us here, the one concerning water and buoyancy. This was launched in the summer of 1611, a session that took three days. It started with a dispute about the nature of cold as a quality, but then shifted into one about buoyancy. The major contention arose when the Aristotelians among those present were shocked to learn that, for Galileo, ice was not condensed water, as they had always assumed. They had to admit that the issue was not completely clear in the classic texts. Although Aristotle had indeed indicated that ice was condensed water, his reflections on this point were rather sketchy. For instance, in his *Metaphysics* he discussed the different senses in which the word "is" can be used, and the examples he offers include ice. He writes: "[the word] 'is' has [a] number of senses; for a thing 'is' a threshold because it is situated in a particular way, and 'to be a threshold' means to be situated in this particular way, and 'to be ice' means to be condensed in this particular way. Some things have their being defined in all these ways: by being partly mixed, partly blended, partly bound, partly condensed."[7,8] Aristotle here takes the idea that ice is condensed water as obvious. Why? We find no clear answer in Aristotle's own works, but his followers filled up the reasoning behind this in the following way. He must have started not from the fact that ice floats on water but from the fact that it is colder than water. Since ice is colder than water, it must be water minus something, minus some amount of heat, and this lack leads to a condensation. It is water with a deficiency, as it were, not with something extra. And as regards the question why ice floats, Aristotelians considered this fact as just one example of buoyancy in general. For them, buoyancy is a matter of shape only. It had nothing to do with density. On this issue, they were certainly following their master who had explained this point quite carefully. In his book *De Caelo*, he argued that shape matters because the determining factor in buoyancy is the difference that the various materials we consider show as regards penetrability. For instance, air is more penetrable than water, and water is more penetrable than earth (see Figure I.1). He adds: "the reason why broad things keep their place [e.g. a plank of wood afloat on water] is because they cover so wide a surface, and the greater quantity [i.e. the water] is less easily disrupted. Bodies of the opposite shape sink down because they occupy so little of the surface, which is therefore easily parted."[9] It is good for us to recall here that, in Galileo's times, Aristotelians used to feel obliged to defend Aristotle, be it on buoyancy or geocentrism, or any other issue, not only because his positions were justified, as indeed they thought they were, but also because they considered these various positions important individual bricks that held an entire worldview in place. For them, removing one brick could have devastating consequences that would destabilize the entire conceptual scheme.

What was Galileo's reaction to this? For him, Aristotelians were seeing the entire issue the wrong way round. They had started from the observation that ice is colder than water and had sidelined the fact that ice floats on water. What they should have done was to start from the fact that ice floats on water. For Galileo, since ice floats on water, it must be rarified water, not condensed water. And as regards buoyancy, Galileo resorted to another ancient source: Archimedes. While Aristotle had developed a shape-theory of buoyancy, Archimedes had developed a density-theory, according to which a thing in water experiences a buoyant force equal to the weight of water displaced. Galileo did not deny that shape matters. He conceded that the shape of a body affected the speed with which it sinks or rises, but was convinced that shape does not affect *whether* it sinks or rises.

Up to this point, the debate seemed well balanced. Both sides presented interesting insights, and both had a heavyweight from Ancient Greece as support. The decisive factor came when Galileo's main opponent, Lodovico delle Colombe, devised a simple but spectacular and decisive experiment. He did not want to resort to Aristotelian deductive reasoning or anything like that. He appealed instead to direct evidence, just like Galileo. He made all the participants gather round the demonstrating table and he showed them how a sphere of ebony, whose density is higher than that of water, sinks when placed on water, while a thin piece of the same material remains afloat even with some weights on it. So the determining factor was shape, not density – full stop.

Galileo must have been quite astounded by this, but he did not give up. He tried to come up with some way of explaining this experiment in his own terms. This was not easy at all, because according to his worldview there should not be any special effect at the surface of a liquid which does not arise elsewhere within liquid. In other words, his view of liquids ruled out what we now call surface tension. He took therefore another line of argument and tried to bring in the relevance of wetness, but this lead to no convincing conclusion. Since the dispute itself became noisy and inconclusive, the meeting was brought to a close, and the main protagonists left with the intention of producing a full written version of their position. Galileo, encouraged to proceed with this by his patron, Duke Cosimo II, took his task seriously, and produced his written text within a year. For him, maintaining the duke's favour was obviously important. We notice once again how science was dependent on patronage to an extent that is hard for us to accept today.

Galileo's written version, entitled *Discourse on Bodies in Water* and published in 1612, was based on Archimedes's classic work *On Floating Bodies*, which had emphasized hydrostatics. Archimedes had offered an account of buoyancy that had been intended to explain the situation once equilibrium is reached. In other words, he had described the state of affairs when a body is stationary and floating, or when it has sunk and lies at the bottom. He had said nothing about the *process* of rising to the surface or of sinking; his view had been limited to *statics* as opposed to *dynamics*. Galileo therefore saw a way of breaking new ground by delving into hydrodynamics. This was a risky business, because in claiming the right to give an account of motion, he was encroaching into the philosophers' domain – yet again. Resorting to the model of the lever, he wanted to explain the downward motion of a sinking body and the corresponding upward rise of the water surface, two motions with different speeds. And he did this by resorting to the model of a lever with different arm-lengths, a lever that makes a short swing on the short side and a quick swing on the long side. He adopts therefore a mechanical view of the world – and this was seriously at odds with the Aristotelian worldview, at least in two senses.

First of all, Aristotelians had always believed that each of the four elements had its own specific motion: for instance earthly bodies move down because they have heaviness, while fiery ones move up, because they have lightness. Heaviness and lightness were for them real attributes belonging to things according to their nature. Each object or material will therefore have its share of overall heaviness or lightness in proportion to its constitution from the elements. From these fundamental, elemental motions, therefore Aristotelians offered the explanation of all motion. As regards the specific case we are dealing with here, the case of sinking or floating, the shape of the body, they used to say, was not the determining factor but only a *causa per accidens*, an explanation of secondary importance. The floating object needs to be understood in terms of its own inherent constitution in terms of the elements, the proportion of which determines the object's intrinsic quantity of heaviness and of lightness. Galileo was dissociating himself entirely from this kind of explanation. He was proposing a worldview in which buoyancy was the result neither of an innate upward trend (lightness as an attribute) nor of an effect of shape. For him, it was the result of the body's downward motion being counterbalanced by a counterforce. The implication here was that bodies, be they predominantly earthy or predominantly fiery, have only one type of motion: downwards. The Aristotelians were not amused.

From water to the stars: a reinterpretation of Galileo's style

Secondly, the fact that water shows a kind of skin at its surface was perfectly in line with the Aristotelians' broad view of liquids in general. For them, water, being a continuum, has a tendency to preserve its cohesion and integrity, as their master had expressed quite clearly in his work *De Caelo*: "Since there are two factors, the force responsible for the downward motion of the heavy body and the disruption-resisting force of the continuous surface, there must be some ratio between the two. For in proportion as the force applied by the heavy thing towards disruption and division exceeds that which resides in the continuum, the quicker will it force its way down; only if the force of the heavy thing is the weaker, will it ride upon the surface."[10] On this issue, Galileo had a problem. For him, water was made up of corpuscles with no intrinsic difference between them. It did not matter whether these corpuscles were at the surface or within the interior of the liquid. This view therefore, as mentioned above, ruled out any idea of surface-tension. How could Galileo then account for the impressive demonstration of his opponent Delle Colombe? To account for the intriguing floating chip of ebony, he had no choice but to resort to an explanation that was considerably extravagant. He proposed that, as the chip is lowered onto the surface, the observable slight depression of the water surface as it floats makes the chip associate itself with

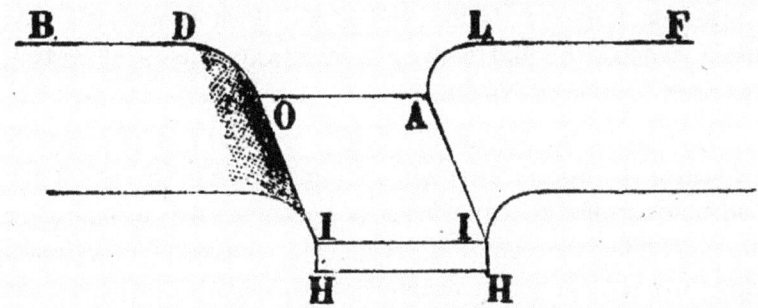

Figure I.1. Galileo's original diagram for his defence against Lodovico delle Colombe.[11]

a layer of air above it. In this way, the composite object, layer of air and layer of ebony, will have a specific weight less than that of water (see Fig. I.1). Was he introducing, through the back door, some occult forces here, some "magnetic virtue of air" as his opponents were quick to remark? These are his words:

> But if it [the ebony chip as it presses down onto the water surface] has already penetrated and is, by its nature, denser than water, then why does it not proceed to sink but stops and remains suspended within that small cavity that had been produced by its weight? I would say: because, as it moves down until its [upper] surface arrives at the water level, it loses a part of its own weight, and it then proceeds to lose the rest of its weight as well by descending deeper even below the water surface, which produces a ridge and a bank around it. It loses weight as it descends in such a way that it drags down to itself the air above it, by adherent contact. This air proceeds to fill up the cavity produced by the little water ridges, in such a way that, in this case, what really descends and is located in water is not just the ebony chip, or the iron chip, but the composite of ebony and air, from which there results a solid [*solido*] which does not exceed water in density as does ebony on its own, or gold on its own.[11,12]

This is the best Galileo could come up with as he tried to reason things out from within his system. I think it is fair to say that, as an explanation, it looks farfetched and *ad hoc*. What it shows is a strong determination on his part to save his overall worldview at all costs. He was ready to go even that far.

So, all in all, we can say that debate on water and buoyancy that had started *viva voce* in 1611 and then dragged on in writing for more than four years had no clear winner.[13] As historians now recognize, one important thing we see in this debate is the emergence of a growing gap between two very different professional identities: on the one side, we have professional philosophers, the Aristotelians, whose principles are derived from acknowledged philosophers; on the other side, we have a specimen of a new species of intellectual, a mathematician-philosopher, who seemed to violate the disciplinary boundaries that had been well established and respected for hundreds of years.

2. Comparing with the astronomy debate

Let us draw a quick comparison now between this debate and the one on the solar system. As is well known, the main story of the solar-system debate, in short, was this. With the use of the telescope, Galileo discovered new evidence in favour of the heliocentric view that had been promoted mathematically by Nicholas Copernicus about fifty years beforehand. Galileo therefore started to defend the idea that Copernicus's view was not a mere mathematical shortcut to obtain quick predictions of planetary positions, but was a true description of how things are. In the ensuing debate, which involved Aristotelians yet again, Galileo was challenged to explain some pretty glaring instances of counterevidence to his proposals. And this is the crucial point where this solar-system debate shows some remarkable similarity with the buoyancy debate. In both cases, Galileo had to deal with counterevidence that seemed obvious and convincing. In both cases, he made proposals that were unconventional and therefore somewhat suspicious.

Instead of going into all the intricate detail of the solar-system debate, let us consider the crucial points only. One obvious element of counterevidence for the proposal that the Earth is in motion is direct experience. We simply have no sensation of movement. In line with this, as common sense suggests, if the Earth were in motion, there should be some detectable displacement during the falling of an object, because, by the time the object hits the ground, the Earth would have moved a little. But nothing of the kind is observed. Here we have, therefore, a serious challenge to anyone who wants to argue that the Earth moves. For Galileo, however, this kind of argument was not the most worrying. He rose to this challenge in a spectacular way by establishing the basic principles of relativity. He proved that, for two reference frames in uniform motion, no such displacement should be expected.[14]

The real worrying element of counterevidence was the lack of stellar parallax. If the Earth were really in motion through space, then the nearby stars should show some displacement with respect to the distant stars. Our view of the night sky would be somewhat like what we see from a moving train: nearby trees shifting across the distant background. But no such effect is evident in the night sky. So again, Galileo had a problem. He tried to use his telescope, but it was all in vain.[15] The only way he could respond to this problem was to adopt what had already been suggested by some commentators before him, namely that the absence of stellar parallax was due to the fact that all stars were infinitely far out in space.[16,17] This suggestion, of course, did solve the problem. It was however *ad hoc* and embarrassing – embarrassing because it went against Galileo's own idea that Aristotle had made a mistake in assuming that there is an essential difference between the sub-lunar universe and the rest. For Galileo, the entire universe should be homogenous with a uniform distribution of stars throughout.

So here we see a clear common feature with the previous debate, a common stylistic feature involving the way science was engaged in. In both cases, Galileo faces an insurmountable problem but sticks to his guns; he does not shy away from defending himself by walking on stilts, as it were: by producing auxiliary hypotheses that, because of their *ad hoc* nature, apparently drain his position of its convincing power.

3. Conclusion

What conclusion can be drawn? There is of course much more that can be said about all the major points highlighted above. The little that has been mentioned however is enough to justify the following three points. First, we need to accept that the practice of science rarely involves clear-cut crucial experiments that decide an issue at one go. What has been highlighted in both debates confirms the idea, proposed by philosopher Imre Lakatos, that science does not develop according to naïve falsificationism but according to a more complex process involving auxiliary hypotheses.[18] These auxiliary hypotheses can have various degrees of plausibility or acceptability, depending on how they fit in with background beliefs that are shared by both the proponent and the opponent of the theory. The early stages of the new scientific paradigm inaugurated by Galileo were vulnerable. There was no knock-down argument on either side. It is true that, in both debates, Galileo's view did eventually turn out to be correct. At that time, however, his case had some obvious weaknesses, even on his own terms. Secondly, a few words about the Church. Although the way the Church handled Galileo during the solar-system debate remains an embarrassment, especially because of its official declaration that heliocentrism was heretical, which it certainly is not since it is not even theological, the arguments mentioned above can nevertheless help us understand why the case was so intriguing, and why some Aristotelians and theologians were not immediately won over by Galileo's arguments.[19,20] And finally, a word about Galileo's genius: as we know, time proved Galileo right in both debates. This shows that he was a man of genius: he had a way of seeing ahead, a way of seeing beyond what can be expressed by reasoned argument and experiment. We see him sometimes groping in the dark, especially in formulating auxiliary hypotheses, but in fact he was groping in the right direction.

Postscript

Dor Ben-Amotz has considered my contribution to the *Aqua Incognita* conference worthy of further reflections, which he kindly added as a postscript to the paper he published in this volume together with Blake

M. Rankin. I would like therefore to take this opportunity to add some clarifications, especially because some readers may easily be mislead by D. Ben-Amotz's comments.

First, it was good of D. Ben-Amotz to remind us that not everything the Jesuits did in their long history was perfect, and that in the theological debates concerning Galileo some of them wrote against Copernicanism and were in favour of Galileo's condemnation.

Ben-Amotz's basic point however seems dubious, because he seems to imply that I embarked on my analysis of Galileo's use of auxiliary hypotheses out of a hidden desire to rehabilitate in some sense those Jesuits in the past who wrote against Galileo. This is, of course, a biased reading of my paper. One cannot argue that, since Jesuits have had defects in the past, and no doubt still have some today, no Jesuit today should discuss the way Galileo responded to criticism. Of course, I readily admit that, in the ill-fated theological debate between the Church and Galileo, the role of the Jesuits, and indeed of the entire Church, is a very important issue that deserves careful and impartial study. The very complex process has been studied openly and at length by professionals, some of them Jesuits, for many decades.

The four Jesuits most closely involved in this event were probably Robert Bellarmine, Christopher Clavius, Orazio Grassi, and Melchior Inchofer. The first one was in charge of the Holy Office, otherwise known as the Roman Inquisition. He had the final say as regards the case at its beginning. In an important letter to Paolo Foscarini, he explained his position very clearly: he remained cautious and conservative, because there had not yet been a definite proof in favour of the Copernican model (such as stellar parallax). He remained convinced, moreover, that no such definite proof will ever be available. Bellarmine argued that, when a mathematical model works well, we cannot automatically say that it represents the world as it is. His overly conservative attitude may indeed be condemned from our present perspective, but his account of science, which philosophers now call an instrumentalist view of science, is certainly not illogical. Some prominent physicists endorse such a view of science even today. Bellarmine was behind the first admonition of 1616, which obliged Galileo to refrain from publicly defending the heliocentric theory, but did not forbid him to conjecture it. Bellarmine died in 1621, and was not involved in the second summoning of Galileo by the Inquisition, which ended with the 1633 condemnation. Christopher Clavius was a prominent mathematician and astronomer at the Roman College, best known for his reform of the calendar still in use today. He was delighted to welcome Galileo at the College and confirm his observations. After Galileo's first admonition, however, Clavius was obliged by his superiors to refrain from any public support of Galileo, and to retain all teaching within the precincts of Aristotelianism. Orazio Grassi, mathematician, astronomer, and architect, had a direct clash with Galileo regarding the nature of comets. For Grassi, comets where real objects above the moon because they showed parallax, while for Galileo they were optical illusions. In *The Assayer*, Galileo used all his wit against Grassi in a way that, for some readers, amounts to unjustified ridicule. The fourth Jesuit in my list, Melchior Inchofer was commissioned by Pope Urban VIII to write a theological justification of the 1633 condemnation. In his final document, he argued that, in the book *Dialogue on the Two Chief World-Systems*, Galileo did not remain neutral as he had promised but sided with the heliocentric view. Inchofer moreover went on to argue that believers were obliged to hold the geocentric view as a matter of faith. Although, given the juridical system within which the event took place, the ultimate responsibility for Galileo's condemnation should be seen as resting on Pope Urban VIII himself, we should also note that the one who wrote most explicitly against heliocentrism and made it look heretical was Inchofer.[21]

This brief sketch is enough to show how the theological debate, with its political and cultural overtones, together with the Jesuits' involvement within it, was a highly complex event, and simple characterizations of it can be misleading. As D. Ben-Amotz rightly pointed out, Galileo was instrumental in reminding theologians that the Bible should not always be taken literally. In other words, Galileo recalled what St Augustine had said about a thousand years earlier. This is an important point, and should undoubtedly be counted in Galileo's favour. Biblical interpretation however was not the topic of my conference paper. Ben-Amotz seems to suggest that, since I did not mention anything about the theological mistakes of the Church, I was deliberately misleading the conference participants. Such a criticism however would be unfair, because my paper was focused on the debate on water and buoyancy – as required by the conference organizers – and that particular debate raised no theological questions whatsoever. For readers to have a better overall picture of the Jesuits' engagement with the natural sciences, it may help to recall that they certainly cannot, as a group, be considered anti-scientific. On the contrary, the Society of Jesus, in spite of its many imperfections, has produced in the course of its four hundred year history many eminent scientists and mathematicians. Without wanting to sound triumphalistic, I may just recall that, by 1773, of the world's 130 astronomy observatories, thirty were operated by Jesuits. There are now thirty-five lunar craters named in honour of Jesuit scientists.

Secondly, D. Ben-Amotz seems to think that, for me, Galileo was a genius because he showed moments of irrationality. This was certainly not my conclusion. A mature and responsible history of science is one that accepts the complexity of any human enterprise whatever it is: there are no perfect Jesuits, there are no perfect scientists. It accepts that scientific development involves both explicit and tacit knowledge, and that there can be both moments of success and moments of failure. Examining the mistakes of scientists is not a way of discrediting them, but a way of appreciating their greatness, which is evident also in the way they pushed ahead, in spite of their limited resources, through horizons that, for them, were pretty dark.

And thirdly, D. Ben-Amotz adds that Galileo's genius lay in his defence of the idea that the scientific method, based on 'visual certainty', should trump all other claims to knowledge. No doubt, Galileo's innovations regarding the scientific method are admirable to the highest degree, and have contributed greatly to civilization. The simple association of the scientific method with 'visual certainty' however needs serious qualification. Observation certainly plays a key role in the methods of the sciences, but, as every philosophy undergraduate knows, scientific observation is inevitably loaded with prior theory, and, as a consequence, demonstration and interpretation are inevitable and not always straightforward. Had science been simply a matter of visual certainty, we would still be holding that the sun moves and the earth stands still. Galileo's point was that, if a scientific claim clashes with a theological one, and the scientific claim is based securely on observation and demonstration, then the theological claim should give way, in the sense that the biblical interpretations or philosophical positions on which it was based should be revised. This is a perfectly valid point. The majority view within Christianity today is that, if the sciences are engaged in properly, and according to moral norms, they will never really conflict with religious faith, because both the material world, studied by these sciences, and divine revelation, studied by theology, derive from the same God. The genuine scientist who labours humbly to discover the truths of nature is in fact being led by the hand of God. This point can be expressed in other words: 'Science can only be created by those who are thoroughly imbued with the aspiration toward truth and understanding. This source of feeling, however, springs from the sphere of religion. To this there also belongs the faith in the possibility that the regulations valid for the world of existence are rational, that is, comprehensible to reason. I cannot conceive of a genuine scientist without that profound faith. The situation may be expressed by an image: science without religion is lame, religion without science is blind.' These are the words of Albert Einstein.[22]

References and Notes

1. Crombie, A. C. *Styles of Scientific Thinking in the European Tradition: The History of Argument and Explanation Especially in the Mathematical and Biomedical Sciences and Arts*, London: Gerald Duckworth & Company, **1995**.
2. This is discussed in Ref. 3, chapter 3; the following paragraphs owe a lot to Biagioli's excellent study. Other useful sources: Refs. 4-6.
3. Biagioli, M. *Galileo Courtier*, University of Chicago Press, **1993**.
4. Drake, S. *Cause, Experiment and Science: A Galilean Dialogue Incorporating a New English Translation of Galileo's "Bodies That Stay atop Water, or Move in It"*, University of Chicago Press, **1981**.
5. Palmieri, P. *Arch. Hist. Exact Sci.* **2005**, *59*, 189-222.
6. Straulino, S.; Gambi, C. M. C.; Righini, A. *Am. J. Phys.* **2011**, *79*, 32-36.
7. *Metaphysics* 1042b 25-28 (my translation). The crucial Greek word here is πεπυκνῶσθαι; this is derived from πυκνόω, which means to make close, to condense. Some translators, like W. D. Ross (see Ref. 8), miss the important nuances by translating πεπυκνῶσθαι as 'solidified' instead of 'condensed'.
8. McKeon, R. (ed.), *The Basic Works of Aristotle*, New York: Random House, **1941**.
9. Aristotle, *De Caelo*, 313b 14-16, see Ref. 8.
10. Aristotle, *De Caelo* 313b 16-22 (see Ref. 8).
11. Galilei 1612, p. 98 (my translation), see Ref. 12.
12. Galilei, G. *Discorso intorno alle cose che stanno in su l'acqua o che in quella si muovono* (1612), in: *Edizione Digitale delle Opere Complete di Galileo Galilei*, Firenze, IMSS 2009, vol. IV, *http://pinakes.imss.fi.it:8080/pinakestext/home.seam?conversationId=1514*
13. The other main texts directly related to this debate include the following. On the side of the Aristotelians: Lodovico delle Colombe, *Discorso apologetico d'intorno al Discorso di Galileo Galilei* (1612); and Vincenzo di Grazia, *Considerazioni sopra 'l Discorso di Galileo Galilei* (1613); on the side of Galileo: Benedetto Castelli, *Risposta alle opposizioni del S. Lodovico delle Colombe e del S. Vincenzo di Grazia contro al Trattato del Sig. Galileo delle cose che stanno in su l'acqua* (1615). This latter author was a friend of Galileo's, but we have now clear indications that this book had been written by Galileo himself.

14. He shows this in "Day Two" of his *Dialogue Concerning the Two Chief World Systems*.
15. Graney, C. M. *Physics in Perspective* **2008**, *10/3*, 258-268.
16. This had been proposed by Thomas Digges (c. 1546-1595) in his translation of Copernicus's book *De revolutionibus*; see Ref. 17, p. 207.
17. Ravetz, J. R. "The Copernican Revolution", in: *A Companion to the History of Modern Science*, R. C. Olby, G. N. Cantor, J. R. R. Christie, and M. J. S. Hodge, (eds.), Routledge, **1996**, pp. 201-216.
18. Lakatos, I. "Falsification and the Methodology of Scientific Research Programmes", in: I. Lakatos and A. Musgrave (eds.), *Criticism and the Growth of Knowledge*, Cambridge University Press, **1970**, pp. 91-196.
19. For more on how Galileo's ideas where incorporated slowly within the school curriculum during the century following his death, see Ref. 20.
20. Caruana, L. "The Jesuits and the quiet side of the Scientific Revolution", in: T. Worcester SJ (ed.), *The Cambridge Companion to the Jesuits*, Cambridge University Press, **2008**, pp. 243-260.
21. One recent study is E. McMullin (ed.), *The Church and Galileo*, Notre Dame IN: Notre Dame University Press **2005**. The literature on this issue is vast.
22. From a talk that was eventually republished as part II of 'Science and Religion' in A. Einstein, *Ideas and Opinions*, New York: Wings Books, **1954**, pp. 41-49.

II
The conservation of wall paintings: the conflictual relationship with water

Piero Baglioni, David Chelazzi, Rodorico Giorgi

Department of Chemistry "Ugo Schiff" and CSGI, University of Florence, 50019 Sesto Fiorentino (Firenze), Italy.
baglioni@csgi.unifi.it

Introduction

Enzo Ferroni wrote that the verb "to conserve", according to its etymology, also means "to inhibit alteration, damage or corruption ... The mere storage of an object inside a case can not ensure its preservation in the sense that it will not be altered. The recognition of the value of a work of art is sufficient to justify the need of storing it. But, in order to avoid its degradation a deep knowledge of the materials constituting the object, and of their stability and reactivity, is strictly required".[1]

Stability and reactivity of materials are concepts that inevitably recall the importance of scientists in the field of works of art conservation.

The devastating flood that occurred in 1966 in Florence gave a strong impetus to research in the field of conservation and restoration (see Figure II.1). In those years the Institute of Chemical Physics, mainly in the person of Enzo Ferroni, gave a decisive contribution to the recovery of the XIV century mural paintings by Taddeo Gaddi in the refectory of Santa Croce church. The detachment of the paintings, in fact, could not be carried out due to the huge amount of nitrate salts

The conservation of wall paintings: the conflictual relationship with water 19

Figure II.1. The Florence's flood in 1966: on the left, some pictures of the Arno river; The picture on the right shows the interior of the Santa Croce convent: in the bottom, the wall painting by T. Gaddi (14th century).

impregnating the walls. Ferroni accepted a public invitation addressed by Umberto Baldini (who headed the restoration following the flood) to anyone who could make a contribution in order to solve the tragic problems of conservation, and proposed the use of tributyl phosphate (TBP) as a sequestering agent of nitrates. This solution allowed completing the detachment of paintings and their restoration, so to that they can be still admired today.[2]

This intervention was followed by the restoration of the XV century paintings by Fra Angelico in the Friary of San Marco (Florence), when an intense partnership was set between Enzo Ferroni and Dino Dini, a conservator and acute experimenter who contributed to the full development of the barium method, which consists of a pre-treatment with an ammonium carbonate solution followed by consolidation with barium hydroxide, so to counteract the dramatic effects of paintings sulphation. In the followings, the XV century paintings in the San Francesco church in Arezzo (Legend of the True Cross's by Piero della Francesca), and the Brancacci Chapel's paintings, by Masaccio, Masolino and Lippi, in Florence were restored.[3,4]

Ferroni's scientific and cultural legacy is now fully expressed by the research activity of the CSGI consortium that he has served as Honorary President until his death. The development of his insights, often sparked by intense discussion with his students, led to the formulation of innovative materials based on modern nanoscience such as microemulsions and micellar systems for removal of detrimental polymer coatings, and nanolimes for the consolidation of wall paintings.[5-7]

1. Wall paintings

Mural paintings represent the interface between the wall and the environment; it is the micron-sized "skin" which in direct contact with the atmosphere and constitutes the discontinuity region where all the physic-chemical properties change. Moreover, looking at the paint layer as the real 'work of art', it is worth noting that it does not exist as a single isolated system, but only as a part of the building with ist specific physic-chemical and mechanical properties.[8]

In order to clarify this aspect is important to recall which are the main characteristics of mural paintings. In particular, the fresco technique will be described. The Italian word 'fresco' attributed to a plaster means that is still wet. The fresco technique requires the application of pigments dispersions when the aerial mortar is still wet. Under this condition, the setting of the mortar allows the incorporation of pigment grains into the newly formed crystalline structure.[9,10]

Before the application of pigments the wall must be prepared. This is done with the sequential superposition of mortars with specific properties and physic-chemical characteristics. The first layer is constituted by a mixture of lime and sand in the ratio 1:3 by volume, where the grains of sand show a very wide distribution with an average size of hundreds of microns, with also some grains millimetre-sized. This layer is called 'arriccio'. This archaic Italian word recalls the image of 'curls'. In fact this layer, after setting, looks very rough and develops a quite large surface area that favour a good adhesion of the second layer of mortar to be applied on top. The mortar composition used for the second layer contains a larger amount of lime (1:2 lime/sand by volume) and very fine sand grains with a narrower size distribution. After setting, this mortar provides a much more regular surface that can be worked to obtain a very flat surface to be painted. This layer after setting is called *intonaco*. The setting process consists of two steps: the first is the drying of the mortar, and the second is the carbonation of calcium hydroxide due to the CO_2 in the air. The first process requires some hours, while a complete carbonation may require several weeks, or even some months, depending on the environmental conditions and the thickness of the layer. Obviously, with the advancement of carbonation, the penetration of CO_2 is hindered by the newly formed compact layers, and reaction occurs according to a kinetic driven by gas diffusion rate.

The third layer is constituted by the pigments, which are applied as a diluted aqueous suspension where some calcium hydroxide is added to provide some more binding material. Being applied when the mortar is still under setting the pigments remain entrapped in correspondence of the surface by the calcium carbonate (as a calcite) that is forming with time. The term fresco is used as opposite to the more diffuse technique known as *secco* painting. The Italian word 'secco' means 'dry'. In this context it indicates that colours can be also applied when the *intonaco* is completely formed, that is lime has completely changed to carbonate and the surface is perfectly dry. In such case, pigments can be bound to the plaster only thanks to some binding media, which typically belong to the class of proteins (for example, casein, animal glue, or egg (yolk, albumen, or a mixture of the two)). This choice is very common in the European tradition; in other cultures, vegetable gums or, more rarely, linseed and poppy-seed oil are also used.

2. The role of temperature and humidity in the natural weathering of materials

Works of art materials exchange energy and matter with the environment and these processes lead to irreversible weathering. In particular, thermo-hygrometric fluctuations (temperature and relative humidity), which may occur daily, are the main factors that rule the degradation process due to the presence of saline solutions within the porous structure of walls and stones. In order to better describe the behaviour of salt solutions confined in porous media, it is necessary to recall the definition of some fundamental parameters of a solution at equilibrium (at a given temperature), namely:

- *Pressure of water vapour.* It is the pressure due to water molecules in gaseous phase, which depends on the water content in the considered 'volume' of atmosphere.

- *Saturated vapour pressure.* When the content of gaseous water per unit volume reaches the finite amount that can be contained, water vapour has reached the saturation point. Any further increase of gaseous water is accompanied by liquid condensation.

- *Absolute humidity.* It is defined as the ratio between the mass of gaseous water and the volume that contains it.

- *Relative Humidity (R.H.).* It is defined as:

$$RH = \frac{P}{P_S} \cdot 100$$

where P is the pressure of the gaseous water and P_S is the saturated vapour pressure at a given temperature. According to the Clausius–

Clapeyron equation the vapour pressure of gases does not increase linearly with temperature. This means that closed spaces (where R.H. is for instance 65%), which can not exchange air with the external environment (therefore air composition remains almost constant) may reach very high humidity, up to saturation, as a consequence of a temperature decrease. In fact, in these conditions, water condensation may take place when temperature variation is about 10-12°C. This is what usually occurs inside historical buildings (i.e. churches, palaces, and so on). However, thermo-hygrometric fluctuations occur also in the interior of museums where air-conditioning is not kept continuously constant but is changed according to the tourists flux.[11]

The vapour pressure of salt aqueous solutions is always lower than that of pure water. According to the Raoult's law, water molecules show a lower 'tendency' to evaporate since they strongly interact with the ionic species of the solute and therefore behave as being "diluted" by the presence of salts. As a result, the relative humidity (RH) of the air in equilibrium with the saline solutions is less than 100%. The greater the concentration of the salt, the lower the relative humidity. The maximum amount of salt that can be dissolved in water, or in other words the minimum amount of gaseous water in equilibrium with the solution, is precisely defined for each salt. This limit is defined as the "relative humidity of equilibrium", and represents, for a given salt, the R.H. of the air in equilibrium with the saturated salt solution. Table 1 shows the equilibrium relative humidity values of some of the most common salts that can be found on mural paintings and stones surfaces.

Table 1. Equilibrium Relative Humidity (%) of salt solutions at 20 °C.

Salt	Equilibrium R.H.
calcium sulphate	99.9
potassium sulphate	98
potassium nitrate	94
sodium sulphate	93
magnesium sulphate	90
potassium chloride	85
ammonium sulphate	81
ammonium chloride	80
sodium chloride	75
sodium nitrate	75
ammonium nitrate	66
calcium nitrate	56
magnesium nitrate	53
magnesium chloride	34
calcium chloride	33

When a porous wall or stone is impregnated with a diluted saline solution, a decrease of the environment relative humidity results in slow drying and consequently in an increase of the salt concentration. When R.H. reaches the equilibrium value, the saline solution will be saturated. Any further decrease of R.H. leads to further evaporation of water and salt crystallization occurs. When humidity regime reaches higher values, dried salt crystals start to adsorb water from the air, and eventually turn into solutions. R.H. fluctuations may have in some cases a daily frequency (more often these process follow a seasonal regime) and the cyclic salt crystallization process exerts a strong mechanical stress within the porous structure of walls and stones (see Figure II.2).[12]

Every time R.H. drops below the equilibrium value, crystallization

The conservation of wall paintings: the conflictual relationship with water   25

Figure II.2. Salt degradation effects due to saline solutions moving by capillarity from the ground (wall paintings in the interior of a church, Cremona, Italy)

takes place; dissolution follows when R.H. value is higher than the equilibrium value. The only way to inhibit this process is to avoid that ambient R.H. values coincide with the salt solutions equilibrium R.H.. If the environmental R.H. is always above or below the equilibrium value of the saline solutions, the latter will always be in the form of solutions or crystals, respectively. The second option is usually preferable. Therefore, using table 1, it is possible to determine the optimal conditions for the conservation of works of art. However, porous works of art materials are usually contaminated by a blend of salts, which may comprise ions such as sodium, magnesium, potassium, calcium, sulphate, nitrate, chloride, and others. When the saline solutions are very diluted, it can be assumed that the behaviour of the different salts is independent of each other. In this case, it would be sufficient to avoid the equilibrium R.H. for each species. Unfortunately, saline solutions are often concentrated. Therefore, it is really difficult to define an ideal R.H. value for inhibiting the cyclic process of salt crystallisation-dissolution, and the best procedure is then to avoid or minimize as much as possible R.H. fluctuations.

Another key factor is represented by salts whose equilibrium R.H. is very low (typically below 50%); these salts are defined as deliquescent (or hygroscopic). A calcium chloride solution, which exhibits an R.H. equilibrium value of 33% at 20°C, represents a classical example of this behaviour. The presence of deliquescent salts is a threat for the conservation of works of art because liquid water adsorbed by these salts may solubilize soluble or poorly soluble salts that, under the same conditions, would be in a 'not active' form.

Figure II.3. Salt degradation effects due to exposure to outdoor conditions (rainfall, vapour condensation, and other). In the bottom left corner, stucco mask of the Templo Major in Mexico city (Aztec civilization); the other pictures show exfoliation and erosion of Angera stone (dolomitic limestone) (Ca'Granda, historical building headquarter of the University of Milan, Milan, Italy).

3. Salt crystallization and degradation: the role of water

Water is undoubtedly involved, both directly and indirectly, in all the major degradation processes that affect cultural heritage materials. Water impregnation of porous matrices can be due to rainfall, capillary sorption from the ground, water vapour adsorption by hygroscopic salts present into walls, and vapour condensation at the dew-point temperature, which is defined as the temperature below which the water vapour in a volume of humid air at constant pressure reach saturation ('dew' indicates the condensed water formed on a solid surface). Table 2 shows the vapour content in saturated air as a function of temperature.

The presence of water is of paramount importance in all the processes of chemical corrosion, biological attack (fungi, algae, lichens)[13] and physical alteration such as freeze-thaw cycles. Moreover, water solubilizes salts, which move through capillarity and spread inside porous materials. As mentioned above, fluctuations of temperature and relative humidity cyclically result in salts dissolution/crystallization processes. For wall paintings, the formation of new crystals inside the pores and/or at the interface between the painted layer and the plaster results either in the detachment of the painted layer or in the cracking and fissuring of the plaster (see Figure II.3).

The supersaturation conditions depend on the type of salt, on the presence of different salts, and on the substrate specific characteristics, including pores shape and pore size distribution, which affect water exchange rates (absorption through capillarity, diffusion and evaporation) and substrate mechanical resistance.[14] This explains the large variety of degradation phenomena that can be observed on mural paintings, such as exfoliation, flaking, and powdering of the painted layer. The same salt can be present as a homogeneous coating (thin and hard) or as a fluffy efflorescence,[15,16] which can be easily removed using a soft brush.

Table 2. Water vapour content in saturated air as a function of temperature.

Temperature (°C)	Water amount (g/m³)
10°	9.5
20°	17
25°	23
30°	30
35°	40

4. Desulphation and consolidation of wall paintings: the Ferroni (or barium) method

Until the end of the 20th century, consolidation and protection of wall paintings and limestone was mainly carried out using synthetic polymers adhesives (also known as resins).[17-19] Starting with the 1960s, the use of polymers was strongly encouraged by eminent international institutions and this technology was enthusiastically adopted worldwide mainly due to the claimed reversibility of these treatments, i.e. the possibility of removing the coating films with the same solvents used for their application.[20] Back then, only few academic institutions alerted on the misuse of polymers, and among them a key role was played by the department of Chemistry of the University of Florence. In fact, polymeric coatings alter the physico-chemical properties of the treated substrates (wall paintings, stones), which can result in severe degradation of the treated artefacts as evidenced by a large number of case studies.[21-23] Moreover, conservators realized that the natural weathering of polymeric materials could dramatically decrease their solubility in solvents, hindering the removal of detrimental coatings and leading to irreversible damage.[24]

On the other hand, the use of materials that exhibit physico-chemical properties similar to those of the artistic/historical substrates (i.e. "compatible" materials) maximizes the durability of restoration interventions. In the first half of the 20th century, inorganic materials for the consolidation of wall paintings and stones included fluosilicates, sodium (and potassium) silicate, and calcium or barium aluminate solutions. However the use of these materials produced poor results and had several limitations.[25-27]

In this framework, Ferroni pioneered the use of compatible and effective consolidation methodologies for the preservation of mural paintings. In fact, at the end of the 1960s he proposed a method that was successfully applied by conservator Dino Dini for the restoration of the wall paintings by Beato Angelico in Florence.

The paintings by Piero della Francesca were then restored with the same method, which has a twofold purpose: the removal of gypsum and the consolidation of the powdered and flaked painted surface (see Figures II.4-5).[28,29]

The method consists in the successive application of ammonium carbonate and barium hydroxide aqueous solutions, using poultices of cellulose pulp.[30,31] The ammonium carbonate solution transforms gypsum into soluble ammonium sulphate, according to the reaction:

$$CaSO_4 \cdot 2H_2O + (NH_4)_2CO_3 \rightarrow (NH_4)_2SO_4 + CaCO_3 + 2H_2O$$

The wet poultice is left onto the surface until dry, so to absorb ammonium sulphate almost completely. Calcium carbonate forms within the substrate pores as a powdery filler with scarce mechanical

Figure II.4. 'The Legend of the True Cross' by Piero della Francesca, 15th century (Arezzo, Italy). The pictures show the same detail before and after the intervention.

Figure II.5. 'The Resurrection' by Beato Angelico, 15th century (Florence, Italy). The pictures show the same detail before and after the intervention.

properties. Then, the application of a saturated aqueous solution of barium hydroxide leads to the formation of calcium hydroxide through a double-exchange reaction with calcium carbonate:

$$Ba(OH)_2 + CaCO_3 \rightarrow Ca(OH)_2 + BaCO_3$$

The process is not thermodynamically favoured (DG^0 is + 51 kJ/mol) but the formed calcium hydroxide is progressively transformed into calcium carbonate by CO_2 and this side-reaction shifts the equilibrium towards the products. It is indeed the slow carbonation of the newly generated $Ca(OH)_2$ that provides, as a long-term effect, the mechanical strengthening (i.e. the consolidation) of the paintings.

Moreover, the application of barium hydroxide transforms residual ammonium sulphate into barium sulphate that, being insoluble, does not undergo to solubilisation-recrystallization cycles:

$$Ba(OH)_2 + (NH_4)_2SO_4 \rightarrow 2NH_3 + 2H_2O + BaSO_4$$

Carbonation of barium hydroxide contributes to a smaller extent to the consolidation effect. The saturated barium hydroxide solution is obtained by dispersing crystalline barium hydroxide powder in water (5% w/w).

To conclude, aqueous saline solutions, which are in principle harmful for the conservation of mural paintings as described above, may provide durable consolidation effects when correctly applied.

References

1. Ferroni, E., *Proc. Symp. on the Restoration of Works of Art*, Florence, 2-7 November 1976. Firenze: Edizioni polistampa, **1981**, 169-178.
2. Ferroni, E.; Dini, D. *Atti della XLI Riunione della SIPS Società Italiana per il Progresso delle Scienze,* Siena, 23-27 September 1967, vol. 2, SIPS Ed., Rome, **1968**, 919-932.
3. Maetzke, A.M. in *Piero della Francesca – La Madonna del Parto. Restauro e iconografia*, Marsilio, Venice, **1993**, 23.
4. Ferroni, E.; Baglioni, P.; Sarti, G. *Quaderni del Restauro 10: 'La Cappella Brancacci. La scienza per Masaccio, Masolino e Filippo Lippi'*, ed. R. ZORZI, Olivetti, Milan (**1992**), 151-161.
5. Baglioni, P.; Giorgi, R. *Soft Matter* **2006**, *2*, 293-303.
6. Baglioni, P.; Chelazzi, D., eds. *Nanoscience for the conservation of works of art*, RSC Publishing, London, **2013**.
7. Baglioni, P.; Chelazzi, D.; Giorgi, R.; Poggi, G. *Langmuir* **2013**, *29*, 5110-5122.
8. Mora, P.; Mora L.; Philippot, P. *Conservation of Wall Paintings*, Butterworths, London, **1984**.
9. Cennini, A.C. *Il libro dell'arte, Manuscript of 1437*, eds. G. Milanesi and C. Milanesi, Le Monnier, Florence, 1859. New editions by: F. Brunello and L. Magagnato, Neri Pozza, Vicenza, 1971; Thompson, D. V. The *Craftsman's Handbook: "Il Libro dell'Arte" by Cennino Cennini*, Dover Publications, NewYork, **1978**.
10. Procacci, U.; Guarnieri, L. *Come nasce un affresco*, Bonechi, Florence, **1975**.

11. Price, C.; Brimblecombe, P. *Preventive Conservation Practice, Theory and Research: Preprints of the Contributions to the Ottawa Congress*, 12-16 September 1994, A. Roy and P. Smith (Eds.), IIC, London, **1994**, 90.

12. Charola, A.E.; Weber, J. *Proceedings of the 7th International Congress on deterioration and conservation of stone, Lisbon, June 15–18*, ed. J. Delgado Rodriguez, F. Enriquez and F. Telmo Jeremias, Lisbon: Laboratorio nacional de engenharia civil, vol. 2, **1992**, 581.

13. Piqué, F.; Dei, L.; Ferroni, E. *Studies in Conservation* **1992**, *37*, 217-227.

14. Dei, L.; Baglioni, P.; Mauro, M.; Manganelli Del Fà, C.; Fratini, F. *Langmuir* **1999**, *15*, 8915-8922.

15. Arnold, A.; Zehnder, K. *J. Crystal Growth* **1989**, *97*, 513-521.

16. Arnold, A.; Zehnder, K. *Proceedings of the symposium organized by the Courtauld Institute of Art and the Getty Conservation Institute*, London, July 13-16 1987, Cather, Sharon Editor, Marina del Rey: Getty Conservation Institute, (**1991**), 103-136.

17. Horie, C. *Materials for Conservation: Organic Consolidants, Adhesives and Coatings*, **1987**, London: Butterworth-Heinemann.

18. Feller, R.L. in *Accelerated Agings – Photochemical and Thermal Aspects*; The Getty Conservation Institute: Los Angeles, CA, **1994**, 63-90.

19. Lazzari, M.; Chiantore, O. *Polymer* **2000**, *41*, 6447–6455.

20. Giorgi, R.; Baglioni, M.; Berti, D.; Baglioni, P. *Acc. Chem. Res.* **2010**, *43*, 695–704.

21. Orea, H.; Magar, V. *Preprints of the 13th Triennial Meeting ICOM Committee for Conservation, ICOM-CC*, Rio de Janeiro, 22-27 September **2002**, (eds. R. Vontobel, James and James), London (England) 176-182.

22. Espinosa, A. *Proceedings of in situ Archaeological Conservation, 6-13 April 1986, Mexico City*; Hodges, H. W. M., Ed.; INAH: Mexico City, **1987**, 84-89.

23. Riederer, J. ICOM-*Committee for Conservation, 7th Triennal Meeting, Copenhagen*, ICOM, Paris, **1984**, 21-22.

24. Dei, L.; Baglioni, P.; Mauro, M. *Preprints of the Conference – Reversibility: Does it Exist?*, Oddy, A. & Carroll S. Eds., London, 8-10 September **1999**, 73-80.

25. Spoto, G. *Acc. Chem. Res.* **2002**, *35*, 652-659.

26. Franchi, R.; Galli, G.; Manganelli Del Fa, C. *Studies in Conservation* **1978**, *23*, 23-37.
27. Matteini, M.; Moles, A. *Preprints ot the ICOM Committee for Conservation – 5th Triennial Meeting*, Zagreb, 1-8 Oct. 1978, International Council of Museums: Paris, **1978**, 78.15.5/1-10.
28. Ferroni E.; Malaguzzi-Valerj V.; Rovida G. *Proceedings of the ICOM Conference*, Amsterdam, **1969**.
29. Baglioni P.; Carretti E.; Dei L.; Giorgi R. *Self-Assembly*, B.H. Robinson Ed., IOS press, **2003**, 32-41.
30. Giovannone, C.; Ioele, M.; Santopadre, P.; Santamaria, U. *Bollettino ICR* no. 6-7 (**2003**), 59-71.
31. Matteini, M.; Moles, A. *Preprints of the ICOM Committee for Conservation – 7th triennial meeting*, Copenhagen, 10-14 September 1984; Froment, Diana de (Editor). International Council of Museums (**1984**), 84.15.15-84.15.19.
32. Matteini, M. *Proceedings of the symposium organized by the Courtauld Institute of Art and the Getty Conservation Institute*, London, July 13-16 1987, Cather, Sharon Editor, Marina del Rey: Getty Conservation Institute, (**1991**), 137-148.

III

The state of water in living systems: from the liquid to the jellyfish[1]

Marc Henry

UMR 7140, Chimie Moleculaire du Solide,
Institut Le Bel, Strasbourg, France.
henry@unistra.fr

Introduction

Here is a little story that may have happened: Seven people were discussing about water at coffee time. "Numbers lying behind all reality, the water concept should be analyzable through numerology", said the mathematician. "I agree", said the biologist, "water is One, the scene action for life apparition as proteins, enzymes and DNA". "I agree and disagree", said a renowned alchemist, "water is the philosophical stone and should be Two (fire and water), saying One miss an essential point". "I agree and disagree", also said the astronomer, "With two hydrogen and one oxygen atoms forged in the stars water should be Three. That's the reality and why bothering with completely useless philosophical statements"? "I agree and disagree" said the physicist, "with ten protons, ten electrons and eight neutrons water is definitively not 1, 2 or 3 but 28 (still a number) otherwise you cannot understand why tradition claims that the moon is full of water, you should always listen at philosophy", "Water is obviously a number and philosophy is nice, but water is 64, the number of hexagrams in the I-Ching", said

---

[1] Reprinted with permission from *Cellular and Molecular Biology* **2005**, *51*, 677-702. DOI 10.1170/T678.

the theoretician. "But you don't say why 64 said the mathematician"? "Oh you crazy guy can't you see that within 10 protons and 8 neutrons you have 54 quarks plus 10 electrons equal 64 elementary particles"! The seventh scientist, a chemist finally said: "no, no, no you are all wrong as water owing to its hydrogen bonding capabilities is able to make any kind of polymer $\{H_2O\}_n$ it should be any number you want between one up to infinity, and so the mathematician was right after all". "But that's the Ouroboros"!!!, realized in chorus this wise assembly. After this discovery of the ultimate truth hidden within the H_2O formula, they all agree that water was a very complex and mysterious thing and asking about who will write the paper, they all said: "the chemist of course"!!!

"All is water, all is one". Was Thales of Miletus (625-566 BC, see Figure III.1) talking to the wind by trying to sum up its philosophy about the origin of things (living or not) in Nature? In order to dwell more deeply into this rather enigmatic sentence, it is worth noting that according to most great religions water is not only just the most convenient vector for bringing inorganic matter to life, but **water is life**. Not convinced?

Take the French word for water *eau* for instance. It comes from the eleventh century word *ewe* and in some local French dialects we find the word *eve* giving full sense to the French word *évier* (a kitchen sink). Now, *Eve* being our common ancestral mother, the link between water and life should be obvious, at least for French people. But notice that *Eve* is a Latin form of *khava* or *khawa* ('v' → 'w' phonetic shift), a Hebrew verb meaning *to live* and that in Sanskrit we have the similar word *khivah* meaning *alive, living*. Adam's choice for naming his wife *Eve* (*Gen.* **3**:20) was because She was literally a life-giver. But the singular future tense of *khawa* is *akwa*, literally meaning *I shall live*, establishing a connection with the Latin word *"aquae"*, a root word related to water in many languages around the world. Figure III.2 shows how by phonetic alteration of the Proto-Indo-European (PIE) roots **au(e)* or

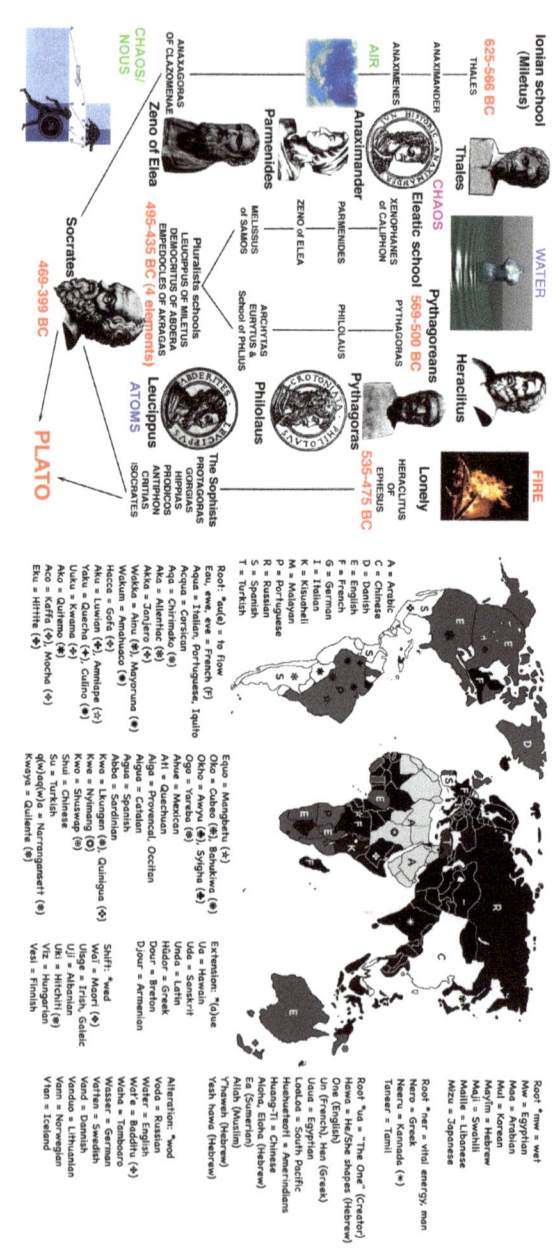

Figure III.1. An overview of the development of antique Greek philosophy (left) and a world trip for words meaning *water* and *Creator* (right).

Figure III.2. An illustration (left) for Surah 21, verse 30 of the Koran, stating that God has made from water every living thing. Also shown, the first two verses of the Bible (right) in Greek (septante LXX translation at the top) and Hebrew (bottom) languages together with an engraved illustration from Ovid.[1]

(a)ue meaning *to flow* (an obvious property of water), the association between life and water cannot be considered as fortuitous.[2] *Eve* is also not very far using the other PIE root **mw* for water as words such as *mouillé, milk, mother, mama, maman...* apply perfectly to *Eve*. A most significant linguistic observation is the **au* → **ua* conversion, showing that by reversing the way water is flowing, it is possible to reach *The One*, i.e. the mighty *Creator* of all things.

Now, opening textbooks devoted to biological facts gives exquisite details about the role of proteins, lipids, carbohydrates, vitamins, ATP, DNA, RNA, etc ... In deep contrast with the above analysis, the first role is not devoted to water and claiming in biology *water is life*, is like talking to the wind... This paper was written to replace water at its central place in science and to try understanding why, even today, the detailed molecular structure of such a common liquid is still elusive. Why, when talking about liquid water structure at the dawn of the XX-Ist century, every scientist has his own theory that is vehemently rejected by other people working in the field? As detailed in the first part this was clearly not the case in the past. The second part will highlight the work of Prof. Ross Aiken Gortner,[3] a scientist who was interested in the exact status of water in the jellyfish[4] and who was talking to the wind by trying to relate life phenomena to the so-called "bound" water concept. The third part introduces another wind blower, Edwin Thompson Jaynes, and from his work the possible reasons for dismiss of Gortner's ideas and the exaggerated importance of diffusion theory by modern biology will be investigated. Keeping in mind that both parts of our brain should be involved in order to satisfy the *libido sciendi* of human beings, the discussion and conclusion try to show that philosophy and science are two equivalent ways of understanding what is hidden behind the water concept.

Water and life in antique civilisations

Two PIE roots **ak'* and **(a)ue* may be found in the Latin word *aquae* derived from the Semitic expression *akwa* meaning *I shall live* in Hebrew. Adding the knowledge that the first root means *sharp*, it leaks out that living things should flow like water by making meanders and should bite like glacial water. But this applies also to snakes, whose tongue is reminiscent of the delta shape of river mouths, explaining the equivalence in all traditions between water, Mother Goddesses (*Eve* for Christians) and serpents (*Eve*'s tempter). But this fundamental triad is by no means limited to Indo-European cultures as it is found in Aztec civilizations the word *atl* that sounds clearly as the PIE root **ak'*. As expected *atl* may be indifferently translated as *water* (think to the Atlantic Ocean) or *arrow* (a sharp object) and, displaying obvious resonance with the Latin *(a)quae* expression, we have the Aztec word *coatl* whose meaning is *snake*.

Christian and Semitic civilizations

Going back for the moment to Christian and Semitic civilizations that refer to *Eve* as a life-giver (*Gen.* 3:20), it is worth noting the explicit link between water and life quoted in the Koran:

> Do not the Unbelievers see that the heavens and the earth were joined together (as one unit of creation), before we clove them asunder? We made from water every living thing. Will they not then believe?"[5] (Figure III.2).

Still more backwards in time, at the very beginning of the creation (*Gen.* 1:1-2), a very well known reference to water is found (Figure III.2). Translation of these two verses in modern languages is surely one of the most debated questions among philosophers and scientists:

> 1. In the beginning God created the heavens and the earth. 2. Now the earth was formless (void) and empty (wasteland), darkness was over the surface of the deep, and the Spirit of God (mighty wind) was hovering (moving, sweeping) over the waters.

Brackets refer to expressions used by different translators stressing difficult Greek and Hebrew expressions and where religious believes may have altered the original meaning. For instance the sentence *ahoratos kai akataskeuastos* and its Hebrew equivalent *tohwou wavohwou* have been quite problematic. Sticking as much as possible to the Greek original meaning would leads to **invisible and without ordering scheme** whereas it perspires **given up to the chaos** in the Hebrew case. In both cases, this means that the world at its very beginning was not empty or void nor it was a formless wasteland. It was just full of something that displayed no visible structure owing to its restless erratic movement. But, looking at the ocean under a stormy weather gives exactly the same feeling! In coherence, the next words refer to the *abyssou* (literally *bottomless sea*) or to the *téhowm* (*deep*) with the additional Greek precision that this chaotic matter cannot be seen *ahoratos*, a clear reference to the absence of light experienced after diving deep into the ocean. Obviously, this inability to see is intrinsic to deep waters having at least 400m of depth and not related to the complete darkness – *skotos* and *roschèk* – existing above the abyss. The last part of the second verse is also quite difficult to translate faithfully, but it suggests that the very secret for life apparition – *theou* and *èlohim* – is lying in the wind – *pneuma* and *rouach* – blowing – *epephereto* and *mérachéfét* – above primeval waters – *hydatos* and *hamayim* –.

In fact, the first verse is also entirely devoted to water through the Greek word *ouranon* and of its Hebrew equivalent *haschamayim*: here the usual English translation *Heavens* hides completely the literal meaning that is *energy hidden in primeval waters* (Hebrew) and *water in the sky* (Greek one). Similarly, translating the Hebrew word *'èréts* by *Earth* is very misleading because the precious indication of dryness (think to the word *arid* in English and *aride* in French) is lost. Here, a slight divergence between Greek and Hebrew versions' may be noted as the word *gên* refers to food producing ability and not to dryness.

Scientists sticking to water as H_2O, a tri-atomic molecule made from electrons, protons and neutrons, themselves made of quarks and so on may be quite skeptic about the usefulness of this analysis. It is however worth noting that words have been created for sharing ideas and concepts among human beings and not for assigning unambiguous names to material objects. Translated in modern language, the first verse is telling us that, as first principles for understanding the world – *En archêi* and *Bere'schit* – , any kind of intelligence – *theou* and *èlohim* – has to distinguish – *epoiêsen* and *bara'* – between mobile things exemplify by air, water or fire (Heavens) and those characterized by their inability to flow (Earth or inertness of dense matter). Then, comes the second verse stating that the erratic and chaotic movement of dark matter was not yet ruled by any life giving principle that was nevertheless present hovering above the deep space-time abyss. Obviously, this is perfectly correct from a scientific viewpoint and the exclamation of Thales of Miletus is alsoquite correct if we understand, **The whole universe is flowing like liquid water**. Moreover, fire and water sharing the same fluid-like behavior the choice of Heraclitus of Ephesus (Figure III.1) to consider fire as a primary element is also correct as everything is in the process of flux (*panta rei*).

Focusing on immaterial attributes of liquid water (ability to flow) instead of its material composition (H_2O), allows unraveling the deep wisdom contained in antique writings in deep coherence with other creation myths met around the world. A wonderful text found inscribed on the interior of coffins belonging to the middle kingdoms of Egypt (2250-1580 B.C.) is worth quoting (*Coffin Texts* '714). It describes how the world came into existence according to Egyptian culture:

> I was The Primeval Waters, he who had no companion when my name came into existence. The most ancient form in which I came into existence was as a drowned one. I was (also) he who came into existence as a circle, he who was the dweller in his egg. I was the one who began (everything), the dweller in the Primeval Waters. First Hahu emerged from me and then I began

to move. I created my limbs in my "glory". I was the maker of myself, in that I formed myself according to my desire and in accord with my heart ...[6]

The parallel with Genesis is striking even in the word ***Hahu*** referring to the wind (***pneuma*** and ***rouach***) that began the separation of the waters and raised the sky in Egyptian's mythology. The explicit reference to the ***Ouroboros***, the mythical serpent biting its tail and dwelling in the egg-shaped primeval waters (Figure III.3) is also worth noting. In echo, it is explained in the ***Book of the Dead*** (chapter XVII) that the birth of light from the waters, and of fire (***Râ***) from the moist mass of primeval matter (***Nun***), formed the starting point of all mythological speculations, conjectures, and theories of the Egyptian priests: *"I am the Great God who created himself. Who is he? The Great God who created himself is the water – it is the Abyss, the Father of the Gods"*.[9] At about the same time (2000 BC) was written the Babylonian epic of creation named ***Enuma Elish'*** from myths of Sumerian people living in Mesopotamia between 5000 BC and 2000 BC (Figure III.3):

> When on high, heaven was not named, below, dry land was not named, Apsu, their first begetter, Mummu (and) Tiamat, the mother of all of them, their waters combined together. Field was not marked off, sprout had not come forth, when none of the gods had yet come forth, had not borne a name, no destinies had been fixed; then gods were created in the midst of heaven ...[10]

Besides the distinction between midst and dry things as first principle (see Genesis' first verse), the occurrence of three entities named ***Apsu*** (fresh water from Sumerian ***ab-zu*** at the origin of the Greek word ***abyssou***), ***Tiamat*** (sea water from Assyrian ***tiâmtu/tâmtu*** meaning *sea*) and ***Mummu*** (violent wind, equivalent of the Egyptian ***Hahu***) is worth noting. The central figure in this triad is obviously ***Tiamat*** or ***Khi-â-mat*** equating *life-matrix/mother* in many languages. According to this myth, life's recipe involves a motion principle (***Mummu***) acting on a Hermaphrodite watery abyss (*Apsu-Tiamat*, equivalent of the Hebrew *tohwou wa-vohwou*).

Other European and Arctic civilizations

Obviously, it may be argued that such coincidences should not be very surprising as Muslim, Greek, Hebrew, Egyptian and Sumerian peoples were all living in a very limited geographical area, close to the strands of the Mediterranean Sea (Figure III.3). But climbing right up to the north, texts entitled ***The Prose Edda or Younger Edda***, written by the Icelandic poet and politician Snorri Sturluson around 1220 AD may be quoted. Interest in these poems comes from their consistent narratives of many of the plot lines of Norse mythology that were scattered among a number of manuscripts which Snorri had access to, but which are now lost (Figure III.3):

> In the beginning there was no earth or heaven, no sand nor sea nor cooling waves. There was only Ginnungagap, a great void. In the north there was Nilfheim, and from Nilfheim's spring flowed eleven rivers, known as Elivagar. As the rivers flowed south, they cooled and hardened into ice. In the south, there was the world of Muspelheim, a fiery world. The northern part of Ginnungagap became filled with the ice and hoar frost from the Elivagar. When that ice formed and was firm, a drizzling rain arose from the venomous rivers and poured over the ice where it cooled into rime, and one layer of ice formed on top of the other throughout Ginnungagap. The southern part of Ginnungagap was lit by the sparks and glowing embers which flew out of Muspelheim. Where the heat from the south met the coolness in the north the ice was thawed and it began to drip and by the might that sent the heat, life appeared in the drops of the running fluid and this fluid formed into the likeness of a man. He was given the name Ymir.[11]

Distinction between wet and dry things, presence of an abyss (***Gin-***

nungagap, here void of matter), water everywhere (rain, river, ice) and life apparition from a *running fluid* resonate in perfect coherence with other myths. Leaving the north for Eastern Europe, Altaic civilizations are met stating that: *"In the beginning, when there was nothing but water..."* whereas the same old song: *"In the beginning, when there was yet no earth, but water covered everything ..."*[12] may be heard in Central Asia Mongolia. Water is still there in Siberia at the extreme northeastern tip of Asia, near the Arctic Ocean and the Behring Strait after listening to this little tale told in October 1900 by Rike´wgi, a Maritime Chukchee man living at Mariinsky Post:

> The Creator lived with his wife. There was nothing, no land, no mountains, only water and above it the sky; also a little piece of ground, just large enough for them to sleep on at night. Creator said to his wife, "Certainly, we feel downcast. We must create something to be company for us". – "All right"! They each took a spade and started to dig the ground, and to throw it in all directions. They dug a ditch so large and deep, that all water flowed down to it. Only the lakes remained in deep hollows, and the

Figure III.3. A map (top right) for earliest Mediterranean civilizations centered on the three rivers: Nile, Tigris and Euphrates. The male/female god ***Nwn*** with its Egyptian spelling and the yin/yang ***Ouroboros*** enclosing the Greek sentence ***hen to pan*** meaning ***All is One*** (drawing from a book De Chrysopoeia or Gold-Making by an early Alchemist, Cleopatra, during the Alexandrian Period). From the ***Enuma Elish'*** epics, ***Ea*** (Babylon) or ***Enki*** (Sumer), the god of water, putting his father ***Apsu*** (fresh water) in a trance in order to kill him and the god ***Marduk*** slaying the dragon ***Mummu-Tiamat*** (sea of chaos) consort of ***Apsu***. In Norse mythology, the watery abyss is named ***Ginnungagap*** and life has appeared, as a giant named ***Ymir*** after melting of the northern frozen ice by the heat emanating from the southern part of the gap.[7] Bottom: A map of Siberia and the picture found on the upper surface of a shaman's drum showing, enclosed in an egg-shaped envelope, a kind of ***Vitruvian Man*** and the raven-shaman holding the drum in the bottom left part of the egg. Also shown, an alchemical emblem[8] showing that the raven is for European alchemy a symbol for the ***Materia primeria***.

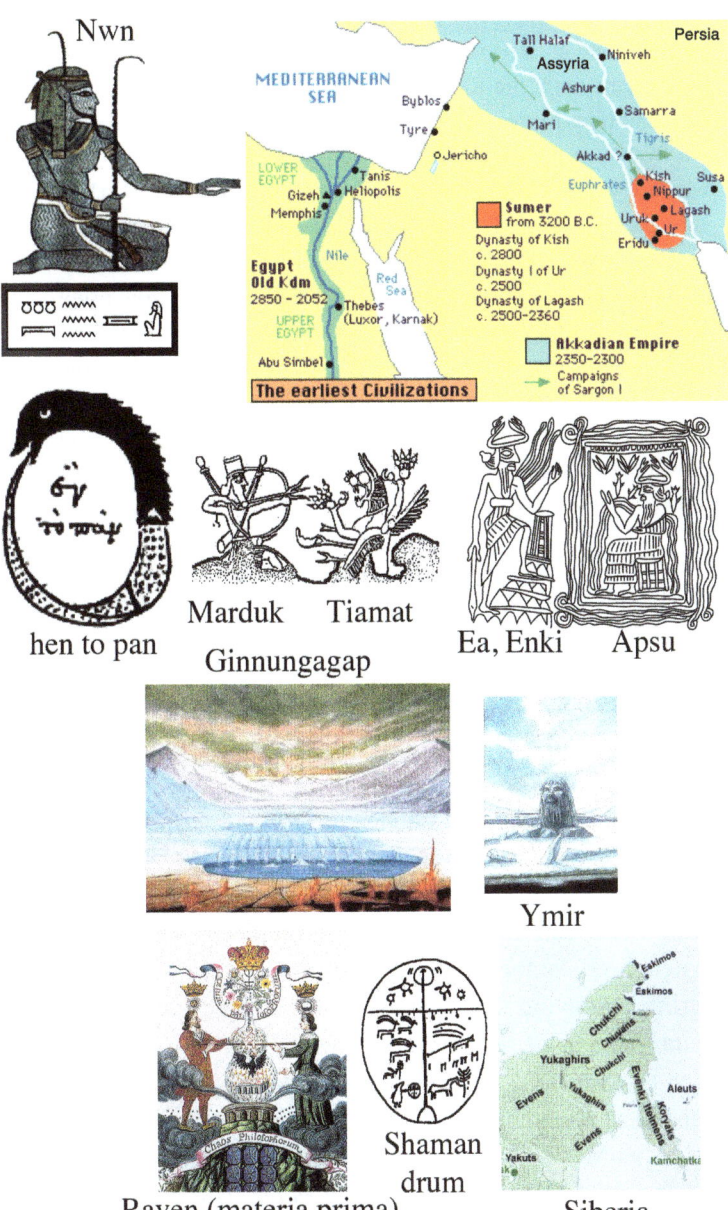

rivers in clefts and ravines. The large ditch became the sea. After that, they created various animals and also men. Then they went away. Only the Raven they forgot to create. They left on their camping-place a large outer garment (ni′glon). Raven came out from it in the night time. He went to visit the Creator. "Oh, who are you"? – "I am Ku′urkil, the self- created one". – "How strange! Self-created! I thought I had created everything, and now it appears that you are of separate origin". – "Yes, yes! I am Ku′urkil, the self- created one". – "All right! – Here, you, bring a few pieces of fly-agaric. Let him eat them, and be full of their force"! Raven ate the fly-agaric. "Oh, oh, I am Ku′urkil! I am the son of the ni′glon. I am Ku′urkil! I am the son of the ni′glon". – "Ah, indeed! And I believed that you were self- created. And now it appears that you are the son of the ni′glon, you are one of mine, created by me, you liar!"[13]

The reference to a deep (the large ditch) filled by flowing water and from which all living things have emerged is not really a surprise, but the end of the tale is truly amazing owing to its wonderful imaged statement of Gödel's incompleteness theorem,[14,15] also called the liar paradox, stating that binary logic is defeated as soon as one element is raised at the level of the whole ensemble (cf the mythical *Ouroboros* dwelling in its egg). The use of fly-agaric mushrooms, **Amanita Muscaria**, for solving the paradox statement, *I am the self-created one*, of the Raven is very elegant, as if the Raven is a material living thing, eating fly-agaric will provoke an *out of body experience* whereas such a separation between mind and body is completely impossible for a self-created God. As the glibly Raven cannot agree to participate in an experiment proving that it was not self-created nature it tells the truth, *I am the son of the ni′glon*, showing that it was a liar, and if the Raven is really a liar... The name **Ku'urkil** may also tentatively be decomposed as **Ku** sounding like **(a)kwa** (*vide supra*), **'ur** (close to the Hebrew word **'ôr** meaning **skin**) and **kil** (close to the Hebrew word **kol** meaning **whole**) reading **skin-coated water**. This suggests that the

Raven was indeed the very first created human being (Shaman), the Inuit counterpart of the Western *Adam*. Connection with Hebrew language is established by the word *ni'glon* that refers to a garment made of reindeer skin. Similar considerations apply to the Siberian Koryak Great Raven **Kutkinnaku** or **Quikinna'qu** and to **Kutq** or **Kutkhu**, the Kamchadal Raven who brought the earth down from the sky with the help of his sister and fixed it immovably in the sea.[16] Connection with Mediterranean myths involving wind action during life apparition may be found in the fact that in most non- European civilizations the raven is used as a symbol for winds and storms. Moreover, in European alchemy the Black Phase or *Melanosis* occurring during the transformation of the *Chaos Philosophorum* (*Materia Prima* or *First Matter*) into the *Coelum Philosophorum* (the perfected *First Matter* or *Lapidis Philosophorum* – Philosopher's Stone) is always symbolized by a raven playing here the role of the mythical self-referenced *Ouroboros*.

The status of water in Indian civilizations

Leaving the frosty regions of the Arctic Ocean, and moving towards warmer countries, the Hindu civilization that have flourished along the Indus River in India from about 2500 B.C is reached. A quite familiar echo originating in other civilizations that had emerged earlier along the Tigris-Euphrates and the Nile rivers may be heard here: *"Water, thou art the source of all things and of all existence!"* (Bhavishyottarapurana 31.14). The *Rig- Veda*, a term meaning *Royal Knowledge*, is the earliest surviving collection of hymns to the gods and ritual texts composed and handed down orally between 1500 and 500 B.C. by the Aryan priests, the Brahmins, a period commonly called the *Vedic Age*. In the hymn called *Creation* (X.129) the ritual story (Figure III.4) involving a self- breathing chaotic watery abyss is found:

    1. THEN was not non-existent nor existent: there was no realm

of air, no sky beyond it. What covered in, and where? And what gave shelter? Was water there, unfathomed depth of water?

2. Death was not then, nor was there aught immortal: no sign was there, the day's and night's divider. That One Thing, breathless, breathed by its own nature: apart from it was nothing whatsoever.

3. Darkness there was: at first concealed in darknew this All was indiscriminated chaos. All that existed then was void and formless: by the great power of Warmth was born that Unit.[18]

From the Sukla part of the **Yajur-Veda** (Satapatha- Brahmana XI.i.6) a self-referenced egg is involved for life apparition: *"Verily, in the beginning, this (universe) was water, nothing but a sea of water. The waters desired, "How can we be reproduced"? They toiled and performed fervid devotions, when they were becoming heated, a golden egg was produced."*[19] Finally, in a collection of laws attributed to **Manu**, the legendary first man, the Adam of the Hindus, universe is found sleeping immersed in darkness with a shining Self-existent will (**Svayambhu**, himself) creating life from waters:

8. He, desiring to produce beings of many kinds from his own body, first with a thought created the waters, and placed his seed in them.

9. That (seed) became a golden egg, in brilliancy equal to the sun; in that (egg) he himself was born as Brahman, the progenitor of the whole world.

10. The waters are called narah, (for) the waters are, indeed, the offspring of Nara; as they were his first residence (ayana), he thence is named Narayana.

11. From that (first) cause, which is indiscernible, eternal, and both real and unreal, was produced that male (Purusha), who is famed in this world (under the appellation of) Brahman.

12. The divine one resided in that egg during a whole year,

then he himself by his thought (alone) divided it into two halves.

13. And out of those two halves he formed heaven and earth, between them the middle sphere, the eight points of the horizon, and the eternal abode of the waters."[20]

Later, in Vedic literature the god *Narayana* – the one who dwells upon the waters – is identified with *Vishnu*, an eternal and all-pervading spirit thought as the preserver of the universe while the two other major Hindu gods **Brahma** (a creative spirit appearing after the growing of the lotus from Vishnu's navel) and **Shiva** (the destroyer of the universe) ruling the overall universe evolution are two other complementary aspects of this watery god. The Great Raven, central figure of Shamanism, appears in India as **Garuda** (the vehicle of *Vishnu*), a giant sized eagle responsible for winds and thunder and which often is shown as a winged human-shaped figure having a beak- like nose. *Vishnu* is usually pictured with the seven-headed snake (Figure III.4) and this denotes *Adi Sesha* or *Ananta Nag* (the timeless or ageless snake). The link between water and time is further glorified in the later *Atharva-Veda* (composed about 1000 B.C. and written down about 200 B.C.) through an hymn (XIX.54) stating that time was the main creator of the primeval waters and of other elements such as fire, wind and earth:

1. From Time (Kâla) the waters did arise, from Time the brahma (spiritual exaltation), the tapas (creative fervour), the regions (of space did arise). Through Time the sun rises, in Time he goes down again.

2. Through Time the wind blows, through Time (exists) the great earth; the great sky is fixed in Time. In Time the son (Prajâpati) begot of yore that which was, and that which shall be.[21]

The status of water in Asian civilizations

All so far studied creation myths were not very explicit in explaining how primeval waters came first into existence and just above the answer was suggested that water, source of all living things, was a self-created child of time. But this notion of an original entity full of nothing able to give birth to matter is also encountered in Taoism, a philosophy developed by Chinese civilization that flourished about 1700 B.C. along the banks of The Yellow River (***Huang Ho***). Accordingly, in chapter 25 of the ***Tao te Ching*** attributed to Lao-tzu (580-500 B.C.) the creative principle of the Tao is introduced:

> There was something undefined and complete, coming into existence before Heaven and Earth. How still it was and formless, standing alone, and undergoing no change, reaching everywhere and in no danger (of being exhausted)! It may be regarded as the Mother of all things. I do not know its name, and I give it the designation of the Tao (the Way or Course). Making an effort (further) to give it a name I call it The Great ...[22]

The legendary unintelligibility of the Tao-te-Ching comes from the fact that this treatise is somehow like the ***Ouroboros*** father of chaos with no definite head or tail nor logical order. Thus taking a look at chapter 8 entitled ***The way of water*** the link between water and Tao is stressed:

> The highest excellence is like (that of) water. The excellence of water appears in its benefiting all things, and in its occupying, without striving (to the contrary), the low place which all men dislike. Hence (its way) is near to (that of) the Tao ...[22]

At last from chapter 42, entitled ***The Transformation of the Tao***, the creative power of the Tao (and by direct inference of water) is stated:

> The Tao produced One; One produced Two; Two produced Three; Three produced All things. All things leave behind them the Obscurity (out of which they have come), and go forward to

embrace the Brightness (into which they have emerged), while they are harmonized by the Breath of Vacancy.[22]

The ubiquitous Yin/Yang drawing (Figure III.4) sums up this conception of creation with **One** being the whole drawing, **Two** the black and white colors, **Three** the two little circles of opposite colors and the S-shaped line. The living aspect of the symbol is linked to the fact that the orientation of the S-shape does not matter, nor the way of coloring the two Yin and Yang regions. The **Breathing of Vacancy** arises from continuous rotation of the S-shaped curve and from the perpetual motion transforming the Yin domain into the Yang one (and *vice versa*) through the two circles connected by an invisible communication channel (*the Tao*) lying in a third dimension, definitively outside the plane defined by the 2D-symbol. A quite striking parallel with Genesis is also possible by identifying **One** with **Elohim**, **Two** (as light/darkness) with the watery abyss pair **Heavens/Earth** and the necessary addition of **Pneuma** (Breath of Vacancy) for getting **Three**, thus restoring the harmony broken by the emergence of duality. Thus while for Hinduism or Taoism, **Zero** (**Kâla, Tao**), appears logically before **One** (**Brahman**, watery abyss), the reality lying before **One** (God) is carefully occulted in western myths, a situation obviously related to the fact that the number zero was unknown to the Egyptians and that for Babylonians it was only a location mark having no tangible reality.[23]

Besides Taoism, Confucianism and Zen Buddhism are the two other main philosophies in Asian civilizations. **Master Dogen** (1200-1253), founder of the Japanese Soto Zen lineage and of Eihei-ji monastery, was the author of two most famous texts entitled **Shobogenzo** and **Eihei Shingi**. Reference to water and life may be found in chapter 3 of the **Shobogenzo** entitled **Genjo-Koan** (***The Realized Universe***):

> Know that water is life and air is life. The bird is life and the fish is life. Life must be the bird and life must be the fish. You can

नासदासीन नो सदासीत तदानीं नासीद रजो नो वयोमापरो यत |
किमावरीवः कुह कस्य शर्मन्नम्भः किमासीद गहनं गभीरम ||
न मृत्युरासीद अमृतं न तर्हि न रात्र्या अह्न आसीत्प्रकेतः |
आनीदवातं सवधया तदेकं तस्माद्धान्यन न परः किं चनास ||
तम आसीत तमसा गूळमग्रे.अप्रकेतं सलिलं सर्वमा इदम |
तुछ्येनाभ्व अपिहितं यदासीत तपसस तन महिनाजायतैकम ||
कामस तदग्रे सम अवर्तताधि मनसो रेतः परथमं यदासीत |
सतो बन्धुम असति निरविन्दन हर्दि परतीष्याकवयो मनीषा ||
तिरश्चीनो विततो रश्मिरेषाम अधः सविदासी.अ.अ.अत |
रेतोधाासन महिमान आसन सवधा अवस्तात परयतिः परस्तात ||
को अद्धा वेद क ड्ह पर वोचत कुत आजाता कुत इयं विस्र्ष्टिः |
अर्वाग देवा अस्य विसर्जनेनाथा को वेद यत अबभूव ||
इयं विस्र्ष्टिर्यत अबभूव यदि वा दधे यदि वा न |
यो अस्याध्यक्षः परमे वयोमन सो अङग वेद यदि वा नवेद ||

第八章

上善若水。水善利萬物而不爭，處眾人之所惡，故幾于道。
居善地，心善淵，與善仁，言善信，政善治，事善能，動善時。
夫唯不爭，故無尤。

Figure III.4. Reproduction (left) in Sanskrit language of the hymn X.129 of the Rig-Veda explaining the formation of the universe from a self-breathing chaotic watery abyss. Below, Lord *Vishnu* symbolizing the infinite ocean (primeval waters) from which the world emerges laying down on the timeless or ageless snake, *Kāla*, shown at right. An excerpt of chapter 8 (right) from the Tao-te-Ching of Lao-Tzu summarized by the ubiquitous yin/yang symbol. Also shown, an illustration of how the world was created according to Shinto religion[17] and a japanese transcription of the koan *Mu soku Wu, Wu soku Mu.*

Vishnu and Kāla

Kāla

Yin/Yang

Lao-Tzu

Takami-Musubi et Kami-Musubi

Mu soku Wu Wu soku Mu

go further. There is practice-enlightenment, which encompasses limited and unlimited life ...²⁴

The first very compact sentence of **Master Dogen** resonates in harmony with the first two verses of the bible as it involves life, water and air (the *pneuma*). In chapter 14 entitled *Sansui-gyo* (**Mountains and Water Sutra**) this very interesting text is found:

> We should study that occasion when the water of the ten directions is seen in the ten directions. There is a study of water seeing water. Water practices and verifies water. Hence we can say there is a practice of water speaking water... It is not only that there is water in the world, but there is a world in water. It is not just in water. There is also a world of sentient beings in clouds. There is a world of sentient beings in the air. There is a world of sentient beings in fire. There is a world of sentient beings on earth. There is a world of sentient beings in the phenomenal world. There is a world of sentient beings in a blade of grass. There is a world of sentient beings in a staff. Wherever there is a world of sentient beings, there is a world of Buddha ancestors. You should thoroughly examine the meaning of this ... Even in a drop of water innumerable Buddha lands appear.²⁴

The Shinto Japanese creation story from the ***Kojiki***, Japan's oldest chronicle, compiled in 712 CE by ***O No Yasumaro*** is also worth quoting. In deep contrast with Zen Buddhism, a mighty rainy pivot deity (***Ame-no-Minaka- Nushi***) able to split into male (***Takami-Musubi***) and female (***Kami-Musubi***) growth principles arises from a shapeless and formless watery chaos haunted by a medusa loafing upon the face of the waters:

> Before the heavens and the earth came into existence, all was a chaos, unimaginably limitless and without definite shape or form. Eon followed eon: then, lo! out of this boundless, shapeless mass something light and transparent rose up and formed the heaven. This was the Plain of High Heaven, in which material-

ized a deity called Ame-no- Minaka-Nushi-no-Mikoto. Next the heavens gave birth to a deity named Takami-Musubi-no-Mikoto, followed by a third called Kami-Musubi-no-Mikoto. These three divine beings are called the Three Creating Deities. In the meantime what was heavy and opaque in the void gradually precipitated and became the earth, but it had taken an immeasurably long time before it condensed sufficiently to form solid ground. In its earliest stages, for millions and millions of years, the earth may be said to have resembled oil floating, medusa-like, upon the face of the waters ...[25]

The status of water in other civilizations

Crossing the Pacific Ocean towards America, Australia or Africa does not alter the central role of water for life apparition. It is however not possible to review here all the numerous oral myths of Native American speaking of the adventures of the singing **Coyote**, relating how **Earth-Maker** began to cry after thinking about what he should do, telling the creative power of **Cyclone**, the wind blowing into water haunted by **water-beetle** and how **Spider Woman** has created human beings from mud and saliva. In Central America, the feathered serpent is found dwelling in water as told in the Mayan hieroglyphic manuscript **Popul Vuh**:

> Only the sea alone is pooled under all the sky ... Whatever there is that might be is simply not there: only the pooled water, only the calm sea, only it alone is pooled ... Only the Maker, Modeler alone, Sovereign Plumed Serpent, the Bearers, Begetters are in the water, a glittering light ...[26]

As shown in Figure III.5, the feathered serpent is also a major deity of Aztecs or Toltecs, but he was rather created from the earth Goddess **Coatlicue** (She with a skirt of snakes) consort of the water God **Tlaloc**. Water is also present in a hymn to **Pachacamac** (literally *He who gives animation to the universe*), the Great Inca Spirit and consort of **Pa-**

chamama (literally *Earth-Mother*), the mother-spirit of mountains, rocks, and plains. Interestingly enough, the same word *pacha* has three usual meaning: *world* (visible things), *time* (things happening in succession) and *property* (things connected with persons such as clothes).

Australian aborigines have lived on the vast island of Australia for over forty thousand years and are one of the oldest surviving cultures in the world dating back – by some estimates – to 65,000 years. For them, earth was originally an expanse of water, colored as the rainbow, and all life was held inside a female Rainbow Serpent, its symbol being the rainbow bridging Heaven and Earth.[29] The Australian Aboriginal people believe the universe has two aspects – the physical world in which we live and another connected world from which it is derived called the **Dreamtime** (Figure III.5). This expression usually refer to the *time before time* or *the time of the creation of all things*, while **Dreaming** is often used to refer to an individual's or group's set of beliefs or spirituality. The rainbow snake is also known in Haiti under the name **Damballah** (Sky-serpent Loa or wise and loving father).

Closing this little world trip by Africa, it is worth quoting the fascinating Dogon creation myth telling the primordial error of the Creator *Amma* raping the female excised earth giving birth to *Jackal*, the deluded and deceitful Son of God. Later, when water, the divine seed, entered the womb of the earth to create, *as is natural and right*, inseparable male and female twins called **Nummo** were born:

> ... Their bodies were green and sleek all over like the surface of water. From head to loins they were human; below that they were serpents. Their red eyes were wide open like people, their tongues were forked like serpents. Their arms were flexible and without joints and their bodies covered with the short, green hairs of vegetation and germination. These spirits were formed of God's divine essence, made of his seed, which is simultaneously the ground, the form, and the substance of the world's life force. They are the motion and persistence of all things. The Pair is present in all water; they are water—the water of seas,

of coasts, of torrents, of storms, and of the spoonfuls we drink. Without Nummo, it would not have been possible to create the earth. For the earth was molded clay and water is in all things that have life. Even stones have this life force, for there is moisture in everything ...'[30]

Water and the jellyfish
The facts according to R.A. Gortner

In this section, the amazing unanimity among all human civilizations for linking water to life apparition will be critically studied using science's quantitative viewpoint. In this scientific world ruled by quantum physics, relativity theory and thermodynamics principles, it is not allowed to invoke a merciful male Creator blowing his pneuma onto female watery matter. The best way to start is to quote a 1930 paper by R.A. Gortner devoted to the state of water in living systems:

> As I write these lines there lie before me two sheets of paper. One is the photograph of a large Medusa (jellyfish) from the Atlantic Gulf Stream, which was photographed immediately after being removed from the water and being placed upon the open pages of a magazine. The Medusa, as removed from the water, weighed in excess of 500 grams. In the photograph one can read the distorted print through the more or less transparent outer portion of the umbrella, but the central portion of the Medusa which measured approximately 10 x 12 cm, was sufficiently dense and opaque to prevent the print underneath from showing in the photograph. The other sheet of paper is the opened pages of the magazine upon which the Medusa had been allowed to dry after being photographed. These pages simply appear as though they have been wetted and then dried. No noticeable film is discernible on the surface of the pages. The print is clear-cut, and even exposing these pages to ultraviolet light results in extremely slight fluorescence. The weight of these pages exceeds by less than 0.45 gram the weight of the pages before the Medusa was

dried upon them. Less than 0.10 percent of dried residue from the large Medusa including the salts, etc..., in the adherent sea water and all of the inorganic constituents of the living organism. Such observations raise the very pertinent question as what constituted the more important part of the original living organism; the organic portion, the proteins, fats, extractives, etc..., or the inorganic portion, the salts, the calcium, magnesium, iron, copper, chlorides, etc..., which together total but a fraction of a percent of the weight of the organism, or the water which comprised more than 99.9 per cent of the animal?..."[4]

For a non-scientist there can be only one answer: Water! But, I bet that for well-trained biologists working at the dawn of the XXI century, the answer would rather be: proteins, or may be DNA, i.e. organic matter! But then how to explain to non-biologists people, the same that are giving you money for doing research, that a bag full of water (at least 99.9%) is able to move, to look for food and to reproduce itself being ruled by less than 0.1% of its constituents? Obviously, the quantitative arguments so useful in science to decide where the truth is lying are here not very helpful. Worse, with just one or two pages devoted to the physics and chemistry of water and with at most 10 lines about a quite mysterious **hydrophobic interaction** in major biological textbooks, it becomes very hard to sustain the obvious fact that the jellyfish is living because its major constituent is water H_2O. But may be that listening to R.A. Gortner's answer could be helpful:

> To my mind there can be only one answer: All are equally important. All are a part of a living substance, which we call the protoplasm, and to concentrate our study only upon the organic portion and the inorganic salts would be analogous to attempting to constructs without plans and without an architect a great cathedral from a mass of rubbles shaken down from a cliff by a gigantic earthquake. The water in the Medusa was as much alive as were proteins, the fats, the lipids, or the carbohydrates, and this living water must have been sharply differentiated in some

way from the great mass of water surrounding the living Medusa in the Gulf Stream. The problem of the nature and degree of this differentiation is a problem which requires for its solution all of the skill and ingenuity of the colloid chemists of the future. When they will have only partially solved these problems, I firmly believe that they will have already inaugurated a new day in biochemistry, physiology, and medicine, for the water relations in the living organism lie at the foundation of problems concerning both health and disease ...[4]

This text was written in 1930 and 75 years later absolutely no progress has been made towards the achievement of this program. R.A. Gortner, in full agreement with all the antique wisdom, points in the right direction and our modern medicine works in the other opposite direction, ridiculing even this naive idea that we can heal people just by resting on water physicochemical properties. Huge amounts of money are now flowing around us for elaborating new sophisticated drugs that will concern at most 30% of our body, whereas nothing is left for getting a better understanding of what we are really made of, i.e. 70% of water. Why a top-quality scientist whose authority was acknowledged, not only in biology but also in chemistry and physics was blowing to the wind by stating that understanding the nature and structure of water should be at the very heart of any modern medicinal research? In his 1929 book dealing with biochemistry, chapters on amino acids, polypeptides, vitamins, carbohydrates, fats, oils, tannins are found coexisting with 290 pages devoted to the colloidal state of matter.[31] Just after its publication, this book has been enthusiastically reviewed:

> To reader of « The Physical Review » the illustrations here afforded of the scope and power of generalizations long familiar should rove intensely interesting. Who that measures vapor-tensions will not enjoy the picture of the cactus on the wall in desert Arizona ten months without rain and sheltered from the dew,

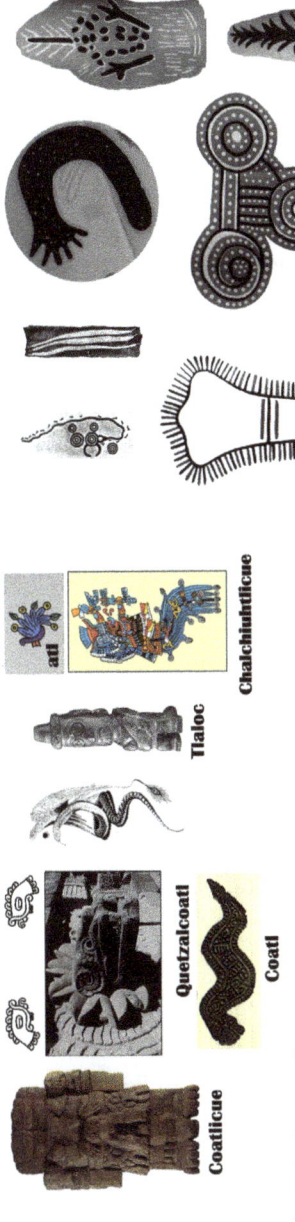

O Pachacamac!
Thou who hast existed from the beginning,
Thou who shalt exist until the end,
powerful but merciful,
Who didst create man by saying,
"Let man be,"
Who defendest us from evil,
and preservest our life and our health,
art Thou in the sky or upon the earth?
In the clouds or in the deeps?
Hear the voice of him who implores Thee,
and grant him his petitions.
Give us life everlasting,
preserve us, and accept this our sacrifice.

It was in the Lalai, the beginning, the Dreamtime, that Wandjina roamed the unformed earth. The most revered Wandjina, the Rainbow Serpent, created waterways, mountains, and valleys on the land by "writhing" its body, then gave birth to many life forms. The birds in the sky, the fish in the sea, the mammals on the land, all of the reptiles, and even humans descended from the Rainbow Serpent...

Figure III.5. Some important Aztec deities (left) with a hymn to *Pachacamac*, Creator god of the incas. Also shown, paintings depicting totemic inhabitants of the Australian landscape in the dreamtime (right). They constitute part of a text, the meaning of which is esoteric knowledge only known to Aboriginal elders.[27,28]

or the story of the three-months sojourn of another cactus over sulfuric acid, the account of drought-resisting cereals in Utah, winter-hardy wheat, and the drying-up of earthworms, or the human- interest paragraphs on the granary-weevils' life without a drink, the new oedema diagnosis, and the rigor of our bodies after death? All these and more are brought into the chapter on Gels to illustrate imbibition and the distinction between bound and liquid water; and not as incidental illustrations merely, for they are examined in the light of tables, formula and charts, and correlated with freezing- points, osmotic-pressures and absorption in the infra-red. Therefore, if any Reader thinks he know more physics though less biology than Gortner, let him go to it; the door is opened wide ... It would be a shame to let mere biochemists monopolize the reading of this book.[32]

Seventy-five years later, a Google's search using *outline of biochemistry* Gortner keywords shows that a building named Gortner's labs exists in the College of Biological Science (University of Minnesota, USA) and that a Gortner Avenue is located in St Paul, Minneapolis. Alas, only 13 hits correspond to papers making direct reference to this rather unique textbook and about 10 links points to various booksellers, one of them being an antiquarian!!! Other links (about 5) pointed to public libraries (mostly in Brazil) allowing borrowing the unique available copy (usually the 3rd edition published in 1949) by "students" lost in the rare books section of their library. Obviously, something went wrong and it perspires that the only culprit was the nature of the object that was the focus of Gortner's research: water as a matrix for life.

The general discussion at the meeting

The general discussion between international experts that have followed the reading of Gortner's paper at the Faraday meeting is worth quoting to understand this disastrous situation. For instance Dr. W. Cramer recalled the water content found in tissues such as liver (72.3%), muscle (75.8%), heart (77.9%), kidney (77.2%), spleen (78.8%), testis (83.0%), embryos (90% if 15th day-old and 84% if 20th day-old) and mouse carcinomas (80-86%). For the investigated cases (except embryos) the ash content as percent of fresh tissue was at least 0.88% and at most 1.6%, showing that the overall organic content was in the range 13-19% for carcinomas and 16-26% for other tissues.[33] A most important point was the observation of a direct correlation between the growth rate of the cells and the amount of **bound water**, the larger the water content, the higher the growth rate. Obviously, this is a convincing quantitative experimental proof of the basic equation water = life, corner stone of all philosophical systems in the world. But, by the way, who was Dr. W. Cramer who was convinced that *the state of water in the cell is of fundamental biological importance* and do you even know his name? All we know from the discussion is that he was working in London in 1930 and, as searching on the Internet gives negative results, he also was talking to the wind. Further analysis of the comments shows that most of them were quite in favour of the position defended by Gortner or at least neutral (bringing only precisions on minor points or reporting other kinds of experimental results related to Gortner's approach). For instance, the famous colloid chemist Prof. H.R. Kruyt (Utrecht University, Utrecht, The Netherlands) was telling:

> I think that Gortner has made important progress with respect to our knowledge of hydration, by working out his method of estimation of bound water... From the investigations of Gortner and his pupils we can see the importance of bound water in several biological processes ...[34]

and was even speaking of a *Gortner-effect*. From Dr. Lecomte du Nouy (Paris, France) who was a dependable, practical and highly productive biophysicist, inventor of the elegant Du Nouy Tensiometer for measuring liquid surface tension and of the Du Nouy Micro-Viscosimeter for studying serum, we may quote: *"The very interesting and highly suggestive paper of Professor Gortner certainly raises one of the most puzzling and fundamental problems of biology ... "*[35] But there were also negative reactions to Gortner's paper and the most noteworthy was coming from Prof. Archibald Vivian Hill (Figure III.6), who was awarded with the 1922 Nobel Prize in physiology or medicine for his discovery related to the production of heat in the muscle:

> I should not like to advance any general criticism of Professor Gortner's paper. He has thought and worked on the subject much more than I have. I feel, however, that he has possibly over-estimated the importance of a study of the state of water in biological systems, so far at least as concerns tissues and fluids of animals. There is an obvious mistake in his statement that Medusa contains only one part in 500 of solid material. Like all other marine invertebrates, this animal must certainly contain the usual constituents of seawater in its fluids, which would make up at least sixteen parts in 500. I have little doubts that the water in the animal, if studied by physicochemical methods would be found to be almost en entirely free ...[36]

This is the heart of the problem as echoed by Dr. Cramer: *"Is the water content of a living cell dependent entirely on osmotic forces or is it dependent also on the imbibition of the lyophilic colloid of the cell content?"*.[33] Here, the term **imbibition** refers to Gortner's **bound water**, i.e. water that is so intimately interacting with the colloidal matter that it displays properties quite different from that of ordinary tap water. On the other hand we have **osmosis**, a phenomenon first evidenced in 1877 by Wilhelm Pfeffer,[37] describing the natural water movement occurring under a concentration gradient in presence of a semi- permeable membrane that admits passage of water but not of solute. By osmotic forces Dr. Cramer is evoking the possibility that water in the cell behaves like ordinary liquid **tap water.** Convinced that water was an essential ingredient for life sustaining, Gortner has devised in the years 1920-30 a very clever method (freezing-point depression shown by a given biological fluid containing 10 g of water and 0.01 mole of sucrose) for getting reliable information about the state of water in a living cell. The basic idea was that if part of these 10 g of water were completely restructured by the underlying biological matter, it would not be available as a solvent for sucrose. Obviously, Gortner's method has been criticized on quantitative grounds[38,39] and the Gortners' have replied to these criticisms by ameliorating the method.[40] But, referring to the three underlined expressions found in A.V. Hill criticism, we are on another ground because he admits that he is not an expert in the field and uses nevertheless authoritative statements giving the uncomfortable feeling that we are leaving experimental evidence for religious belief. Responding to so much arbitrary affirmations and certitudes coming from a Nobel Prize winner is very difficult and the only possible way is to stick to experimental facts as shown in Gortner's reply:

> ... Professor Hill suggests that there must be some obvious mistake in my statements in regard to the solid content of a Medusa. In the illustration cited in my opening paragraph, I was unfortunately forced to depend upon the statement of the collector as to the original weight of the Medusa, although I made the weight of the dried material. On the other hand, many years ago, while associated with the Station for Experimental Evolution of the Carnegie Institution of Washington, I had occasion to dry down small forms of salt-water Medusae and in a number of instances found to my surprise that the dry matter content was considerably less than 1 per cent of the original weight as taken from the open sea water. This would indicate that the salt content of the water in the umbrella of the jellyfish must in a good many instances be less than the salt content of the seawater surrounding the organism.[3]

This is a very kind way of recalling to his prominent challenger that, owing to the presence of membranes, the jellyfish is definitively not a mere bag full of sea-water but that it is a living organism able to play very cleverly, at little energy costs, with electrolytes. Yet another Hill's criticism was that Gortner used 1M sucrose solutions that may perturb significantly (through adsorption for instance) the initial partition between "free" and "bound" water. At first sight this objection may seems quite sound but after all this is a general problem in science and measuring without making perturbations is physically impossible.[41] Furthermore, the suggestion of Hill to replace the 1 M cane sugar solution by a 0.1 M NaCl solution is very strange, first because 0.1 M is by no means a dilute solution and second because Hill should have known that, owing to the very large compression of the double layers, increasing the ionic strength at such levels would have been very harmful to any kind colloidal matter, alive or dead!!! The suggestion of moving from the freezing-point depression method towards the thermal method of vapour pressure depression is also not very clever, as increasing temperature should be even much more destructive for the fragile colloidal matter than lowering it. Gortner's answer to these objections was:

> I am heartily in agreement with Professor Hill's suggestion, that, at least for warm-blooded animals, a method should be devised, which would differentiate between the free and bound water at the normal temperature of the animal, but as yet no method, with the possible exception of the vapor pressure method suggested by Professor Hill, has been available. As Professor Burton points out our problems here at Minnesota and Newton's problems in Western Canada have been concerned to a very large extent with the phenomenon of winter-hardiness, and the measurements which have grown out of winter-hardiness studies demonstrate rather conclusively that the binding of water by lyophilic colloids is an important, if not the determinant factor in cold resistant forms.[3]

Again, all Gortner's arguments are based upon strong experimental evidence that may be easily verified by any farmer trying to make plant growing under cold conditions, and the selected freezing-point depression with sugar and not ionic salts sticks most closely to what is observed in nature. But A.V. Hill was attacking not only Gortner's views but also other experimental evidences:

> The belief in the existence of a large amount of bound water in muscle seems to have originated largely from the work of Overton, published in 1902; he found that muscles immersed in hypotonic solutions did not swell to anything like the degree expected, on the assumption that all the water contained in them is imprisoned within semi-permeable membranes. His experiments can be confirmed, but their explanation I believe to be quite different, i) the slowness of attainment of equilibrium, ii) and most important, the loss of semi-permeability in a considerable fraction of the fibres when immersed in the hypotonic solution, and iii) possibly, a progressive swelling or shrinkage of the tissue owing to colloidal changes of other kinds.[36]

The underlined sentence illustrates well the latent conflict that always exists between science (*Think first then measure for validation*) and religion (*Believe even if experimental facts are against you*).

Concerning the attacked scientist, Overton, Hill gave just his name, without judging even worth quoting a full reference. Moreover, his alternative proposed explanations are not very convincing because i) equilibrium can never be attained in any kind of experiment (but ideas are instantaneous), ii) loss of semi-permeability preventing water flowing would have mean closing of the pores (but by what substance?) and iii) the expression *owing to colloidal changes of other kinds* is just a convenient way to admit that no other good arguments were available. But, by the way, who was exactly this poor Overton who was not able to interpret his experiments correctly? First, his full name was Sir Charles Overton the very first scientist who, having evidenced that lipophilic molecules were easily absorbed by a frog mus-

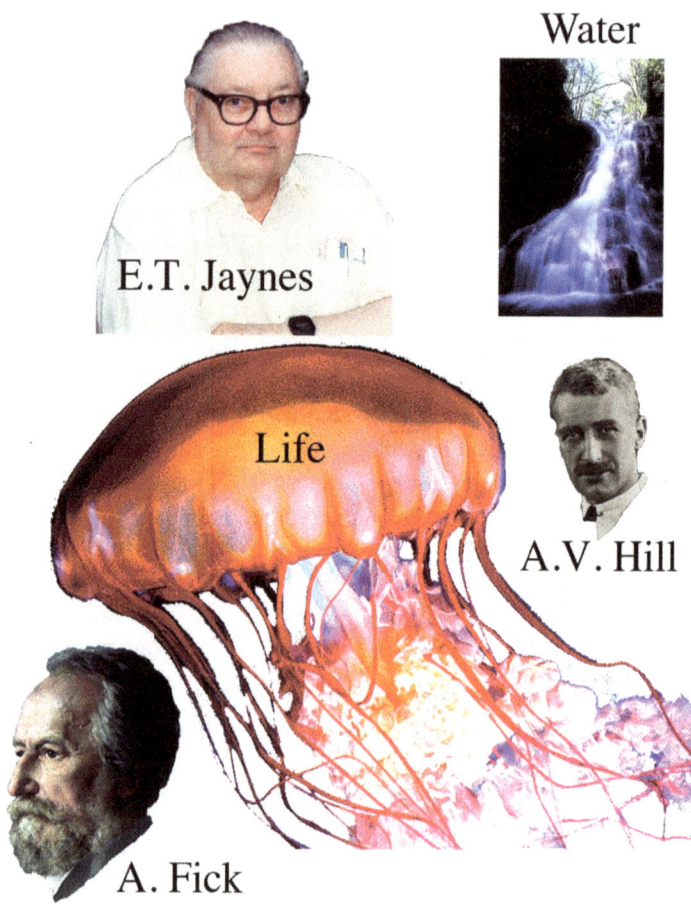

Figure III.6. Some prominent scientists and the jellyfish posing the quantitative problem of the state of water in living systems.

cle, has emitted the supposition that most cells should be surrounded by a semi-permeable lipid membrane.[42] Here, it is worth quoting the Nobel lecture given on December, 11, 1963 by Allan Lloyd Hodgkin about the ionic basis of nervous conduction stressing three times the importance of Overton's work:

> ... There were several early attempts to provide a theoretical basis for the reversal (of the E.M.F.), but most of these were speculative and not easily subject to experimental test. A simpler explanation, now known as the sodium hypothesis, was worked out with Katz and Huxley and tested during the summer of 1947. The hypothesis, which owed a good deal to the classical experiments of Overton, was based on a comparison of the ionic composition of the axoplasm of squid nerve with that of blood or sea water ... If all the external sodium was removed the axon became reversibly inexcitable, in agreement with Overton's experiment on frog muscle ... Nevertheless, if they are to be of any use to the animal, nerve fibres must be equipped with a mechanism for reversing the ionic exchanges that occur during electrical activity. The necessity for such a system was foreseen by Overton in 1902 when he pointed out that human heart muscle carried out some 2.4 109 contractions in 70 years, yet as far as he knew contained as much potassium and as little sodium in old age as in early youth ...[43]

But, of course this is the kind of things people are saying after having being rewarded and honored by a Nobel Prize. The reality is quite different as confessed recently by Andrew F. Huxley that had shared the same 1963 Nobel Prize for medicine with A.L. Hodgkin and Sir John Eccles. In 1939, Hodgkin and Huxley were quite surprised by the experimental findings showing that in both crabs and squids nervous fibers the action potential of the membrane exceeded its resting potential by 40-50 mV, ruling completely out the idea that electrical activity was depending on a breakdown of the membrane (Berstein's theory). Here is the end of the story as told by A.F. Huxley himself:

This was contrary to Bernstein's theory, but it was not altogether a surprise to Hodgkin as he already had hints (unpublished) of this discrepancy from external measurements on fibres of crabs and lobsters. We could not pursue the problem further because war was obviously imminent, and we left Plymouth two days before Hitler invaded Poland. We published the result in a letter to Nature (1939) with no discussion or explanation. In a full paper (1945), we gave four possible explanations, all wrong. It was also in 1945 that we began discussing the explanation that turned out to be correct, namely that the increase in permeability was highly specific for sodium ions, which were thereby enabled to diffuse inwards carrying their positive charge. This was confirmed experimentally by Hodgkin and Katz (1949). If we had known the paper of Overton (1902), I am sure we would have reached that conclusion immediately in 1939.[44]

Remembering the absence of reference and the negative opinion expressed by A.V. Hill, the missing by Hodgkin and Huxley of Overton's fundamental contribution is understandable. Obviously, this does not mean that A.V. Hill was not a good scientist, as there is no doubt that in his expertise domain (physiology of the muscle) he was one of the best. The problem is rather that he was speaking outside his expertise domain and was too much confident in the ability of the mechanistic approach to explain the emergence of life from inert matter. As he was also Nobel Prize winner, he has used his scientific authority to definitively orient all future research in the direction of viewing the cell as a little semi-permeable bag filled with ordinary water and holding electrolytes and substances in solution. If this view is perfectly correct for a dead cell, a living cell is quite different, being able to move matter against concentration gradients and resisting rather than aiding naturally occurring diffusion processes.

The diffusion paradigm in biology

This leads us to a crucial question: **what is the real molecular mechanism driving diffusion and osmosis**, or is it possible to derive osmotic forces **from a suitable thermodynamic physical potential**? Quoting the celebrated Vant'Hoff equation[45] established empirically in 1887, $\Pi = RT \cdot \Delta C$ (where Π is osmotic pressure, R the gas constant, T the absolute temperature, and C the concentration difference of non-permeable solutes across the membrane) is of no help. It does not explain why water molecules are moving, but just states what may be observed after a long time when the flux of water molecules across the membrane will be the same in both directions. But, is irreversible thermodynamics able to give a mechanical justification of van'Hoff law? We recall that its links the steady-state volume flux across a membrane, J_v, to its hydraulic conductivity, L_p, through the Kedem-Katchalsky equation:[46]

$$J_v = L_p(\Delta P - s \cdot RT \cdot \Delta C)$$

Here σ is a suitable reflection (or selectivity) coefficient ($\sigma = 1$ or 0 for an ideal semi- or completely permeable membrane, respectively). It is worth noticing that the same coefficient L_p is used for the purely mechanical term ΔP and for the non-mechanical osmotic one.[47] The problem obviously lies in the macroscopic nature of van't Hoff law as opposed to universal mechanical conservation laws (energy, impulsion, angular momentum ...) that apply at any scale. Saying that water move **because** of the presence of a concentration gradient cannot be correct as this notion is a macroscopic concept, useful for scientists but not for molecules, microscopic entities responding only to potential gradients of electromagnetic or gravitational nature according to Newton's law of mechanics. So, why molecules do respond to a concentration gradient by moving from the most concentrated region towards the most diluted one? The answer to this fundamental question may be found in the work of Edwin Thompson Jaynes (July 5, 1922-

April 30, 1998) whose contribution to science was far more important than that of Albert Einstein. A short biography may be found under the URL: *http://bayes.wustl.edu/etj/etj.html* saying that:

> Jaynes retired in 1992 after a long and productive career. Jaynes' contributions to science were of the highest caliber. His work in reformulating statistical mechanics has illuminated the foundations of that theory and enabled extensions to non-equilibrium systems. His dedication to rooting out contradictions in quantum mechanics is legendary. He must have single-handedly sparked more debate in quantum mechanics than any other person in the last 50 years. The verdict on his neoclassical radiation theory is still not in, and may not be for many more years. It may yet prove to be a better description of nature than quantum electrodynamics. He also helped take an interpretation of probability theory from being virtually unknown to a healthy research area that is being applied daily in economics, biology, physics, nuclear magnetic resonance and many other disciplines. His writing helped to clarify the foundations of probability theory in a way never achieved before. He wrote profusely, in a warm and friendly way that enabled one to see complex points as if they were intuitively obvious. He spoke as he wrote. When he criticized someone's work, he always stuck to the facts; he never reverted to name-calling. His friendship was hard to earn, and hard to keep, for he had little tolerance for incompetence. He would undoubtedly be uncomfortable with all of the attention being lavished on him now that he is dead.

If science were really as rational as it claims to be, E.T. Jaynes would be nowadays as celebrated as Albert Einstein himself, as Jaynes was convinced that Einstein was often not very clear by trying to clear up Nature's mysteries and that he was also the only man able to understand the real signification of the second law of thermodynamics.[48] So, how does one calculate a diffusion coefficient for moving matter and not get zero? This apparently simple exercise in kinetic theory that has puzzled generations of physics students is easily solved following Jaynes' arguments:

> Think, for definiteness, of a solution of sugar in water, so dilute that each sugar molecule interacts constantly with the surrounding water, but almost never encounters another sugar molecule. At time $t = 0$ the sugar concentration varies with position according to a function $n(x; 0)$. At a later time we expect that these variations will smooth out, and eventually $n(x; t)$ will tend to a uniform distribution ... Since sugar molecules – or as we shall call them, "particles" – are not created or destroyed, it seems natural to think that there must have been a diffusion current, or flux $J(x; t)$ carrying them from the high density regions to the low ... Our present problem is: how do we calculate $J(x; t)$ from first principles? Maxwell gave the simple kinetic theory of diffusion in gases, based on the idea of the mean free path. But in a liquid there is no mean free path ...[48]

With a clear idea of what is really a liquid medium, Jaynes immediately rejects all previous attempts based on kinetic theory of gases, and acknowledging that a straightforward computation of the flux $J(x; t)$ by summing over all particles in a small region gives a null average velocity drift, he notes:

> Surely, this must be right, for our particle, interacting only with the surrounding water, has no way of knowing that other sugar molecules are present, much less that there is any density gradient. From the standpoint of dynamics alone (i.e. forces and equations of motion) there is nothing that can give it any tendency to drift to regions of lower rather than higher density. Yet diffusion does happen!"[48]

As diffusion theory persists nowadays as a cornerstone of biology and medicine,[49] what can be done and who can help? Albert Einstein, may be? Yes, but according to Jaynes,[48] in order to establish the correct formula for the diffusion coefficient:

... Einstein was forced to invent strange, roundabout arguments – half theoretical, half phenomenological. For example, first estimate how the density $n(x; t)$ would be changed a long time in the future by combining the distributions generated by many different particles, then substitute it into the phenomenological diffusion equation; and from that reason backwards to the present time to see what the diffusion flux must have been. This kind of indirect reasoning has been followed faithfully ever since in treatments of irreversible processes, because it has seemed to be the only thing that works. Attempts to calculate a flux directly at the present time give zero from symmetry, so one resorts to "forward integration" followed by backward reasoning. Yet this puzzles every thoughtful student, who thinks that we ought to be able to solve the problem by direct reasoning: calculate the flux $J(x; t)$ here and now, straight out of the physics of the situation. That symmetry cannot be exactly right; but where is the error in the reasoning? Furthermore, instead of our having to assume a phenomenological form, a correct analysis ought to give it automatically; i.e., it should tell us from first principles why it is the density gradient, and not some other function of the density, that matters, and also under what conditions this will be true. Evidently, we have a real mystery here ...[48]

To solve this deep mystery, it must first be acknowledged that the quest for a physical prediction from the dynamics is a dead end, as *"The equations of motion do not require particles to move about at all"*.[48] Why? Well, because the equations of motion are symmetric in past and future! To break this symmetry, the prior knowledge of the varying density of particles in the past must be taken into account. As information about where the particles were in the past is not a time-reversed mirror image of where they will be in the future, the right question is: *"What is the best estimate we can make about how the particles are in fact moving in the present instance, based on all the information we have?"*[48] It is a problem of Bayesian inference that has nothing to do with dynamics that predicts rightly a zero movement. In other words:

> ... The Gaussian form of the probability distribution expresses an average over the class of all possible motions compatible with the dynamics, in which movements to the right and the left have, from symmetry, equal weight. But of course, our particular particle is in fact executing only one of those motions. Our prior information selects out of the class of all possibilities in a Gaussian distribution a smaller class in which our particle is likely to be, in which movements to the right and left do not have equal weight. It is not the dynamics, but the prior information, that breaks the symmetry and leads us to predict a non-zero flux ..."[48]

The amazing thing is that using just Bayes' theorem and after only six lines of elementary mathematics, the well-known Einstein's formula for the diffusion coefficient was recovered and the predicted average diffusion flux was found to be proportional to the density gradient (Fick's first law of diffusion). Consequently, in diffusion and osmotic processes matter is not moving because of forces arising from a concentration gradient. Instead, as matter can physically move in only one direction at a time, it is this asymmetric information field that gives a higher probability of displacement towards regions of low density. Clearly, the misleading mathematical simplicity of Fick's law and its striking success for explaining the observed behaviour of inert matter without no explicit assumptions other than those of statistical mechanics was giving the illusion to be on safe grounds. But, like many classics of science, Fick's 1855 paper is far more widely acknowledged than read, hiding the fact that Fick's law was derived in 1855 from dubious experimental data and justified using a rather fickly analogy *with the law which Fourier has established for the spread of heat in a conductor; and which Ohm has transferred with such splendid success to the spread of electricity.*[50] It was probably the obsession of Adolf Fick to explain all living phenomena by the principles of physics and chemistry that apply to non-living matter, that renders him blind to the fundamental difference existing between physical gradients (heat, electricity) and information gradients (diffusion, osmosis). The road was thus wide opened for mechanistic materialism having the faith that a rational medicine would emerge only when each pathological process was attributed to a specific physiological cause, itself understood in the language of physics.

The claim by A.V. Hill in 1930 that practically all the water inside the cell was "free" rather than "bound"[51] can thus be traced back to the very conservative mechanistic materialism of Adolf Fick. The inescapable logic that was responsible for the dismiss of Gortner's ideas relating bound water to life sustaining was thus deep- rooted in the failure of the whole scientific community to recognize the shortcomings of diffusion theory. As well explained by Agutter et al.,[49] in order to answer some of the most pressing questions of experimental physiology, agriculture, and medicine, biochemistry was designed to investigate metabolism using the concepts and methods of organic chemistry. Organic chemistry being able to cope only with unstructured solutions, it was mandatory to portray the cell as an unstructured solution and Fick's law was assumed to hold in order to maintain the validity of the model. Despite all the experimental facts showing that a living cell cannot obey to Fick's law, the entrenchment of diffusion theory in biology was nevertheless irreversible owing to this vicious feedback mechanism.

Ontological and epistemological levels

This is why the necessity of putting prior information into the probabilities in order to get non-zero matter fluxes is so important for biology. But conventional orthodox training in physical sciences does not provide any technical means for taking prior information into account, focusing 100% on physical prediction and relying on the belief that probabilities are real physical properties of systems. However, that prior information is often highly cogent, and sound reasoning requires that it be taken into account. In other fields this is considered a platitude; what would you think of a physician who looked only at your present symptoms, and refused to take note of your medical history? Accordingly, using Bayesian inference requires probabilities that are dependent on the current state of knowledge about systems, and every biologist or scientist convinced that equations or models are describing the real world should ear Jaynes' warning:

> We are all under an ego-driven temptation to project our private thoughts out onto the real world, by supposing that the creations of one's own imagination are real properties of Nature, or that one's own ignorance signifies some kind of indecision on the part of Nature ... To appreciate the distinction between physical prediction and inference it is essential to recognize that propositions at two different levels are involved. In physical prediction we are trying to describe the real world; in inference we are describing only our state of knowledge about the world. A philosopher would say that physical prediction operates at the ontological level, inference at the epistemological level. Failure to see the distinction between reality and our knowledge of reality puts us on the Royal Road to Confusion ... As a Bayesian calculation operates on the epistemological level, it gives us only the best predictions that can be made from the prior available information and if in the real world there are extra controlling factors not yet identified, predictions may be wrong ...[48]

Scientists are very often missing the point that a wrong epistemological prediction is a good thing, because it points to the existence of unknown factors and allows learning new things about the real world. Rejecting a Bayesian calculation because it has given an incorrect prediction is like disconnecting a fire alarm because that annoying bell keeps ringing. Realizing that inert matter can move not only under physical potential gradients (attraction, repulsion, sedimentation...) acting at the ontological level (**prediction**) but also in the presence of subjective information gradients (diffusion, osmosis...) acting at a very different epistemological level (**inference**) is of the utmost importance for scientific training in biology. Trying to understand the behavior of living matter without this distinction in mind, there is great danger to slip towards mechanistic materialism as illustrated above for the entrenchment of diffusion theory in the corpus of biological knowledge.

This observation immediately raises the question of the status of thermodynamics' second law in biology, as the general self-organizing power of biological systems seemed to be in conflict with *the tendency to disorder* philosophy derived from this concept. For Jaynes, the starting point is the simple observation that on a hot day when the ambient temperature is above body temperature ($T = 310$ K), animals are still able to move. In more scientific words, the world's most universally available source of work – the animal muscle – is constantly violating the second law because it is able to deliver useful work (efficiency as high as 70%) when there is no cold reservoir at hand. This is obviously quite amazing if the cell has to behave as a thermodynamic machine similar to a heat engine, and even for scientists, it is very difficult to resist the temptation of seeing the hand of the Great Creator behind this matter of fact. But before turning to philosophy, let's see first what E.T. Jaynes has to say on this very controversial subject: *"The obvious first answer is of course that a muscle is not a heat engine. It draws its energy not from any heat reservoir but from the activated molecules produced by a chemical reaction..."*[48]. The actual source of energy is then easily identified in the hydrolysis of adenosine triphosphate (ATP) for which the reported heat of reaction is 0.43 eV per molecule (i.e. 41.4 kJmol^{-1}). Concerning the possible violation of the second law, Jaynes's tells us that there is no need to invoke the Great Creator. Basically, biological systems, far from violating physical laws, exhibit the operation of those laws in their full generality and diversity, knowledge far beyond our present level of understanding of thermodynamics laws'. In other words, the tendency to disorder arguments are too vague to be of any constructive use and shuffled cards, messy desks, and disorderly dorm rooms as examples of entropy increase are more misleading than useful.[52] Only a general statement of the second law, including biological phenomena as a particular case, is needed and all the sophisticated explanations and speculations already published on non-equilibrium thermodynamics must be ignored. Going back to original sources is the right way of proceeding with the muscle's problem:

> Some have thought that it would be a highly difficult theoretical problem, calling for a generalized ergodic theory to include analysis of "mixing" and "chaos". Another school of thought holds that we need a modification of the microscopic equations of motion to circumvent Liouville's theorem (conservation of phase volume in classical Hamiltonian systems, or unitarity in quantum theory), which is thought to be in conflict with the second law. We suggest, on the contrary, that only very simple physical reasoning is required, and all the clues needed to determine the answer can be found already in the writings of James Clerk Maxwell and Joshuah Willard Gibbs over 100 years ago. Both had perceived the epistemological nature of the second law and we think that, had either lived a few years longer, our generalized second law would long since have been familiar to all scientists ... The observed efficiency of muscles may be more cogent for this purpose than one might at first think. Since animals have evolved the senses of sight, sound, and smell to the limiting sensitivity permitted by physical law, it is only to be expected that they would also have evolved muscle efficiency (which must be of equal survival value) correspondingly. If so, then the maximum observed efficiency of muscles should be not merely a lower bound on the maximum theoretical efficiency we seek, but close to it numerically ...[48]

As Jaynes' paper may be downloaded at the previously indicated

URL, it suffices to say that the fundamental keyword characterizing the second law is not **disorder** but **reproducibility** and that the limitation it puts on processes arises from a deep interplay between the epistemological macrostate, representing our knowledge about the system, and the ontological microstate, i.e. the coordinates and momenta of individual atoms, which determine what the system will in fact do. The existence of the second law is the price that must be paid for having an engine working in a reproducible way, as an engine that delivered work only occasionally by chance would be completely useless in engineering and biology alike. Just by making this clear differentiation between the epistemological macrostate and the ontological microstate it was possible to show that Liouville's theorem is the reason for the existence of the second law. It was also possible to elaborate a quantitative theoretical model predicting that the maximum muscle efficiency would be 76.5% (a value very close to what is observed experimentally) provided that the energy of ATP hydrolysis is confined to a single vibration mode in a striated muscle. Had the energy of hydrolysis spread to ten vibration modes before being recaptured by the muscle, the predicted efficiency would have dropped down to only 8%. This nicely explains the relative stiffness and massiveness of the myosin head whose function is first to resist to rapid thermalisation while transferring its energy into the macroscopic sliding of the actin fiber. Another interesting conclusion was that a muscle is able to work efficiently not because it violates any laws of thermodynamics (as often stated by the supporters of the vitalism doctrine), but because it is powered by **tiny spots** of molecular size as hot as $T = 5060$ K, a temperature similar to that at the surface of the sun!!! This astonishing result opens fascinating perspectives in terms of biological evolution and may help to unveil the physical reason why the ATP molecule has been selected as the universal energy carrier in all living cells. Accordingly, ATP hydrolysis giving an energy flash that is just of the same order of magnitude than the average thermal radiation emanating from sun, this molecule should be viewed as a molecular archive recalling us that at the very beginning, the first living cell was probably powered directly by the sun. If so, then a planet with a hotter sun would evolve life with still higher muscle efficiency. Bere'schit bara' 'èlohim 'èt haschamayim wé-'êt ha'èréts..., the ouroboros is closed.

Discussion

« *Ô water! Ô water!* » was saying Chung-nî while Hsü, the disciple of Meng-tzu, was asking to his master: « *What did he find in water to praise?* » and Meng-tzu replied « *a superior man is ashamed of a reputation beyond his merits* ».[53] After traveling around the world to learn that water was the matrix for life and after diving at the roots of diffusion theory to understand why matter is moving, anybody should know, like Chung-nî, what is worth praising in water. The survey of all creation myths has shown that three notions are invariably invoked for life apparition: the magic expression **In the beginning...** acting on a whole nothingness within a watery matrix principle. If this fundamental scheme is particularly obvious and explicit in Taoism, it is also present, very often implicit in other myths. For R.A. Gortner and its jellyfish, life was primarily water informed by tiny amount of organic matter from a quantitative viewpoint and the Faraday meeting of 1930 was marking a change of paradigm for biological sciences. Accordingly, before 1930 the existence of « bound » water supporting life phenomena was a platitude, being experimentally validated and proven, whereas after 1930 this view was completely dismissed by the attacks of molecular biologists who were rather uncomfortable with the coexistence of two kind of water molecules in the cell. With free water obeying to classical physicochemical laws (diffusion and osmosis) that rules inert matter, these lazy scientists were quite happy being in the (false) secure land of dilute aqueous solutions ruled by macroscopic thermodynamics. With the other one, bound water, they were very unhappy because dissolution of electrolytes were very specific escaping general osmotic rules and also because most of matter movements were occurring against concentration gradients with the consequence of the appearance of strongly ordered nanostructures, suggesting a possible violation of the second law of thermodynamics. As experimental techniques were not developed enough for a clear understanding of the nature and structure of this bound water and as there were grave theoretical misunderstandings concerning the nature of diffusion and osmotic processes, the image of the cell behaving as a bag of solution delimited by an inert semi-permeable membrane where matter transport depends on specific biological machinery at the expenditure of cellular energy has become a paradigm and dogma of molecular biology.[49,54]

What is known today about water in the cell

Seventy-five years later, it is clear that this paradigm must be discarded, mainly because it puts the most abundant ingredient of the cell (water) at the back of the stage devoting all the important primary roles to proteins, enzymes, RNA, DNA, etc ... Accordingly, with current analytical and computing techniques, we now know that it is because water molecules interact with proteins and DNA that life is possible. For instance, it was recently possible to study the local water structure of the saccharide-free form of concanavalin A and its bound water, including 167 intact D_2O molecules and 60 oxygen atoms at 15 K to 2.5-Å resolution using neutron diffraction.[55] From detailed X- ray diffraction structures, it becomes possible to study how interfacial water molecules act as key allosteric mediators in a cooperative dimeric hemoglobin demonstrating their crucial role in communication between proteinic subunits[56] or to study the network of ordered water molecules with a polygonal organization that binds three DNA duplexes.[57] Using Inelastic Incoherent Neutron Scattering (IINS) techniques, water-DNA and water- proteolipid membrane systems may be studied over a range of hydration states.[58,59] It is found that below a critical concentration, bulk water is essentially absent, i.e. all the water in the system is interacting with the biological macromolecules.[60] From computer simulation studies we know that the lifetime of the ubiquitous DNA double helix in the vacuum without any water molecules around it is only 50 picoseconds and that this molecule should be heavily hydrated for stabilizing the double-helical structure, support of the genetic code.[61] Moreover, subsequent simulations of DNA in water have revealed that water molecules are able to interact with nearly every part of DNA's double helix, including the base pairs that constitute the genetic code. It is well known that cytochrome c oxidase is a redox-driven proton pump, which couples the reduction of oxygen to water to the translocation of protons across the membrane. The recently solved X-ray structures of cytochrome c oxidase have permitted molecular dynamics simulations and allow establishing the proton pump mechanism, i.e. the transport of the substrates, oxygen and protons, through the enzyme. A large number of water sites are predicted within the protein, which form two channels along which protons can enter from the cytoplasmic (matrix) side of the protein and reach the binuclear center.[62] Taking into account the knowledge[63,64] that liquid water is a schizophrenic medium displaying two coexisting states differing by their local densities, the living cell has acquired new levels of complexity with the beautiful work of Philippa Wiggins showing how water solvent properties may change with its structure[65] thus leading to a very convincing molecular mechanism for the otherwise mysterious hydrophobic interaction.[66] The recent isolation from self-assembled nanocapsules in solution of such microdomains (Figure III.7) and their full characterization from the point of view of their structure and relative H-bond strengths open new fascinating perspectives for the possibility to mimic cation transport through biological ion channels, but also to get general information about transport processes through molecular pores.[68-70] The intrinsic properties of these nanocapsules, which can be even synthesized on a (kilo)gram scale, are also thought to open a new era in the study of water structures that may be formed at the nanometer scale in the presence or absence of electrolytes.[70]

If we go back to our jellyfish, it is clear that R.A. Gortner was right and A.V. Hill was doubly wrong, first on a scientific quantitative standpoint and second on a qualitative philosophical ground. By denying the existence of bound water, Hill was turning his back to millenaries of wisdom claiming that water was the source of all life and existence and this simple knowledge would have been enough to suggest him a more modest position. But, referring to philosophical statements is not the favourite pastime of most biologists as lucidly confessed by A. Szent-Györgyi, another Nobel Prize in physiology and medicine in 1937 for his work on vitamin C:

> One of my difficulties with protein chemistry was that I could not imagine how such a protein molecule can live. Even the most involved protein structural formula looks stupid, if I may say so ... It looks as if some basic fact about life were still missing, without which any real understanding is impossible ...[71]

The diagnostic was right (a protein without his water shell looks stupid), the feeling was good (some basic thing is still missing) and the answer was obvious (no life without water) but it takes several years of hard work to find it. The surprising thing is thus that most biologists do not follow Szent-Györgyi, who was one of the first to recognize the importance of water for a living cell. This has led to the current paradox of seeing scientists the most concerned by the life phenomenon (biologists), focus research on the minor components of the cell, leaving the study of the major component to those who are the least concerned (physicists and chemists). A possible reason for this strange situation was given by Szent-Györgyi himself: « *It is a puzzling question why water with its enormous physiological and pathological importance, has received relatively little attention from biologists. I think that the correct answer was given by Oliver Lodge, who said that the very last thing a deep-sea fish would discover is water* ».[72] Consequently, under the harmful impulsion of A.V. Hill, a paradigm shift has occurred in biology around 1930, explaining why only naked proteins, naked DNA or RNA, naked membranes, naked enzymes are found in most textbooks. Of course, biologists are not the only people to be blamed and it suffices to say that physical scientists are still fighting in literature about liquid water structure, which has intrigued them for centuries. Even in 2005, they still don't know if this liquid should be viewed as a giant polymeric continuous H-bonded network or as this two state (high-density liquid HDL and low-density liquid LDL) substance evidenced by all spectroscopic measurements. At last, chemists lying at the borderline of physics and biology are also to blame when they try to mimic through extensive and costly syntheses un-watered mythical biological architectures. Obviously, working with sophisticated organic and/or inorganic self-assemblies is nice and helps to forget that chemists don't know how to manipulate and control the most flexible and simple building block that Nature has ever created: a mere water molecule H_2O.

Science at the ontological level

About this irresolvable dichotomy existing in liquid water between the one-state polymeric model favored by computers and the schizophrenic two-state model evidenced by spectroscopy it is worth recalling Jaynes' fundamental distinction between an ontological reality (with its perfectly reversibility in time that can only be think) and an epistemological knowledge of this reality (full of irreversible facts that are experienced every day). These two notions (**ontology** vs **epistemology**) having been used with so much different meanings by both scientists and philosopher, a clarification is needed before continuing. French protestant theologian Jean le Clerc coined the word ontology in 1692 from modern Latin *ontologia* in order to combine the notion of « being » (*ontos* genitive of Greek on prepositional of *einai* « to be ») with that of « word, idea, or reason » (from Greek *logos*). An ontological statement is thus something that can be written down (mathematical equation) or expressed verbally (assertion or axiom) and gives an idea or the reason why something should exist or transform. For instance, it is an ontological law of Nature that the total energy $H(q,p,t)$ (the so-called **Hamiltonian**) viewed as a function of position q, associate momentum p and time t should rule the time behavior of any other function $u(q,p,t)$ through the fundamental equation of motion:

The deep reality engraved in this compact equation may be realized by setting u = H, implying that $dH/dt = \partial H/\partial t$. Furthermore, if the system is also at equilibrium ($\partial H/\partial t = 0$) then the total energy should be conserved *(dH/dt = 0)*. This is by essence the first principle of thermodynamics stated by Rudolf Clausius under the form: "*Die Energie der Welt bleibt constant*", or « the energy of the world remains constant ». But it is also possible to set $u = q$ or $u = p$ in the above equation leading effortlessly to Hamilton's equations of motion ruling classical mechanics ($dq/dt = \partial H/\partial p$ and $dp/dt = -\partial H/\partial q$). For systems made of a large number N of particles, it becomes impossible to follow the variation with time of each (q,p)-couple and the solution is to introduce the **Liouville** function $W_N(q_1,p_1,...,q_N,p_N,t)$ giving the probability of that at time *t* the ith particle would be found at location q_i with momentum pi. Then, taking $u = W_N$ leads to the Liouville equation, the very foundation of statistical mechanics, leading after application of the mass conservation law to Liouville's theorem stating that the probability flux W_N in a 6N-dimensional phase space is a conserved quantity $(dW_N/dt = 0)$. The shift from classical to quantum mechanics comes just from the realization that the **Poisson bracket** of the position q with its associated momentum {q,p} was not equal to zero ($[q,p] = ih/2\pi$ in quantum mechanics) as assumed by classical mechanics. Accordingly, the above equation still holds (the so-called Heisenberg's representation involving **commutators** instead of Poisson brackets) with the only change that *u*, *q*, *p*, and *H* are no more functions but differential operators (hence also the move from the classical Liouville function W_N to the density matrix ρ of quantum statistics). Finally, it is also worth noticing that it is possible to derive from a suitable 4D vector potential another Hamiltonian allowing putting Maxwell's equations ruling all electromagnetic phenomena under the same universal mathematical form quoted just above. This kind of universal behavior with a single equation ruling all possible phenomena (mechanical, electrical, quantum-mechanical, statistical, etc ...) is the signature of an ontological reality where time is just a label. This is well shown by the invariance of the fundamental equation under the transformation *t'* = -*t*, implying *q'* = *q* and *p'* = -*p*, meaning that things and facts occur inexorably and in a reversible way provided that no external cause is acting on the system. A direct consequence is that such a world should have no experimental reality, as there is no room for the kind of beings called observers (cf. Gödel's incompleteness theorem).[14,15] At the ontological level (the reality in Science) there is no life, no light to see, no matter to touch, only a purely mental activity is allowed.

Science at the epistemological level

Moving to the field of biology, it should be realized that the following sentence is typically of ontological nature: *There is water in a living cell* and has definitively not the same meaning as this apparently quite similar one: *In a living cell there is water*. Accordingly, it should be clear that the first statement is an abstract law stating that the second concept (life) cannot exists without the first one (water), or more concisely that *water is life*. Reversing this statement (l*ife is water*) immediately shows that something new has happened even if the same verb « is » is employed. If it seems so absurd relative to the first statement (a glass of water is obviously not living), it is because it refers to a sensible world where life and water are easily identified

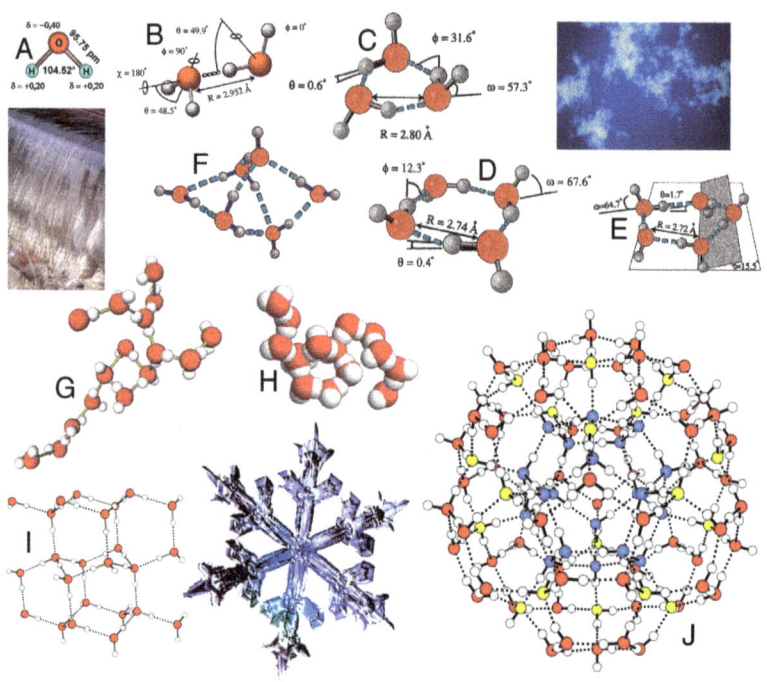

Figure III.7. A short overview of the current scientific knowledge concerning water structure. Ouroboros-type (A-F) in the vapor state.[67] Serpent-type in the liquid state as dynamic (G) and static (H) water clusters evidenced by molecular dynamics studies.[64] Solid state structures in hexagonal ice (I) and in encapsulated $\{H_2O\}_{100}$ (J) nano-assemblies.[68]

through their properties and are no more pure abstract concepts as in the first case. In order to describe correctly this completely new situation Scottish philosopher James F. Ferrier, coined in 1856 another word **epistemology**, from Greek *epi* « over », *histasthai*, « to stand » and *logos*, « word, idea or reason ». Literally, an epistemological statement allows a sensible being named *observer* to **stand over the ideas** giving him practical rules for manipulating the reality by giving him limited access to mental ontological quantities. Knowing these practical recipes, it becomes possible to predict, but with no absolute certainty, how a giving event may affect the sensible world in which the observer is living. The tricky point is that the same root logos is used in both words showing that the same mathematical language has to be used to describe such different aspects of reality, ontological laws of Nature and observers' epistemological recipes. Thus, the correct way of reversing the *water is life* statement is to say that *life is more than water*. At the mathematical level we have move from an implication (causal ontological statement: water $\Rightarrow$ life) to an inequality (conditional epistemological statement: life > water). The heart of the problem is thus the fascination of the human mind for perfect ontological **symmetry** (if $A = B$, then $B = A$) by opposition to the epistemological **asymmetry** (if $A \Rightarrow B$, then $B > A$) of the human brain. This is basically the reason why commutators $[A,B] = AB - BA$ are so important in Science leading to uncertainty relations because if I am « *the son of my mother* », I cannot also be at the same time « *the mother of my son* ». This is also why biochemistry rely so heavily on chirality and why Master Dogen, in his water sutra is saying: "*There is a study of water seeing water. Water practices and verifies water. Here, we can say that there is a practice of water speaking water... It is not only that there is water in the world, but there is a world in water...*".[51] It should be now obvious that he explicitly refer to the existence of two kinds of water, one existing at the ontological level, the immaterial water concept giving life by its structure in space and time, and the other one at the epistemological level, the material thing that scientists represents by the symbol H_2O.

Having identified the problem, we may go back to the second law of thermodynamics under its original statement by Rudolf Clausius: "*Die Entropie der Welt strebt einem Maximum zu*", or « The entropy of the world tends towards a maximum ». The use of the verb « tends » strongly suggests the epistemological nature of this second law while its mathematical expression through a well-known inequality $\Delta S \geq 0$

finishes the demonstration. Realizing that entropy cannot be of ontological nature, being a pure epistemological quantity accessible only through experiments in a sensible world has great implications in biology. First, it allows to identify those reproducible **macroscopic** experiments that can be repeated many times leading always to the same results owing to the existence of two equilibrium states obeying the rule $\Delta S = S_{final} - S_{initial} \geq 0$. Other non-equilibrium states for which entropy cannot be defined, may be observed but cannot be controlled or reproduced. Basically, an increase in entropy is the price to pay for being able to manipulate and control a macroscopic system in a reproducible way. It is a kind of tax perceived by Nature for every muscle contraction or matter transport. The living entity pay the tax and in return there is a guarantee by Nature expressed by Liouville's theorem $(dW_N/dt = 0)$, that the organism will not be projected in an unwanted place or time, nor torn apart by the powerful forces at work in Nature. If the tax is not paid, living is still possible (freedom exists at the epistemological level) but there is absolutely no guarantee that what have occurred in a first trial would be observed many times in the future! This is basically the reason why life cannot be observed with molecules but only with macromolecules. With macromolecules, the number of degree of freedom becomes large enough for the entropy tax to be perceived on a single object (protein, enzyme, nucleic acid) rather than from an uncontrollable chaotic manifold of small molecules.

The importance of information theory

The link existing between this epistemological entropy and **information theory**[73] may also be perceived through the use of Boltzmann's formula $S = k_B \cdot \ln W$, where k_B stands for Boltzmann's constant $(1.380658 \cdot 10^{-23}$ J.K$^{-1})$ and W is the number of different ways of realizing a given macroscopic situation from the definition of an ensemble of possible ontological microstates. As beautifully explained 75 years ago[74] it is worth considering the case of a single molecule placed in a box of volume V in contact with a thermostat T and equipped with an impermeable piston having two compartments of equal size named A (left) and B (right). Assuming that the molecule cannot escape from the box, this system is characterized by only two microstates: molecule in A or molecule in B, that is to say $W = 2$. At the beginning, the piston is assumed to be at the extreme right of the box so that it is impossible to know in which compartment is located the molecule owing to its erratic unpredictable movement, a situation described by assigning an entropy $S_{gas} = k_B \cdot ln\ 2$ to this « gas ». Now, the piston is moved very slowly towards the left until the middle of the box is reached. At that time the molecule is known to be in the left compartment and so $W = 1$ meaning that $S_{gas} = 0$. As this process is typical reversible the work required to overcome the pressure exerted by the molecule should then be $k_B T \cdot ln\ 2$. This shows that in going from an unknown distribution into a known one, work has to be done and as a consequence there is a decrease in entropy of the gas by $k_B \cdot ln\ 2$. Opening a frictionless shutter in the piston would now lead to an irreversible diffusion of the molecule in order to occupy the whole available volume V, with at equilibrium an increase of entropy by $S_{gas} = k_B \cdot ln\ 2$. Consequently, when going irreversibly from a *known* distribution to an *unknown* distribution there is an increase of entropy. In other words, **gain in entropy** always means **loss of information** and nothing more.[73] More particularly, existence of irreversible processes neither implies one-way time, nor has any temporal implications. The second law has nothing to say about the rate or the path followed by a given transformation, statements of the kind $dS/dt \geq 0$ being meaningless and not deducible from the definition of entropy. The reason why so much confusion lurks around the entropy concept probably comes from a very unhappy choice of terminology by Clausius, who forged this word from Greek *tropê*, meaning « round, revolution, turn, circuit, rotation », in order to look and sound like the word « energy » previously coined by William Thompson (future Lord Kelvin) from Greek *en* meaning « in » and *ergon* meaning « action, work ». His idea was probably to condense in a single word Carnot's principle stating that during a large number of Carnot cycles, the heat *dQ* exchanged at temperature T during each cycle was such that $\oint dQ/T \leq 0$ for any heat engine, the equality holding only in the case of reversible engines. Clausius was also appealed by the fact that in Greek the word *tropê* may also be interpreted depending on the context as: « change, transformation, evolution, conversion » leading after combination with the Greek word *en* to the idea of **content of transformation**, a striking parallel with the idea of **content of action** expressed by the word « energy ». But semantic problems then immediately arise by noting that another Greek word *tropos* meaning « direction, kind, fashion » exists. Thus, further association with *en* would suggest existence of a **time arrow** associated with entropy *(dS/dt ≥ 0)*, a statement in flagrant contradiction with information theory. The fact that the Greek word *entropê* also exists meaning « confusion, mind reversal » adds some spicy flavor to this incredible entanglement of different concepts in a single word. This is probably why when mathematician John von Neumann was asked by Claude Shannon what name should he give to his measure of the information content of a signal, Von Neumann replied rather coyly « *You should call it entropy ... [since] ... no one knows what entropy really is, so in a debate you will always have the advantage!* ».[75] It was a very bad joke, but Shannon apparently took him seriously, adding still more confusion by calling entropy what was in fact **neguentropy**, i.e. entropy with a minus sign.[76]

When it is accepted that entropy gain means information loss, a useful link may be established with quantum mechanics. Let's consider again the case of a box containing a single molecule. If the temperature is fixed, the ignorance of its exact location means that W $\propto$ V and that its momentum will vary according to some probability distribution characterized by its width Δp reflecting the ignorance of its exact velocity. For a 3D-box, having the same extension Δx along the three direction, it is possible to write $V = (\Delta q)^3$ and the measure of this lack of information concerning position and momentum may be reflected by a phase volume accessible for the molecule $W \propto (\Delta q \Delta p)^3$. This phase volume is clearly of epistemological nature owing to the absence of equality sign and reference to observable quantities: size of the box Δq and an uncertainty Δp that can be measured. Yet, application of Boltzmann's formula is impossible because the logarithm function is an abstract mathematical function that should only apply to abstract (i.e. unitless) quantities. However, as $ln\ 0$ cannot correspond to a physical situation the existence of a minimum action unit $\hbar$ is a necessary condition for a correct definition of the state of minimal entropy *(S = 0)* characterized by the situation *W = 1*. It is worth noting that this normalization is necessary only because we are interested in a material box having physical walls enclosing a molecule (epistemological level) and not at all by the abstract associated problem where x and p are mere numbers attached an hypothetical particle submitted to a potential energy U such that $U_{inside} = 0\ and\ U_{outside} = \infty$ (ontological level). Staying at the epistemological level allows then to compute the entropy as $S = k_B \cdot ln(W/\hbar^3) = 3k_B \cdot ln(\Delta q \Delta p/\hbar)$ with as expected *S = 0* when $\Delta q \Delta p = \hbar$. Starting from this state of maximal information and letting the molecule free to move at temperature T until equilibrium is reached would correspond to a variation in entropy $\Delta S = 3k_B \cdot ln(\Delta q \Delta p/\hbar) \geq 0$, or as $ln\ x$ is a monotonic function of *x:* $\Delta q \cdot \Delta p \geq \hbar$. From this rigorous derivation of **Heisenberg's inequalities** from entropy considerations, it should be clear that classical and quantum mechanics are not separate disciplines with the latter more correct than the former. The only difference lies in their position in collecting information, with classical mechanics claiming **theoretically infinite information** that is available only at the ontological level in observer's mind without tax. By contrast, quantum mechanics claims only **concrete finite information** that is available from measurements at the epistemological level after having paid the tax. The problem with quantum mechanics is that both level of reality are completely mixed. Accordingly, only epistemological wave functions ψ are written in equations whereas interpretative words between those equations use only the language of ontological particles. This situation has led to the troublesome **wave-particle duality** that may just reflect the inability of scientists to decide what they are talking about. This derivation is also enlightening because it allows to state that without Heisenberg's inequalities, macroscopic processes mediated by macromolecules would not be reproducible and thus that no life would be possible. This is a much more positive attitude than the usual disappointment after learning the impossibility to measure simultaneously both position and momentum.

Conclusion

Several points are worth emphasizing after this study of the status of water in the living cell. First, the recognition of two levels of reality (ontological by the mind and epistemological by the body) is of crucial importance for realizing that matter does not move under the influence of energy, but rather under the influence of information. This is because energy is an ontological concept that remains the same whether matter is moving or not, while there is **always** an irreversible loss of epistemological information after diffusion has occurred. The concrete thing that is happening during spontaneous diffusion is that irreversible information loss leads to an increase in entropy from which heat is generated allowing extraction of work used for moving inert matter. In other words, it is because equilibrium distributions always correspond to minimum information (or maximum entropy) that molecular things have tendency to finish by being mixed up. The great error is to assimilate minimum information states and hence maximum entropy to **disorder**. This subtle point may be most easily perceived by considering a small RNA fragment carrying the information AGCTAGTGCATCTAGC corresponding according to C. Shannon to an information content: $I = -16 \cdot ln(1/4)/ln\ 2 = 32$ *bits*. As there is a material support (here the sheet of paper or the sugar/phosphate backbone in the case of a real cell) the minimum entropy tax that has to be paid for writing or reading this information should be $S = 4.4 \cdot 10^{-22}$ $J \cdot K^{-1}$, corresponding to a dissipated heat of at least 0.8 eV. Keeping the same length as in the sequence AAAAAAAAAAGGCCTT shows that there is, as expected, less information relative to the first gene: $I = - 16 \cdot [5 \cdot ln(5/8)/8 + 3 \cdot ln(1/8)/8]/ln\ 2 = 25$ bits and consequently, less heat (at least 0.6 eV) would be dissipated for manipulating it. The key point is that the sequence AGATACATAGAACAAA contains also the same 25 bits of information and has thus exactly the same thermodynamic entropy, despite the fact that it is obviously more disordered than the one displaying the same number of A,G,C and T symbols.

In fact both messages carry **subjective information** through the interpretation attached to each sequence that has nothing to do with thermodynamic entropy considerations. Consequently, assimilation of entropy to disorder lead to the disastrous belief that at a macroscopic scale, things are able to move spontaneously leading after waiting a time long enough, to a state where disorder is maximum! This is a very perverse interpretation of the meaning of the second law that opens full the road for the belief in the existence of an omnipotent God, who by using its divine finger, forces things and facts to proceed into the **good direction**, that is less disorder. It is worth stressing that the reality is quite different and first that second law limitation on macroscopic processes $(\Delta S \geq 0)$ is easily understood in objectively meaningful terms, in both biology and physics, as the price to pay for **reproducibility**. Second, the bigger the object the more reproducible would be its behavior, as if it is too small (molecule), Brownian motion would make its behavior erratic and irreproducible. But, if it becomes too big, frictional forces, resulting from additive van der Waals attractions, with other macroscopic objects gravitation and, at a still larger scale, gravitation (i.e. space-time curvature) will exert non-negligible external forces (not really God's finger). Consequently, any initial movement tends to be irremediably stopped and external work has to be done in order to keep such macroscopic objects moving. This means that if disorder, a concept relevant only at a macroscopic scale, arises it is because earthquakes, hurricanes or ATP molecules burning in muscles that **forces** objects to become distributed according to the most probable configuration.[45] Obviously the macroscopic human brain does not easily perceive such a reversal in causality, with macroscopic objects moving under the influence of energy and not of entropy and molecules moving under the influence of entropy and not of energy. This is still more difficult for the human mind ruled by an ontological determinism requiring either the existing of an external force (God's finger) or the absence of such forces (so nothing can change). This is why it is so useful to discuss matter movement neither in terms of energy nor entropy but in terms what's is happening to **information**. Considering only information, there is no need to reverse the way of thinking (cf. the Greek word *entropia*) on going from the macroscopic scale to the microscopic one, as now inert matter should move as soon as information is either created (energy dominant) or erased (entropy dominant). The good thing is that this fundamental rule apply not only to **objective information**, an epistemological concept related to entropy and evaluated by Shannon's formula, but also to **subjective information**, an ontological concept related to energy that can be only mentally evaluated or manipulated. This new attitude opens very interesting perspectives in biology despite the trivial fact that living organisms does not violate anymore the second law by their mere existence. It clearly shows that observed concentration gradients, osmotic effects or ATP-activity are not the **active cause** for the sustaining of life, but rather the **passive consequence** of the constant information processing occurring within the living cell. It should then become obvious to see that water with its highly structured percolating H-bond network, is the physical medium onto which information is constantly written, read or erased, insuring that any short-range matter movements would induce long-range information transmission. This is why there is so much water in the cell and why the brain is almost pure water, as it must processes both material objective information and mental subjective information. This is also why macromolecules are so densely in a cell packed in order to prevent any long-range matter movement or diffusion that would erase irreversibly part of the vital information that is needed for keeping the cell into a healthy state.

But it is in medicine that this change of paradigm should have the most fruitful consequences, by making realize that the key-lock mechanism (drugs' administration) or the chirurgical intervention (using instruments or radiations) are just two active ways of healing a cell that is no more able to process information correctly. Relying on passive entropic forces mediated by water, is also quite valuable and should not be neglected nor treated scornfully. In fact, the most difficult point for any physician facing a disease should be to recognize when to act actively by adding new information and when to wait passively that entropy makes its natural job of erasing the wrong information at the origin of the illness. Obviously, dealing only with objective information this kind of medicine whether active or passive remains entirely in the domain of epistemology. Moving to the level of ontology, immaterial subjective information should also be taken into consideration opening quite new, but also quite dangerous, perspectives for medicine. Because there is a complex interweaving between an epistemological body and an ontological mind in Man, it is quite difficult to treat both aspects separately and this may well be the real challenge for any kind of medicinal approach.

The above conclusions also help to reconcile Science with Philosophy. As stated above, the big problem with scientists is that they oscillate in permanence between an epistemological standpoint by making experiments and an ontological viewpoint by interpreting their observations. The previously reported discussion between R.A. Gortner and A.V. Hill about the state of water in the jellyfish is a good example.

A.V. Hill was clearly speaking using ontological terms (**I believe that** ...) while R.A. Gortner was responding with epistemological arguments (**I observe that** ...). This was the very reason why they could not find a common ground of understanding. Among other famous examples, the dispute opposing Albert Einstein (an ontologist) to Niels Bohr (an epistemologist) is worth quoting. This is a very dangerous game, because any crisscrossing between the two levels is immediately sanctioned by a paradox leading to infinite discussions between two irreconcilable factions. A simple look at the abundant scientific literature shows that paradoxes are indeed not very rare: Gibbs paradox and Maxwell's daemons in thermodynamics, Zeno's paradox in mechanics, Schrödinger's cat and Einstein-Podolsky-Rosen (EPR) paradox in quantum mechanics, time paradoxes in astronomy, Liar's paradox in mathematics, Darwinian evolution and genetic paradoxes in biology, etc... By contrast, philosophers are more immune to paradoxes because they never make experiments staying as much as possible at the ontological level. They are nevertheless doomed to the same fate as scientists as soon as they introduce some epistemological considerations in their way of thinking. It is thus quite satisfying to see that both Science and Philosophy agree on the concept of information and on the ontology/epistemology dualism. This was well shown in Master Dogen's water sutra with, "*water seeing water or water speaking water*". Information was also present but kept hidden in Thales of Miletus motto, *All is water, all is one*, that may be tentatively translated as, *Life is information*, in Science's language. This is why the creator says, "*Let's eat fly-agaric*", in the tale of the glibly Raven *Ku'urkil*, pretending to be an ontological self-created creature where as it was only the epistemological "*son of the ni'glon*". And what is **Tao** if not a perfect equilibrium ($0 = 1 - 1$) between a **conservative** ontological information principle (+1), called Yang in Taoism or energy in Science, and a **dissipative** epistemological information principle (-1), called Yin in Taoism and entropy in Science? Obviously, it is because Yin

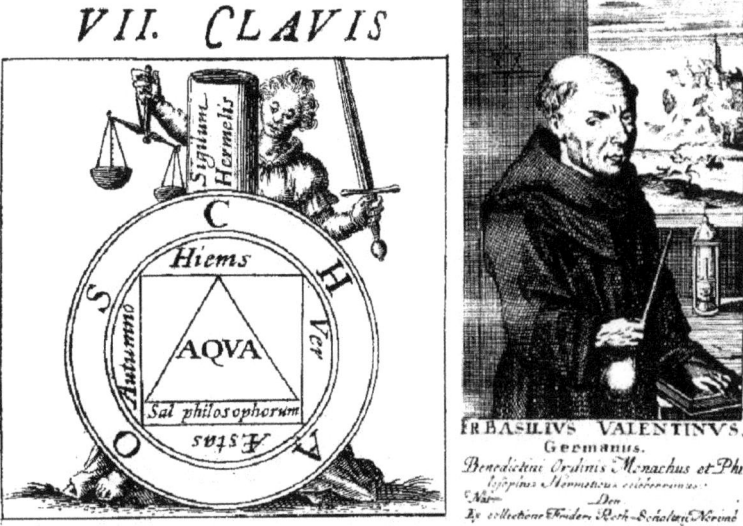

Figure III.8. The seventh key of Basile Valentine, the mythical Alsacian monk (right) thought to have lived in Ehrfurt during the XVth century.[77] This beautiful emblem centered on water recalls the two levels of comprehension of the universe, ontological by the sword and epistemological by the scale, the triangular shape of the water molecule, the existence of random information (chaos) and the concept of sensible time through the succession of the 4 seasons (square) involving a double duality Fire (Hot/Cold) and Water (Dry/Moist) united by the symbol of the Ouroboros (circle).

erases information that Yang is obliged to move $[1 + (-1+1) = (1-1) + 1]$, as if Yin is absent Yang can only stay immobile $(1 + 0 = 1)$. The exchange of round brackets in the previous expression symbolizes what is called the **breathing of the vacancy** in Taoism, a concept called **pacha** by Incas and **time** by Science. So Tao (0), the *random information*, creates one (1), the *oriented information*, One creates Two (1-1), the **neutralized information** symbolized by the Ouroboros, Dogon's Nummo or the egg and Two creates Three $[1 + (-1+1) = (1-1) + 1]$, the *living information*. Three creates every living thing because Yin,

Yang and Tao defines altogether a dynamic, through a regular alternation of 0 and 1, named **God** in most religions and **big bang** in Science. But more generally, the familiar introduction of most myths: *"In the beginning ..., or Bereschit ..., or En archê ..., or In principio ..."* has also clearly two levels of interpretation. At the ontological level or Aborigines' dreamtime, it may have the sense « In principle » or « By equivalence », suggesting symmetry, eternity, immobility, potentiality and thus reversibility. Keeping the same words but just switching at the second epistemological level automatically gives the usual sense of the beginning of something, implying symmetry breaking, creation, time flowing, movement, realization that is to say irreversibility. Everything becomes clear and limpid after taking conscience of the intrinsic duality of all reality. That's the message coming from water study (Figure III.8) and that's a real shame for mankind that Science persists to waste time in resolving paradoxes that arises because both levels of interpretation are confused into a single concept named **reality**.

References

1. Ovid, *Les Metamorphoses*, Panckoucke, Paris, **1667-1771**.
2. Ruhlen, M. *The Origin of Language*, John Wiley & Sons, New York, **1994**, pp. 107.
3. Mann, F. D. *Perspect. Biol. Med.* **1976**, *20*, 142.
4. Gortner, R. A. *Trans. Faraday Soc.* **1930**, *26*, 678.
5. Yusufali, A. *The Qur'an Translation 5th Ed.*, Tahrike Tarsile Qur'an, New York, **2000**, surah 21, "al-anbiyah", verse 30.
6. Rundle Clark, R. T. *Myth and Symbol in Ancient Egypt*, Thames & Hudson, London, **1959**, p. 74.
7. Beale, N. *Encyclopédie de la Mythologie*, Cotterell, A. (ed.), Celiv, Paris, **1996**, p. 193, p. 250.
8. Hauck, D. W. *The Emerald Tablet. Alchemy for personal transformation*, Penguin Arcana, London, **1999**, cover picture.

9. Budge, E. A. W. *The Papyrus of Ani, British Museum*, London, **1895**, pp. xciii-ci.
10. Jastrow Jr. M. *The Civilization of Babylonia and Assyria*, J.B. Lippincott, Philadelphia, **1915**, pp. 428.
11. Sturluson, S. *The Prose Edda: Tales from Norse Mythology*, J. I. Young translator, University of California Press, Berkeley, **1964**, Gylfaginning.
12. Holmberg, U. *The Mythology of All Races, 4. Finno-Ugaric Siberian Mythology, Marshall Jones*, Boston, **1927**, pp. 317.
13. Bogoras, W. *The Jessup North Pacific Expedition*, Vol. VIII. I. Chuckchee Mythology, Boas, F. (ed.), Memoir of the American, Museum of Natural History, Leiden, **1910**, p. 154.
14. Gödel, K. *Monatsh. Math. Phys.* **1931**, *38*, 173.
15. Gödel, K. *On Formally Undecidable Propositions of Principia Mathematica and Related Systems*, Meltzer, M. Translator, Dover Publ., New York, **1992**.
16. Czaplicka, M.A. *Shamanism in Siberia*, Clarendon Press, Oxford, **1914**, Chap. XIII-3.
17. Hosoda, T. *Kamiya no Masagoto Tokiwagusa*, Kyoto, **1827**.
18. Griffith, R. T. H. *The Hymns of the Rig-Veda*, London, **1889**.
19. Eggeling, J. *The Satapatha-Brahmana according to the Text of the Mâdhyandina School, Sacred Books of the East*, Vol. 44, Müller, M. (ed.), Clarendon Press, Oxford, **1900**, p. 12.
20. Bühler, G. *The Laws of Manu, Sacred Books of the East*, Vol. 25, Müller, M. (ed.), Clarendon Press, Oxford, **1886**.
21. Bloomfield, M. *Hymns of the Arthava-Veda, Sacred Books of the East*, Vol. 42, Müller, M. (ed.), Clarendon Press, Oxford, **1897**.
22. Lao-tzu, *Tao te Ching, Sacred Books of the East*, Vol. 39, Müller, M. (ed.), Legge, J. (Transl.), Clarendon Press, Oxford, **1891**.
23. Seife, C. *Zero. The Biography of a Dangerous Idea*, Viking Penguin, Toronto, **2000**, Chap. 1.
24. Nishijima, G. W. *Master Dogen's Shinji Shobogenzo: 301 Koan Stories*, Leutchford, M. and Pearson, J. (eds.), Windbell Publ., **2003**, koans 3 and 14.
25. Brians, P.; Gallwey, M.; Hughes, D.; Hussain, A.; Law, R.; Myers, M.;

Neville, M.; Schlesinger, R.; Spitzer, A.; Swanin, S. *Reading about the World, Vol. 1*, Harcourt Brace Custom Publ., 3rd ed. New York, **1999**, Japanese Creation Myth.

26. Tedlock, D. *Popol Vuh: The Definitive Edition of the Mayan book of the Dawn of Life and the Glories of Gods and Kings*, Simon and Schuster, New York, **1985**, p. 72.
27. Spencer, B.; Gillen, F. J. *The Northern Tribes of Central Australia*, London, **1904**, p. 740, Figure 312.
28. Spencer, B.; Gillen, F. J. *The Native Tribes of Central Australia*, London, 1899.
29. Tacon, P. S. C. *Austr. Canad. J. Native Stud.* **1989**, *9*, 317.
30. Griaule, M. *Conversations with Ogotommeli*, Oxford University Press, Oxford, **1975**.
31. Gortner, R. A. *Outlines of Biochemistry*, Wiley and Sons, New York, **1929**.
32. Miller, W. L. *Phys. Rev.* **1930**, *35*, 566.
33. Cramer, W. *Trans. Faraday Soc.* **1930**, *26*, 686.
34. Kruyt, H. R. *Trans. Faraday Soc.* **1930**, *26*, 689.
35. Lecomte du Noüy, P. *Trans. Faraday Soc.* **1930**, *26*, 693.
36. Hill, A. V. *Trans. Faraday Soc.* **1930**, *26*, 687.
37. Pfeffer, W. *Osmotische Untersuchungen*, Engelman, Leipzig, **1877**.
38. Grollman, A. *J. Gen. Physiol.* **1931**, *14*, 661.
39. Moran, T.; Smith, E. C. *Trans. Faraday Soc.* **1930**, *26*, 695.
40. Gortner, R. A.; Gortner, W. A. *J. Gen. Physiol.* **1934**, *17*, 327.
41. Szilard, L. *Zeits. Physik* **1929**, *53*, 840.
42. Overton, C. E. *Viertel. Naturforsch. Ges. Zürich* **1899**, *44*, 88.
43. Overton, C. E. *Pflüger's Arch. Ges. Physiol.* **1902**, *92*, 346.
44. Huxley, A. F. *J. Physiol.* **2002**, *538.1*, 2.
45. van't Hoff, J. H. *Z. Phys. Chem.* **1887**, *1*, 481.
46. Kedem, O.; Katchalsky, A. *Biochim. Biophys. Acta* **1958**, *27*, 229.
47. Dainty J. *Adv. Bot. Res.* **1963**, *1*, 279.
48. Jaynes, E. T. *Maximum Entropy and Bayesian Methods*, Skilling, J. (ed.), Kluwer Acad. Publ., Dordrecht, **1989**, pp. 1.

49. Agutter, P. S.; Malone, P. C.; Wheatley, D. N. *J. Hist. Biol.* **2000**, *33*, 71.
50. Fick, A. E. *Ann. Phys. Leipzig* **1855**, *94*, 59.
51. Hill, A. V. *Proc. Roy. Soc. London A* **1930**, *127*, 9.
52. Lambert, F. L. *J. Chem. Educ.* **1999**, *76*, 1385.
53. Mencius, *Chinese Classics*, Vol. 2, James Legge, J. (translator), Oxford University Press, Oxford, **1895**, chap. 12.
54. Mentré, P. *L'eau dans la Cellule*, Masson, Paris, **1995**.
55. Blakeley, M. P.; Kalb-Gilboa, A. J.; Helliwell, J. R.; Myles, D. A. A. *Proc. Natl. Acad. Sci. USA* **2004**, *101*, 16405.
56. Royer Jr., W. E.; Pardanani, A.; Gibson, Q. H.; Peterson, E. S.; Friedman, J. M. *Proc. Natl. Acad. Sci. USA* **1996**, *93*, 14526.
57. Soler-Lopez, M.; Malinina, L.; Subirana, J. A. *J. Biol. Chem.* **2000**, *275*, 23034.
58. Michalarias, I.; Xiuli Gao, X.; Robert, C.; Ford, R. C.; Li, J. *J. Mol. Liq.* **2005**, *117*, 107.
59. Ruffle, S. V.; Michalarias, I.; Li, J.-C.; Robert, C.; Ford, R. C. *J. Am. Chem. Soc.* **2002**, *124*, 565.
60. Mentré, P. *Cell. Mol. Biol.* **2001**, *47*, 709.
61. Gerstein, M.; Levitt, M.; *Scientific American* **1998**, November issue, 101.
62. Hofacker, I.; Schulten K. *Proteins: Structure, Function, and Genetics* **1998**, *30*, 100.
63. Robinson, G. W.; Cho, C. H. *Biophys. J.* **1999**, *77*, 3311.
64. Stanley, H. E.; Buldyrev, S. V.; Giovambattista, N.; La Nave, E.; Scala, A.; Sciortino, F.; Starr, F. W. *Physica A* **2002**, *306*, 230.
65. Wiggins, P. M. *Physica A* **2002**, *314*, 485.
66. Wiggins, P. M. *Physica A* **1997**, *238*, 113.
67. Keutsch, F. N.; Saykally, R. J. *Proc. Natl. Acad. Sci. USA* **2001**, *98*, 10533.
68. Henry, M.; Bögge, H.; Diemann, E.; Müller, A. *J. Mol. Liq.* **2005**, *118*, 155.
69. Müller, A.; Henry, M. *C. R. Chimie* **2003**, *6*, 1201.
70. Müller, A.; Bögge, H.; Henry, M. *C. R. Chimie* **2005**, *8*, 47.
71. Szent-Györgyi, A. *Science* **1941**, *93*, 609.
72. Szent-Györgyi, A. *Cell-Associated Water*, Drost-Hansen, W. and Clegg, J.S. (eds.), Acad. Press, New York, **1979**, preface.

73. Rothstein, J. *Science* **1951**, *114*, 171.
74. Lewis, G. N. *Science* **1930**, *71*, 569.
75. Tribus, M.; McIrvine, C. E. *Sci. Am.* **1971**, *225*, 179.
76. Brillouin, L. *La Science et la Théorie de l'Information*, Masson, Paris, **1959**.
77. Maier, M. *Tripus Aureus, hoc est, Tres Tractatus Chymici Selectissimi*, Paul Jacob for Lucas Jennis, Frankfurt, **1618**, VII clavis.

IV

Water, salt and oil: an exploration of the foundations of molecular forces

Barry W. Ninham[§,], Pierandrea Lo Nostro[‡,#]*

§: Department of Applied Mathematics,
Research School of Physical Sciences and Engineering,
Australian National University, Canberra, Australia 0200.
barry.ninham@anu.edu.au

‡: Department of Chemistry "Ugo Schiff" and CSGI,
University of Florence, 50019 Sesto Fiorentino (Firenze), Italy.
#: Enzo Ferroni Foundation, 50019 Sesto Fiorentino (Firenze), Italy.
pln@csgi.unifi.it

Our commemoration of the celebrated dispute on water 400 years ago, with Galileo the key player, suggests that we might profitably take a longer view on issues in physical chemistry than usual. Here are some key events.

1.1 Early ideas

The Library of Ashurbanipal: Nineveh 700 BC.

The first scientific paper we know of was discovered by David Tabor in the Ashmolean Museum in Oxford.[1] It dealt with the spreading of oil on water. The "paper" was in cuneiform on clay tablets, part of the vast library of the Assyrian biblical king. Although devoted to the same topic that interested Ben Franklin, the Assyrian Priest's interest was in necromancy, a black art to facilitate communication with the dead. It was used last by Ulysses in the *Odyssey* to facilitate a visit to Hades around the same time.

We have lost the art, but not interest in what we now call hydrophobic forces.

1.2 Newton, and Bošković

Another event was Newton's comment in Article 31 in the *Principia*; he attempted to measure (molecular) forces between surfaces, but failed saying "Surface combinations were owing". That is, contamination defeated him, and the matter had to wait until the work of Israelachvili, Tabor and colleagues 300 years later.

But Newton did deduce that attractive intermolecular force potentals were $\sim r^{-6}$, ahead of Thomas Young and van der Waals by 200-300 years.

Ruder Bošković, a Croatian priest working in Rome published a system of the world (in Latin) based on molecular forces whose interactions had oscillatory pair distribution functions, anticipating modern statistical mechanics. Mention of his name apparently sent the normally phlegmatic J. Clerk Maxwell into a fury. The Ruder Bošković Institut in Zagreb is named in his honour.

1.3 Berthollet and the foundations

The foundations of physical chemistry can arguably be assigned to Claude-Louis Berthollet, a French scientist who was a member of Napoleon's expedition to Egypt in 1795.

He observed on the banks of the Nile hard shining rocks of soda lime, sodium carbonate.[2] Everyone knows that calcium carbonate (limestone) should precipitate out and sodium chloride stay in solution. The reason for the reversal is that cooperative water structure changes with temperature for reasons not understood. Perhaps selective nucleation is affected by atmospheric gas. (The temperature reaches greater than 60° C in the heat of summer and dissolved gas depends on temperature and salt).

This was the defining moment. Physical Chemistry depended on temperature, and thermodynamics came into existence.

1.4 Hydration, Poisson and Laplace

A concept of key importance for water and forces was a dispute between Poisson and Laplace in 1831, following Thomas Young's great paper on surface tension of 1805. Laplace stole Young's paper without attribution and added it to the appendix of volume 6 to his treatise on the system of the world (on planetary motion). Young's biographer (Professor George Peacock) in the 1850s was apoplectic at Laplace's (French) perfidy. The Laplace-Young assumption was that a liquid retained its bulk properties to an infinitesmal (i.e. atomic) distance from a surface.

Poisson insisted that a surface ought to interact with the liquid and induce a profile of density, and liquid order (in our language dipole moment, hydrogen bonding) extending over several molecular layers. Here then we had hydration. At a solid flat surface, or at the surface of a molecule or protein, hydration is a core concept.

The current state of knowledge on molecular forces and hydrodynamics of colloidal suspensions and the dispute was reviewed by the Rev. Challis of Trinity College (Newtons' College) Cambridge in two papers given at the British Association meeting in 1836. In these he introduced the term Mathematical Physics for "this, the highest Department of Science". In other words mathematical physics originally meant what we mean by colloid and surface science!

Poisson lost out to Laplace and Young because of a mistake of a factor of 2 in his analysis, and due also to too rigorous an application of Ockham's Razor. (The simplest proposition is not always necessarily the correct one, contra Ockham).

1.5 Maxwell to the rescue

Honor was restored by the publication of J. Clerk Maxwell's great paper on Capillary Action in 1876 published in the 9th edition of the *Encyclopaedia Britannica*. He showed that Poisson was right, and calculated the range of the exponentially decaying profile of liquid order at a surface and the size of the forces between surfaces. He introduced many modern concepts like a Landau-De Gennes Hamiltonian and scaled particle theory and anticipated by exactly 100 years the same theory, rediscovered by Marcelja in 1976!

Hydration forces were quantified by theory and direct force measurements confirmed this at the same time around 1976, (and actually at and between molecularly smooth surfaces were indeed oscillatory as Bošković intuited).

2.1 Enter Hofmeister and specific ion effects (1870s)

All chemistry is specific, all salts specific.

An awareness that salt water systems should show some systematics was first anticipated by Poisson in work on viscosities in the 1830s.

Then came Hofmeister, a pharmacologist.

His work on the amount of salt required to precipitate from solution a protein egg white lecithin showed that there is an ordering of effectiveness.[3]

As measured by concentration, the order is:

At fixed cation:

$$CO_3^{2-} > SO_4^{2-} > S_2O_3^{2-} > H_2PO_4^- > F^- > Cl^- > Br^- \approx NO_3^- > I^- > ClO_4^- > SCN^-$$

At fixed anion:

$$NH_4^+ > K^+ > Na^+ > Li^+ > Mg^{2+} > Ca^{2+} > C(NH_2)_3^+$$

Anions were apparently more effective than cations.

The Hofmeister sequence applied to other colloids, and to this day is often wrongly assumed to be a given for all systems.

Soon after, reverse Hofmeister series appeared for other systems. Hofmeister's work remains as a central challenge. It is rediscovered in every decade (It ranks in importance with Mendel's work in genetics, likewise forgotten for decades). The reason is clear. Theories of electrolyte solutions that took root over the next 150 years like Debye-Hückel and double layer and the DLVO theory do not accommodate Hofmeister effects. They were an embarrassment to theorists, and – deviating from theory -could be then dismissed as "specific" ion effects.

Hofmeister remained bemused about whether his effects, a consequence of whatever molecular forces operated between the proteins and salts in water should be attributed to short range effects (hydration), or to some unknown very long range water structure induced by the salts. See Marc Henry's and other papers in this volume.

2.2 Enter biologists

The notion of recognition, of *the geometry of molecules* (lock and key) is comprehensible. But the specificity of *long range forces* that drive molecular recognition, and hence physical association of macromolecules, remained a mystery. The source of reproducible (chemical) energy to drive enzymatic catalysis is a key problem, obviously. To understand such forces was D'Arcy Thompson's plea, and that of the founders of the cell theory of biology, and of the physiologists, from the mid 19th century. It was the main aim of F. W. Ostwald and his students. They included Arrhenius, van't Hoff, Nernst and W. Ostwald, the founder of colloid chemistry. The first three were among the first Nobel prize winners as later was the senior F. Ostwald. His interests included electrochemistry, and enzymes that he recognised were catalysts, both ion specific.

3.1 Lebedev 1894: the notion of cooperativity in forces

So far then, conceptually, pre-quantum mechanics, we had three concepts: water as a continuum between two interacting bodies, action at a distance but the water actually irrelevant; water with very long range "structure"; and short ranged hydration water mediating unknown forces that were ion specific. And then our hydrophobic, repulsive necromantic forces between oil and water.

Lebedev, a biophysicist, friend of Clerk Maxwell and step-father of Deryaguin, the leader of the Russian School for a half a century later, discovered light radiation pressure. He articulated this vision of the problem of how molecules recognise each other, "feel their vibes" as it were: "Of special interest and difficulty is the process that takes place in a physical body when many molecules interact simultaneously, the oscillations of the latter being interdependent due to their proximity. If the solution of this problem ever becomes possible we shall be able to calculate the values of the intermolecular forces due to molecular inter-radiation, deduce the laws of their temperature dependence, solve the fundamental problem of molecular physics whether all the so-called "molecular forces" are confined to the known mechanical action of light radiation, to electromagnetic forces, or whether forces of unknown origin are involved".

This succinct grand conception of the cooperativity of molecular forces bridged the unbridgeable, atoms on one hand and continuum media on the other, long and short range.

Cooperativity and how macromolecules like proteins send out and feel their vibes mediated by water was explicit.

Half a century later the vision was accomplished by Lifshitz (in theory) and Lebedev's stepson Deryaguin (in experiment), later by Israelachvili and colleagues.

3.2 A Gallimaufry of theories.

The textbook theories that we are familiar that had grown up over the last century are necessarily a mish mash.

Quantum mechanics and the quantification of the idea of a chemical bond in chemistry arrived on the scene only around 1940.

The theories for water and electrolytes were (and are still) based at first on electrostatic interactions only, typically extensions of the Born free energy of an ion, Debye-Hückel Law for activities, Onsager theory for interfacial tensions, and the double layer theory for electrolytes at interfaces.

The DLVO theory for colloid stability balanced these electrostatic forces against attractive quantum mechanical forces. Later extensions to include ion size "worked" better but really only were curve fitting with no predictablity.

The assumptions were broadly these : water is a continuum with uniform bulk properties up to an interface and its high dielectric constant alone is enough to reflect its many body molecular properties.

Short range hydration was sometimes included by the arbitrary addition of hydration shells.

Quantum dispersion forces between ions and ions and surfaces were usually not considered. When included, as for the DLVO theory, they were very crudely parametrised.

More sophisticated statistical mechanical theories to allow molecular solvent structure (dipole-dipole) interactions did capture something of hydration and overlap of hydration profiles. But they did no better really than the cruder characterisations like extended Debye-Hückel theory.

The situation was analogous to the description of valence in terms of paired electrons by G. N. Lewis ca. 1900, prior to quantum mechanics. A nice idea that captured some essence, but did not quantify matters.

Theory did not account for specific ion effects, or their influence on water and vice versa. In other words it did not account for the real chemistry of electrolytes in aqueous solution.[4]

Here is a partial list of where Hofmeister effects show up: Born energies, water activity, pH and buffers, dissociation of electrolytes and formation of ion pairs, heats and entropies and partial molal volumes of solution, self-diffusion coefficients of ions in water, and water in electrolytes, viscosity of aqueous and non-aqueous salt solutions, interfacial tensions, adsorption at interfaces, ion exchange chromatography, electrophoresis, surfactant and polymer cloud points and critical micellar concentration, polymer swelling and gelation, protein solubility and denaturation temperatures, degree of protein aggregation, coacervate behaviour, microemulsions, enzyme activity, growth rates of bacteria, optical activity of chiral molecules, water absorption in natural fibers, bubble-bubble coalescence, ion complexation by ligands at the air/water interface, formation of host-guest complexes, transition in surfactant lamellar phases, microemulsions, vesicle formation polymers, colloid interactions, direct force measurements between interfaces, ion binding, membrane and zeta potentials, electrophoresis, self assembly of surfactants and lipids, ion "pumps", solubility of salts in non aqueous solvents.[4,5]

The effects can be dramatic: for example a microemulsion can revert for oil-in-water to water–in-oil with change in counterion. In other cases (like in membrane potentials), the effects provide more confronting challenges.[4] Force measurements between bilayers change by orders of magnitude with change of counterion. The last item, Hofmeister effects in non aqueous solvents deserves note. It shows that the effects are not necessarily attributable due to "hydrogen bonding" of water.[6]

3.3 The simulation conundrum. In favour:

Faced with this frustrating situation, one of the armies of the tribes of the physical chemists have retreated and took a different approach. It began around 1960 with a paper of Barker, Henderson and Watts that took a molecular potential for atom-atom van der Waals interactions for argon. They used this potential to simulate on a computer the entire phase diagram of argon. In fact not quite, as to obtain the interfacial tension theory had to use three body potentials derived from quantum mechanics. In general, for water this is an impossible task. The success of this procedure has to be tempered by the admission that their two body van der Waals potential was wrong by a factor of 2.

If the results of such machine experiments could be so insensitive, the uninitiated would be forgiven for taking a jaundiced view of the whole proceedure.

Nonetheless the approach took off, proliferated without bound with the increase in computer size, and is applied to a vast number of ion specific problems. The sample of water and salt molecules that can be simulated is limited. There are doubts that such small samples can capture enough of the many body interactions in water.

The fact that simulation can not decide the sign of the air-water interfacial potential if such can be defined gives room for pause. But if popularity were the measure of success in science, simulation may win.

3.4 The Alternative View

This is probably closer to Langmuir's view, that water is more like a dynamic giant molecule acting in concert than a collection of single molecules.

Lifshitz theory and Lebedev's view is more like that.

Lebedev's program was implemented, at the behest of Deryaguin, by Lifshitz in a very major advance, extended by Dzyaloshinski,

Lifshitz and Pitaevski to include interactions between planar media separated by another dielectric usually liquid water medium around 1960. The theory was confirmed by measurements of Abrikossova and Deryaguin and promptly forgotten in the West. This was truly a triumph denied Newton. Later many measurements of Tabor and Israelachvili, and Israelachvili and Pashley and colleagues confirmed the theory. The technique involved interferometry to measure distance along the same lines suggested by Rev. Challis in 1836!

Their experiments involved forces between molecularly smooth mica interacting across water, and across electrolytes. All the concepts we have discussed above surfaced. The net force was a combination of electrostatic double layer forces and van der Waals–Lifshitz forces as DLVO theory said it should be. Specific ion effects occurred at close separations showing up in oscillations and other "hydration" effects.

This was agreeable. Between hydrophobic surfaces the forces were larger than van der Waals forces by orders of magnitude and attractive.

But agreement beween experiment and theory was more apparent than real, with sufficient flexibility in experimental uncertainty and inadequate understanding of theory to massage the agreement. Oscillatory forces in clay systems had long been confirmed and were ion specific, a matter of critical importance to soil irrigation.

But some measurements could not be doubted.

Forces between bilayers in salt solutions followed in Hofmeister series in agreement with observation on counterion dependence of self assembled ionic bilayer phases. E.g., the forces with acetate as counterion were 100 times larger than those for bromide with bilayers formed from quaternary ammonium surfactants. (This mimicked the bulk self assembly behaviour, the one forming lamellar phases, the other spontaneous single walled vesicles. Apparently arcane, these things are important to biological self assembly).

All this was very satisfactory.

3.5 The conceptual advance was the more important

The idea of Lifshitz was essentially to decompose the (electromagnetic) interaction between two bodies into a set of allowed Fourier modes, with boundary conditions at the media surface set by the frequency dependent dielectric susceptibilities. Each allowed mode was then assigned an oscillator free energy. The resulting formula was impossibly complicated and it was impossible to use and unravel what it meant for a decade or so. But then it was easy. It employed the *measured* dielectric responses of each bulk medium for all electromagnetic frequencies. These are peculiar to each body and to water. They include in principle all many body interactions and temperature. Different frequencies locked in at different distances. Two concepts were important. The first is to emphasise again that measured frequency dependent dielectric responses contain implicitly *all* information about many body interactions. The second is this. Two bodies at a distance at first sense low frequency electromagnetic vibrations, their own and those of water in the microwave region. At closer distance they sense infrared modes.

As they come closer still, additional forces kick in successively – optical frequencies, ultraviolet and finally in the far ultraviolet at the angstrom level, chemistry takes over if needs be.

More generally the interactions depend on shape and experience a torque that lines them up. Lebedev's space shuttle docking idea for macromolecules was confirmed.

Its deficiencies lay in use of a bulk medium description for the interacting bodies and intervening solvent. It missed on hydration. And need it be said, it missed specific ion effects, included implicitly but apparently far too small.

4.1 Hydrophobic interactions and atmospheric gas.

We encountered hydrophobicity with the Assyrians. Oil and water do not mix. An elephant in the living room. Put our problems to one side for the moment. Some other experiments have been swept under the carpet:

With colloidal particle suspensions degassing can reduce flocculation rates by factors of at least 10. "Hydrophobic" forces are switched off. Hofmeister ordering becomes different with degassing. The same occurs for force measurements. More dramatically oil in water emulsions become stable when degassed! Latex polymerisation does not proceed without gas – implying a connection to chemical reactivity. Nanosecond laser pulses produce plasma breakdown (due presumably to cooperative dipole interactions in water). On degassing no such breakdown occurs.

4.2 Specific ion effects in bubble-bubble interactions.

(See the paper of V. Craig, this volume). If the complication of dissolved atmospheric gas were not enough vexation, the simplest experiment of all is:

If a gas is passed through a glass frit at the base of a column (a fish tank will do) the bubbles fuse and grow larger as they ascend. If salt is added to the column at a concentration of 0.15 M (the concentration of salt in the blood of land animals) the bubbles no longer fuse. The column becomes opaque, filled with densely packed small bubbles. The same occurs for divalent and other ion pairs. They all scale on the same curve with the same critical electrolyte Debye length for all.

The visual effect is astonishing. But more astonishing is the fact that for an equal number of other ion pairs no effect on bubble bubble inhibition occurs! For these electrolytes bubbles continue to fuse up to 6 M.

There is a "combining rule" that predicts which ion pairs behave in one way or another, and indeed for mixtures of salts. Although known and publicised for at least 30 years the phenomenon has been virtually completely ignored as it does not conform to established theory.

A similar phenomenon occurs with isomers of sugars with critical concentrations around one molar.

Bubbles form and foam at the ocean, but on fresh water no foam lasts, a clear demonstration of the phenomenon. It has been used for a century or more in coal flotation in Russia. Since this has been observed by humans from time immemorial it is probably one of the first scientific observations in physical chemistry.

5.1 The pH and buffer problem

We now illustrate more explicitly some practical examples of specific ion effects. Consider pH. Everyone knows that unless we use a buffer, dissolved CO_2 sets pH at around 5 rather than the expected value of 7 for pure water at room temperature.

If we set the pH at 7 using a standard phosphate buffer, we can measure pH as a function of added salt.

The salt ought to be irrelevant.

Instead, the measured pH decreases with added salt. The amount of this decrease follows a Hofmeister series. The variation over the series is almost 1 pH unit. This does not seem too catastrophic. But a change in pH of that order is crucial to assigning pK_as of proteins. Then if we try the same experiment with a different buffer, cacodylate, at the same nominal pH of 7, the Hofmeister series is reversed!

This can not be a bulk electrolyte effect, at least we think so.

Then if we change the cation of the added electrolytes from sodium to potassium, the series reverses again!!

The same effects occur with other buffers. For some, pH increases with salt. All show a Hofmeister series.

Zeta potentials can change in sign with buffers.

Keeping in mind that many biological experiments involve addition of concentrated charged macromolecules, the meaning of pH can be seen to be generally elusive or meaningless.

That pH is meaningless seems to bother people not at all.

It is impossible to comprehend such phenomena with classical theories based on electrostatic interactions alone. The buffer problem is more serious than it appears.

With ionic microemulsions – whose microstructure involves typically microtubes or spheres of diameter 1-3 nm – it is impossible to use buffered water. The microemulsions do not form!

5.2 A biological challenge: restriction enzyme action and Hofmeister effects

The activity of a standard restriction enzyme called Hindi 2, that cuts linear DNA at a particular palandromic sequence is strongly dependent on the nature of background electrolytes.[4,5,7,8] The hydrophobic pocket, the "active site", is hydrophobic, around 3 nm diameter. Its structure depends on a specific ion, magnesium, that can be substituted for by a few other ions e.g., Mn(II). The enzyme joins with the DNA. A dimer of two enzymes associates with and diffuses along the chain until they find the right hydrophobic site by stereochemistry.

The question is how does it cut so precisely and so repeatedly, and: where does the energy come from?

The results indicate that with fixed cation (Na^+), the enzyme activity follows the sequences:

for phosphate buffer:

$Br^- < I^- < NO_3^- < SO_4^{2-} < ClO_4^- < Cl^-, CH_3COO^- < HPO_3^-$

for cacodylate buffer:

$ClO_4^- < SO_4^{2-} < I^-, NO_3^- < Cl^- < CH_3COO^- < (CH_3)_2AsO_2^- < Br^-$

Circular dichroism shows the DNA conformation does not depend on pH (5-8). The enzyme is active for salt concentrations lower than 0.5 M.

Results in summary are these:

1. pH is irrelevant?
2. The buffer anion is crucial to activity.
3. The anion of the background salt is crucial.
4. the cation anion pair is important.
5. Competition of 2, 3, and 4 for the DNA and enzyme surface determines enzyme activity.

We have one other piece of information; addition of vitamin C, a well known free radical scavenger, stops the enzyme dead in its tracks.

All this is completely counterintuitive. We can deduce only that our basic theories must be missing something.

Specific ion effects in biology show up everywhere, e.g, in bacteria as discussed in another paper in this volume.

6.1 Making sense of the mess

The examples above show that textbook theories of physical chemistry and colloid science that inform our intuition are impotent to explain the phenomena they were meant to explain. This situation means that physical chemistry that ought to be the enabling discipline behind biology is irrelevant. In fact it is probably fair to say that it is inadvertently rather malevolent! – in that biologists use the tools of physical chemistry, from pH meters to buffers, NMR, light scattering, X rays, electrophoresis and so on. It is not always so, but the *interpretation* of many measurements using these tools depend on topics that need more work. This confounds things even more.

6.2 Flaws in the theory

Conceptual locks and progress.

The problems we face have only recently been recognised. (We leave aside simulation experiments).

We can illustrate what went wrong by considering the prototype theoretical problem of colloid science – The forces between two planar charged media separated by salt water.

The long range interaction between them is considered to be due to the action of two forces. These are: the repulsive double layer force that scales with the Debye length; and the attractive van der Waals (or Lifshitz) force due to dispersion forces. We do not consider this or its decorations to include hydration in more detail. The idea leads to the DLVO theory.

The intervening water solvent is treated as a continuum bulk liquid for both forces.

But the problem lies in the ansatz that the electrostatic double layer forces and quantum mechanical dispersion forces can be separated. They can not. The double layer forces follow from nonlinear theory (usually a Poisson-Boltzmann distribution). The Lifshitz forces derive from a linear theory. It turns out this is inconsistent and fundamentally wrong. So wrong in fact that the entire venerable theory on which our intuition and measurements are based has to be rebuilt.

The matter is highly technical, and a more detailed discussion is given in the Appendix at the end of this chapter.

The inclusion of dispersion forces on same footing as non linear electrostatic theory is absolutely necessary for a consistent theory and to access specific ion effects. Explicitly, for the dispersion forces acting on ions we have to take account of their density profile self consistently with the coupled many body electrostatic forces.

The same problem occurs with all our other difficulties, most of which ignore dispersion forces entirely, i.e., the source of the very phenomenon we are trying to understand. In restricting forces to electrostatics and hydration, we compound the error further. Hydration (hydrogen bond or dipole orientation and density) is due to not just electrostatic and steric forces but depends strongly on dispersion forces, cooperatively not separately.

6.3 Improved theories, counterintuitive concepts

Once these problems were recognised it was possible to cut through the confusion.

It is probably fair to say that real progress towards a predictive new theory that includes Hofmeister effects is being made.

Ion size and dynamic polarisabilities can and are now being defined consistently via *ab initio* Quantum Mechanics.

Hydration of ions and of "surfaces" is included in *ab initio* extensions.

The theory quantifies Gurney (Collins) potentials, the overlap of hydration shells of ions that give rise to short range forces. It includes cavity energies, quadrupole and octupole extensions. Anisotropy is also included.

The emerging theory is not perfect but seems on the right track.

Hofmeister reversal is accommodated in principle and has been exhibited in practice as a consequence of the theory.

The most important development is conceptual.

When dispersion forces (and resulting hydration due to them) are taken into account, forces between ions of *like* charge can be *attractive*. This is completely counterintuitive. In the older theory e.g. only counterions could be attracted to a surface (or ion) of opposite charge. Now coions complete for the same sites.

This can be seen in the pH and buffer problem.

The interpretation of a pH measurement by a glass electrode rests on a surface potential measurement. The Nernst equation is unquestioned, being a thermodynamic statement. But at some point a distribution function for the electrolyte molecule has to be introduced. Both the double layer (Poisson-Boltzmann) and the activity (extended Debye-Hückel theory have to be involved in the interpretation which ignores ion specific dispersion interactions). The difference in specific ion effects for two examples, phosphate and cacodylate buffers is due to competitive adsorption of buffer anion, electrolyte cation and anion, and hydronium for the gas electrode surface. Similarly for most of the other effects we have discussed like zeta and membrane potentials.

6.4 Inexplicable phenomena. Dissolved gas a dilemma

Besides the specific ion challenge, the other conceptual lock that confronts us is that of dissolved gas.

At one atmosphere of pressure and room temperature dissolved gas, oxygen around 10^{-3} M, is at very low concentration compared with electrolytes we are generally concerned with ≈ 0.15 M in physiology.

Hardly worth considering at first sight. Yet the problems listed have a major or minor dependence on dissolved gas that we do not understand: nanoparticle synthesis, self assembly of lipids and proteins; emulsion stability, emulsion polymerisation, laser cavitation, bubble bubble interactions, colloid stability and interactions, hydrophobic forces, sonochemistry, cloud points of surfactants aqueous dispersions and formation of host-guest inclusion compounds.

All have a dependence on dissolved gas that can be exploited pratically, and all have to do with hydrophobic forces, in one way or another. It is impossible to simulate these effects.

Without any dissolved gas, simulation and dimensional arguments agree on the range of hydrophobic interactions between hydrophobic surfaces. It is due to a surface induced ordering and fluctuations that lower the density of water molecules in the gap. The range is around ~ 3 nm. So the very long range of measured hydrophobic forces makes no sense.

But there is a hint in that the fluctuations that lower the water gap density typically have a size ~ 0.1 A, corresponding to surface induced water dipole ordering.

In Figure IV.1 the slabs represent the hydrophobic surfaces, the double arrow segment such a fluctuation and the circles the dissolved gas molecules. In fact the dissolved gas molecules at $5 \cdot 10^{-3}$ M are about 3 nm apart, much more near a hydrophobic surface. As far as the fluctuation or crack in the water is concerned such a molecule (of diameter ~ 0.4 nm) presents a large surface.

So the range of the fluctuations and force can be extended in range by percolation of the "cracks" from one gas molecule to another. This process can be enhanced or inhibited by different ion pairs which can order adjacent water molecules differently.

This picture is consistent with the well known tensile strength of water, abut 200 times less than theory would predict.

6.5 Restriction Enzyme Kinetics

We can now also try to make sense of our enzyme problem.

Recall that:

The enzyme active site is hydrophobic.

The enzyme attaches to DNA. It diffuses to a hydrophobic nucleotide sequence (stereochemistry).

Then spontaneous hydrophobic cavitation occurs (known from force measurements).

OH radicals (known from sonochemistry) transfer to Cl^- to create an energetic radical which cuts the phosphate bond.

This proposition is a *cooperative* mechanism to harness weak molecular forces (physical) to produce a large energy (chemical). Cavitation is prohibited above about 0.175 M. This is consistent with consistent with bubble-bubble and other counterintuitive specific ion effects

We can check the hypothesis by adding vitamin C, a well known free radical scavenger. This stops the enzyme dead! There will be different mechanisms for ATP–ADP and horseradish peroxidases.

7. Theory and measurement

In summary the interpretation of measurements like pH, buffers, zeta and membrane potentials, ion binding via NMR, light and neutron scattering free energies of molecular transfer ... are all based on an invalid classical theory.

These textbook theories omit ion specific quantum fluctuation (dispersion) forces and ignore dissolved gas.

By so doing they confound the meaning of hydration, their meaning is dubious. And then again simulation compares real water (containing gas) with theoretical water!

The same difficulties underlie experiments that involve interaction of light with molecules in water like circular dichroism and light initiated electron transfer.

Their consideration would take us too far afield.

As opposed to this appalling situation – the recognition that the classical intuition is badly flawed, and that theory needs a complete revamp – there are some positive consequences.

A major source of specific ion effects has been identified, and quantified, (almost) predictively.

The role of dissolved gas has been identified.

And once recognised these and other solute effects, can be

exploited to control forces, crystallisation, crystal growth, reactivity in nanotechnology, self assembly and formulation.

Some quite unanticipated consequences of the removal of what Stephen J. Gould called conceptual locks are likely.

Among these are reexamination of the old Gilbert Ling controversy of the 1950's, the ion pump hypothesis in biology, the meaning of biomembrane potentials and zeta potentials.

And in passing we note that the problem of why ice floats on water is addressed in a paper of Anderson and Ninham.[9]

We have hardly begun to tackle other new concepts in this volume, of M. Henri and D. J. Morré.

The third Aqua Incognita conference will make our musings appear as naïve as those of Galileo and colleagues.

Appendix. technical problems with the theoretical mishmash

It turns out that the DLVO theory that is built on the ansatz of separation of forces is wrong on several counts. With the electrostatic double layer forces the theory is non linear. The ions of the electrolyte experience a force from the charged or even uncharged (image forces) surfaces. This sets up a profile of counterions and coions. The overlap of profiles gives rise to an osmotic force, just as for hydration forces. If we like, the electrolyte is the "liquid" (in the passive water surround), and the coion and counterion density profiles are induced by the charged surfaces. The profiles are a result of surface induced ordering of the liquid just as are hydration profiles proper. On the other hand, for the quantum mechanical dispersion forces, the liquid, including the electrolyte in the gap, is treated as a medium of uniform density up to a molecular distance from the surfaces. Clearly this is not right, conceptually, and not right qualitatively. (We get over the very bad approximation of a pairwise summation of individual molecular dispersion forces by using Lifshitz and its extensions. This includes all many body molecular forces through its reliance on all frequency contributions. These are extracted from the measured dielectric susceptibilities on the interacting media).

But it is not sufficient. Technically, the mishmash violates the Gibbs adsorption equation and charge continuity when an electrolyte is involved because of the nonlinear distribution of ions (see chapter 4 in Ref. 4). In addition, both components of the forces assume that water itself is a continuum medium up to the interacting surfaces (no hydration forces induced by surfaces or ions).

We can make this statement explicit.[10] Consider the simplest prototype colloid problem – Two **uncharged** surfaces separated by a dielectric liquid. The force per unit area between them includes a sum of quantum mechanical terms from different frequencies. It includes also a temperature dependent classical term. If the theory is generalised to allow ions in the medium, this term, which dominates in biological media, takes on a different form. It can be shown that it is exactly the extension to two surfaces of Onsager-Samaris theory. This theory calculates the change in interfacial tension at an interface due to dissolved salt. It is a limiting law, valid if at all to concentrations below 10^{-4} M.

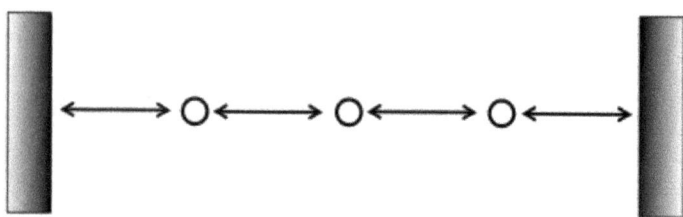

Figure IV.1. The slabs represent the hydrophobic surfaces, the double arrow segment refers to the fluctuation and the circles the dissolved gas molecules.

It does not depend on ion size, and involves only the Bjerrum length. Onsager himself recognised that the theory is missing some unknown forces. No ion specificity is involved. So too with the colloidal particle interaction problem, no specificity is included.

Now if we go further and consider **charged** colloidal particles, the situation is very much worse because of the non linear coion and counterion profiles set up by the double layer.

Ion specificity is missing from the theory. However the dispersion forces acting on an ion can be evaluated from an extension of Lifshitz theory and built into a generalisation of Poisson Boltzmann double layer that does allow for ion specificity.[4] It includes electrostatic and quantum (dispersion) forces on the same footing.

Such a program is being implemented, see e.g., articles of Beck, Duignan and Parsons in this book.

Parenthetically, we remark that the same, identical, problems that have bedeviled physical chemistry are endemic and unrecognised in nuclear and particle physics and quantum electrodynamics proper, and more seriously.

References

1. Tabor, D. *J. Coll. Interface Sci.* **1980**, *75*, 240-245.
2. *http://www.lindahall.org/events_exhib/exhibit/exhibits/napoleon/soda_lakes.shtml*
3. Kunz, W.; Henle, J.; Ninham, B. W. *Curr. Opin. Coll. Interface Sci.* **2004**, *9*, 19-37.
4. Ninham, B. W.; Lo Nostro, P. *Molecular Forces and Self Assembly. In Colloid, Nano Sciences and Biology*, Cambridge University Press, Cambridge, 2010.
5. Lo Nostro, P.; Ninham, B. W. *Chem. Rev.* **2012**, *112*, 2286-2322.
6. Peruzzi, N.; Ninham, B. W.; Lo Nostro, P.; Baglioni, P. *J. Phys. Chem. B* **2012**, *116*, 14398-14405.

7. Kim, H.-K.; Tuite, E.; Nord en, B.; Ninham, B. W. *Eur. Phys. J. E.* **2001**, *4*, 411-417.
8. Parsons, D. F.; Boström, M.; Lo Nostro, P.; Ninham, B. W. *Phys. Chem. Chem. Phys.* **2011**, *13*, 12352–12367.
9. Andersson, S.; Ninham, B. W. *Solid State Science* **2003**, *5*, 683-693.
10. Ninham, B. W.; Yaminsky, V. *Langmuir* **1997**, *13*, 2097-2108.

V
Hofmeister effects in living organisms

Niccolò Peruzzi[◊], Barry W. Ninham[#], Pierandrea Lo Nostro[*,◊,‡]

◊: Department of Chemistry "Ugo Schiff" and CSGI,
University of Florence, 50019 Sesto Fiorentino (Firenze), Italy.

#: Department of Applied Mathematics, Research School
of Physical Sciences and Engineering, Australian National
University, Canberra, Australia 0200.

‡: Enzo Ferroni Foundation, 50019 Sesto Fiorentino (Firenze), Italy.
pln@csgi.unifi.it

Overview

Hofmeister specific ion effects occur everywhere, in bulk solutions of aqueous and non-aqueous solvents and at interfaces (air/water, oil/water, solid/water, polymer/water, etc.). Hofmeister phenomena consist in differences in behavior of electrolytes in solution, usually above 0.1 M (but in some cases even in the micromolar range). Hofmeister found that salts can be ordered in a particular order – the Hofmeister series – according to their effectiveness in precipitating egg yolk albumin from a water dispersion. The specificity of ion effects in biology and biochemistry is universal and essential to life. And their relevance to geology and life sciences is obvious. The challenge to classical physical chemistry is that it does not explain Hofmeister effects. We observe that the most important ions that regulate life are Na^+, K^+, Mg^{2+}, Ca^{2+}, CO_3^{2-}, HCO_3^-, SO_4^{2-}, Cl^-, $H_2PO_4^-$, HPO_4^{2-}, and a group of transition metals like Zn^{2+}, Mn^{2+}, Cu^{2+} at very small concentrations.

In this contribution we will focus on one aspect of Hofmeister phenomena that is almost completely unknown: the effect of different ions on living organisms.

Introduction

The trends discovered by Hofmeister in his pioneering work on the precipitation of egg yolk albumin from water dispersion list effectiveness by concentration in the order:[1]

with fixed cation:

$SO_4^{2-} >$ $HPO_4^{2-} >$ $CH_3COO^- >$ $Cl^- >$ $NO_3^- >$ $Br^- >$ $I^- >$ $ClO_4^- >$ SCN^-

with fixed anion:

$NH_4^+ >$ $K^+ >$ $Na^+ >$ $Li^+ >$ $Mg^{2+} >$ $Ca^{2+} >$ $C(NH_2)_3^+$.

However the order is not always the same; there are re-orderings along the series, depending on the hydrophobicity and surface charge of the substrate, on the solvent and on the concentration. In some cases a total inversion of the series is observed.

Similar observations are found in completely different systems, for example for activity coefficients, viscosity, conductivity, pH measurements, osmotic pressure, aggregation of proteins and formation of amyloid fibrils, activity of enzymes, critical micellar concentration and phase behavior of amphiphiles and microemulsions, and stability of surfactant-based systems, just to mention a few. The interested reader can found several extensive reviews on this subject in the existing literature.[2]

While Hofmeister phenomena in bulk solutions and at interfaces are reported in thousands of papers, the effect of salts on living species is still almost completely unexplored.[3,4] There are two obvious reasons: i) living organisms are difficult to handle, and sometimes it is hard to define a standard system that can be representative of an entire population, and ii) it is not easy to pin down how and where a specific salt affects the entire complicated dynamic mechanism that regulates life. However it is clear that halophilic bacteria and other salt-loving species evolved in peculiar and extreme environments where the concentration of salts is particularly high and prohibitive for most living species. The osmoregulating activity that these special forms of life have developed is indeed another case of a Hofmeister phenomenon, because all salts behave differently.

In this contribution we will discuss the effect of different ions on the vitality of different living organisms, two strains of bacteria (*Staphylococcus aureus* and *Pseudomonas aeruginosa*) and brine shrimps (*Artemia salina*).

Before proceeding with the discussion of the experimental results, we need to recall some issues related to Hofmeister phenomena.

Kosmotropic and Chaotropic species

Chemists are familiar with the terms "hard" and "soft" proposed by Pearson in the Hard Soft Acid Base (HSAB) theory.[5] Here a hard ion is descriptor for an ion with a small radius or a large charge (for example Al^{3+} or F^-) while a soft species possesses a large radius and a small charge (for example Cs^+ or SCN^-). The terms classify ions through their charge density. The distinction is based on a purely electrostatic model for the description of the ionic interactions.

In Hofmeister phenomena, where specific ion effects take place, ions are classified as either "kosmotropic" or "chaotropic". These terms were introduced by Jones and Dole.[6] Solutions of kosmotropes are more viscous and solutions of chaotropes are more fluid than water at the same temperature. For a wide range of concentrations (between 5 mM and 1 M), the viscosity η of an aqueous electrolyte solution can be fitted to the equation

$$\eta = \eta_w \left(1 + A\sqrt{c} + Bc\right) \qquad (1)$$

Here c is the molar concentration of the salt and η_w is the viscosity of pure water at the same temperature. For dilute solutions (when electrostatic interactions dominate), the term in c vanishes and that in $c^{1/2}$ prevails. On the other hand, the term in c dominates in more concentrated solutions and reflects the specificity of the dissolved salt. The ion-specific Jones-Dole coefficient B is negative for chaotropes *($\eta < \eta_w$)* or positive for kosmotropes *($\eta > \eta_w$)*.[7,8]

The etymology of the words kosmotrope and chaotrope recalls the perturbation that these molecules or ions produce in the local structure of water.[8,9]

The overall perturbation of an ion in water is a balance between two opposing effects: the disorder that any ion induces in the structure of the solvent, by interrupting the local hydrogen-bonded tridimensional network, and the re-ordering effect produced by the ionic electric field on the water dipoles. In fact, the strong electrostatic field produced by a small and strongly hydrated (hard) ionic kosmotrope forces the orientation of the permanent dipole moments of the surrounding water molecules and imparts a higher order on local water molecules via ion-dipole interactions. On the other hand, a large and poorly hydrated (soft) chaotrope produces a weaker electrostatic field, perturbs the dynamic quasi-ordered (hydrogen-bonded) array in bulk water and makes the surrounding water molecules more disordered.[10]

However, no general agreement exists on how the water structure is to be defined and on how the extent of hydrogen bonding should be measured or computed.[6] The very language we use, with terms like ion-dipole or dipole-dipole interactions are inadequate because the strong, highly coupled, many body interactions in water imply that it is impossible to dissect matters into individual molecular contributions.

Moreover, when the water structural effects of the ions are discussed, effects beyond their hydration shells are generally implied. Whether and how much an ion affects the solvating water molecules beyond the hydration shell(s) is still an open question.[2]

Another reason why the terms kosmo- and chaotrope should be handled with care is because in some cases the series reverses, i.e. the behavior of a specific ion does not match with the common terminology, especially when the ion sits in the central part of the Hofmeister ranking, as in the case of sodium and chloride.[11]

With these caveats, and because of familarity and convenience, we will retain the classifications of ions as either kosmotropes or chaotropes. A kosmotropic species possesses a large charge, small size and polarizability and is strongly hydrated. Conversely a chaotrope has a small charge, a large radius and polarizability and is weakly hydrated.

The relevance of non-electrostatic interactions

One of the issues that are still debated is related to the interactions that an ion establishes in solution and at interfaces.

Basically, the theories that describe the behavior of ions in solutions refer to electrostatic interactions only. Some unquantified hydration effects are included via effective size or exclusion zones at interfaces. The Debye-Hückel and DLVO theory of colloid interactions – that are valid in very dilute systems – consider the ions as charges that interact with other ions and the solvent molecules only through Coulombic and ion-dipole forces. The solvent effects are usually accounted for as a continuum with the same dielectric properties of the bulk phase. These (electrostatic) models, even with the inclusion of many-body attractive van der Waals forces, do not explain specific ion effects in terms of interactions.[1,2]

Non-electrostatic forces operate at a shorter range than Coulomb interactions, and depend on the polarizability and ionization energy of the intervening species, that is on ion or molecular properties that are strictly related to their electronic configuration, and therefore to their specific nature.[2]

Basically, the polarizability quantifies the "softness" of the electron cloud around the nucleus, and reflects the different response of the molecule/ion under the effect of an external oscillating electric field, a field that includes electromagnetic fluctuations due to its neighbors. As a matter of fact, fluoride ions are very hard, with the lowest polarizability, while iodide ions are very soft due to the presence of several internal electrons that shield the nuclear attractive force from the outer valence shell. Therefore, in principle F^- should be less sensitive than I^- to other ions and dipolar molecules.

Shape is another factor. Halides and most of the cations are spherical, meaning that the interactions they establish are isotropic. This cannot hold for ions such as nitrate, thiocyanate or dihydrogen phosphate, which have a nonspherical structure. Because of the existence of symmetry axes, their polarizability will depend on the direction in which it is calculated. And the interaction between a nonspherical ion and another molecule or interface will change, depending on the direction of approach. The shape issue shows up strongly in specific ion effects.

In conclusion, a consistent theory needs to take into account all of the effects of non-electrostatic forces, many-body interactions, and structural features such as anisotropy. And too of the effects of the (quantum mechanical) non electrostatic forces in inducing hydration.[12-15]

Effect of salts on the growth rate of *S. aureus* and *P. aeruginosa*

S. aureus is a Gram-positive halophilic bacterium that proliferates in osmotically stressful enviroments.[16] These microorganisms possess a bacterial wall with an external coating layer, made up of a complex matrix composed of polysaccharides, lipids, and proteins.[17] *P. aeruginosa* is a Gram-negative opportunistic pathogen quite widely diffused in the environment.[18] This microorganism develops in low salt concentration conditions, and even in distilled water. Although the effect of different ions on bacterial processes is already known,[19] and may have a significant impact on different bio-medical and industrial applications (e.g. in food safety and biotechnology), the number of studies devoted to this aspect is particularly small.

We studied the effect of different sodium salts (fluoride, chloride, bromide, iodide, thiocyanate, phosphate, dihydrogen phosphate, sulphate, selenate, nitrate, acetate, tetrafluoborate, cyanate, carbonate, iodate, bromate, chlorate, formate, succinate, citrate, and maleate) on the growth rate of a strain of *S. aureus* (ATCC 25953) and of *P. aeruginosa* (ATCC 27853) as a function of the salt concentration, according to a standard procedure.[20]

In this contribution we show that the bacterial growth is clearly affected by the salt concentration (ranging between 0.2 and 0.9 M), and by the specific nature of ions according to a Hofmeister sequence. This specificity correlates with corresponding changes of enzyme activity with high salt in, e.g., restriction enzymes,[21] Lipase A (*Aspergillus niger*),[22] horseradish peroxidase,[23] and growth of yeast cells.[3]

The bacteria growth rate was calculated as $Log(C/C_0)$, where C and C_0 are the counts of the bacterial suspension in the presence and in the absence of the salt.[20]

The growth rate sequences are:

for *S. aureus*

Cl^-, CH_3COO^- > succinate > NO_3^- > $HCOO^-$ > Br^- > ClO_3^- > citrate, maleate > BrO_3^- > SCN^-, SO_4^{2-} > I^- > (H_2O) > SeO_4^{2-} > $H_2PO_4^-$ > F^- > BF_4^-, CNO^-, CO_3^{2-}, PO_4^{3-}, IO_3^-

for *P. aeruginosa*

(H_2O) > Cl^- > NO_3^- > CH_3COO^- > Br^- > I^-, SCN^-

Again, the two bacterial strains behave differently. This difference may be related to the different structure of the bacterial walls and membranes. In the case of *S. aureus* some anions (left of water in the series) favour the microorganism, whilst other salts (right of water) inhibit its growth.

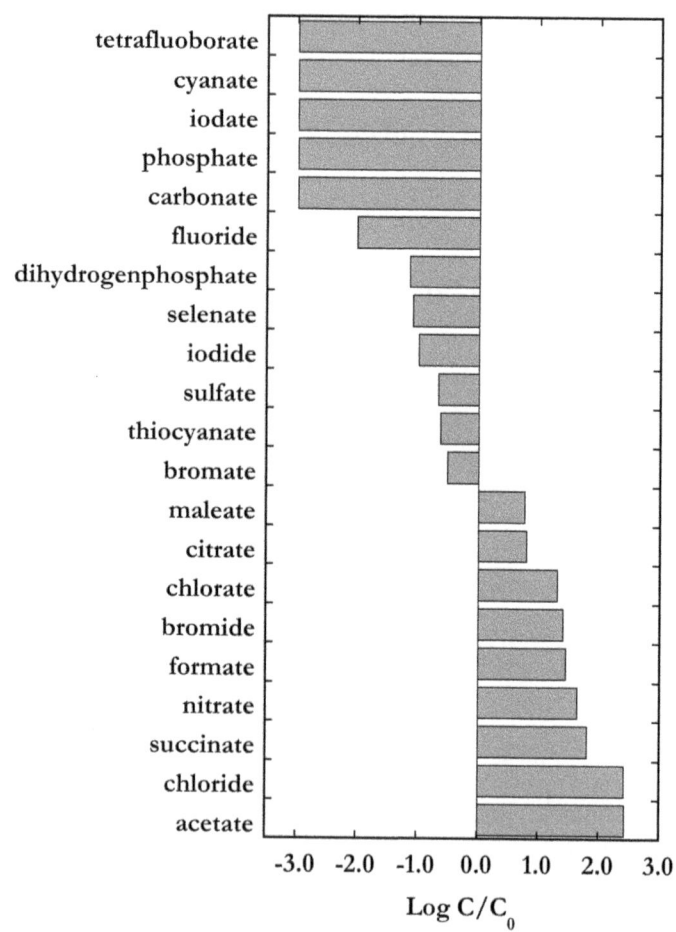

Figure V.1. *$Log(C/C_0)$ of S. aureus* in the presence of 0.9 M aqueous solutions of the investigated sodium salts.

Figure V.1 shows the $Log(C/C_0)$ of *S. aureus* in the presence of 0.9 M aqueous solutions of the investigated sodium salts. The results indicate that the bacteria proliferate in the presence of organic anions (citrate, succinate, maleate, acetate) or in the presence of mild kosmo- or chaotropic ions such as chloride, bromide or nitrate. The growth of

S. aureus is significantly inhibited by strong kosmo- or chatropic anions such as phosphate, carbonate, fluoride, sulphate and iodide, and particularly by those species that produce an alkaline solution.

At the same salt concentration, *P. aeruginosa* growth is always depressed, with the exception of NO_3^-.

Figure V.2 shows the growth rate of *S. aureus* (a) and of *P. aeruginosa* (b) as a function of the background electrolyte concentration. For *S. aureus*, $Log(C/C_0)$ reaches a maximum at about 0.3 M and then decreases. Growth is a factor of 10^5 higher in 0.2 M salts than in pure water, reducing to a factor of 10^3 or below at 1 M. This observation is consistent with the typical halophilicity of *S. aureus*, that grows preferentially in diluted salt solutions, and in some cases even at high concentrations. On the other hand, the growth of *P. aeruginosa* is not favored by the addition of salts: the plot shows that all ions increase the bacterial growth at 0.2 M, with the exception of acetate and thiocyanate, that inhibit the growth also at low concentration in agreement to a previous study.[24] Interestingly, the two trends parallel the specific ion effect of salts on the activity of a restriction enzyme.[21]

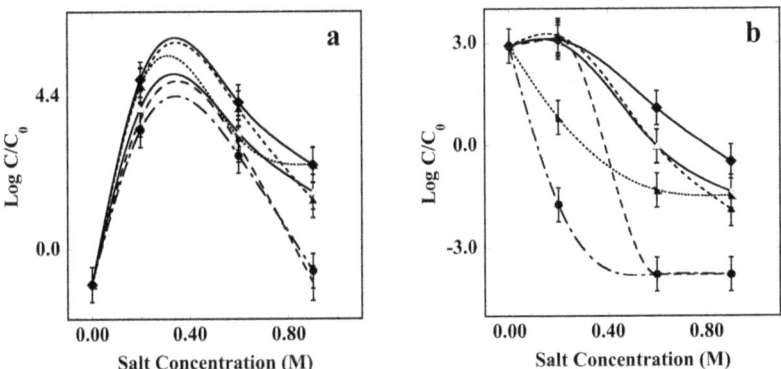

Figure V.2. $Log(C/C_0)$ of *S. aureus* (a) and of *P. aeruginosa* (b) as a function of the salt concentration. ●: SCN^-; O: I^-; ▲: Br^-; ∆: NO_3^-; ♦: Cl^-; ▲: CH_3COO^-. The curves are guidelines for the eye. Reproduced with permission from ref 20. Copyright 2005 IOP Publishing Ltd.

Discussion

There are different mechanisms through which bacteria react to a hyperosmotic shock due to the presence of concentrated salt solutions. The most important are the osmoregulating activity of compatible kosmotropic solutes that are either found in the growing medium or synthesized by the same microorganisms.[25,26] And the folding-unfolding of proteins and enzymes as a response to preferential interactions between the ions and the bacterial membranes.[25,27,28] In particular, in the case of *S. aureus*, it has been proved that chloride protects the microorganism against thermal shock, unlike sulphate and nitrate.[29] And more generally, Cl⁻ is necessary for some halophilic and halotolerant species.[30] Multivalent salts have stronger adverse effects, probably due to their charge and to the larger amount of Na⁺ counterions. Chaotropic ions act at the interface and can permeate across the membrane through the chloride channels,[27] or reduce that membrane lipid order and induce changes in membrane composition.[25] On the other hand, chaotropic anions protect yeast cells against ethanol-induced deactivation.[31]

According to our results on the growth rate of *S. aureus* and *P. aeruginosa*, the largest values of $Log(C/C_0)$ are found for chloride, which is considered to be only a weak chaotrope.[25] Strong kosmotropes such as fluoride and multivalent ions have an adverse effect on the bacterial survival. This effect is reflected by the values of some physico-chemical parameters that are directly related to the kosmotropicity/chaotropicity of the anion: a_W (water activity, calculated for 1 M sodium salts solutions at 25°C from osmotic coefficient data),[32] and ΔS_{II} (the water structure entropy change).[20] In the case of *P. aeruginosa* the growth rate appears to be more sensitive to the availability of free water than for *S. aureus*.

Kosmotropic (small and multivalent such as F⁻ and SO_4^{2-}) anions have large and negative ΔS_{II} values (reduced mobility of water due to strong hydration). Inversely, chaotropic (large monovalent species like I⁻) possess positive ΔS_{II}.[33] In order to infer that dispersion forces are directly involved in these specific ion effects, we compared $Log(C/C_0)$ to the partial molar volume (V_s, see Figure V.3, left plot) and to the polarizability (α, see Figure V.3, right plot), that are closely related to dispersion interactions.[34,35] The plot of $Log(C/C_0)$ as a function of V_s and α show very regular trends, indicating that the *S. aureus* growth rate (full lines) is significantly decreased by the ions with large V_s and large α. The same trend is observed for *P. aeruginosa* (dashed lines).

This effect is an indication that dispersion forces must be involved in whatever mechanism it is by which anions specifically affect the bacterial growth. This argument is strongly corroborated by our previous studies on Hofmeister effects that take place in different systems. In all these cases, where hydrophobic interactions and dispersion forces play the dominant role, we detected a significant dependence on anion polarizability and other physico-chemical parameters that are involved in non-electrostatic interactions.[2]

In conclusion, the results show a dependence of the bacterial growth

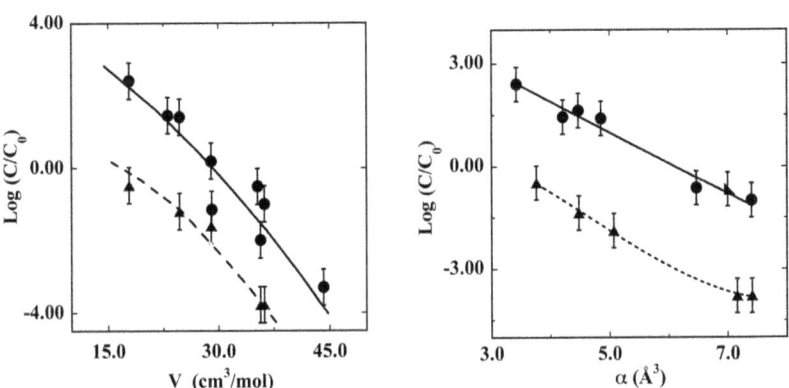

Figure V.3. $Log(C/C_0)$ of *S. aureus* (●) and of *P. aeruginosa* (▲) as a function of the anion partial molar volume V_s (left) and of the anion polarisability α (right). The curves are guidelines for the eye. Reproduced with permission from ref 20. Copyright 2005 IOP Publishing Ltd.

rate on the water structure activity induced by the anions.[36] The reasons that both strong chaotropes and kosmotropes eventually have the same effect on *S. aureus*, can be due to multiple effects. Among these there are certainly the depletion of water molecules from the bacterial outer surface due to chaotropes, the water activity change induced by all ions and especially kosmotropes, the osmoregulating activity,[26] the preferential interaction of some anions with the membrane, and the hydrophobic interactions.[27,28] In the case of *P. aeruginosa* similar trends are recorded, but the toxic effect of all anions, here due to what for this organism is an extremely high salt concentration, causes the shift of the growth rates to lower values.

The second feature that springs forth from these data is related to the kind of forces that ions experience in solution at these non-diluted conditions. The anion dependence of $Log(C/C_0)$ on a_w in the case of *S. aureus* indicates that the bacterial growth reaches its maximum with chloride, and then decreases regardless of a_w (see Figure V.4). In contrast for the case of *P. aeruginosa*, $Log\ C/C_0$ decreases regularly with decreasing a_w, according to the behavior of this bacterium in salt environments.

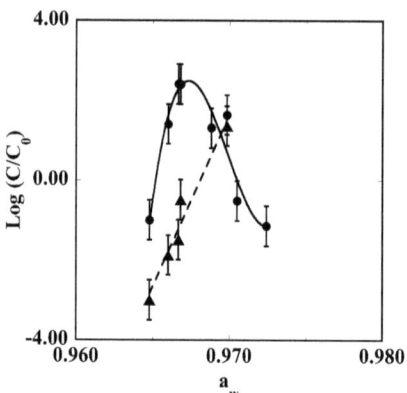

Figure V.4. $Log(C/C_0)$ of *S. aureus* (●) and of *P. aeruginosa* (▲) as a function of the water activity a_w (calculated for a 1 M aqueous solution of the sodium salt at 298 K). The curves are guidelines for the eye.

Effect of salts on *Artemia salina*

Artemia salina (brine shrimps) are branchiopod crustaceans that develop very well in hypersaline conditions, as in solar saltworks for table salt and chemical industries. They are essential for limiting algae bloom and favors the growth of halophilic bacteria.[37]

Adult *Artemia* have an elongated body with two stalked complex eyes, a linear digestive tract, sensorial antennulae and eleven pairs of functional thoracopods. Brine shrimp possess an osmoregulatory system based on trehalose, the capacity to synthesize efficient respiratory pigments to cope with the low O_2 levels at high salinities, and the ability to produce dormant embryos (cysts) depending on the salt concentration in the environment. After 24-h incubation in seawater, these cysts release free-swimming nauplii.

The best environment for *Artemia* are thalassohaline concentrated saltwaters where NaCl is the major component. However there are different species that adapted to sulphate, carbonate, or potassium rich athalassohaline environments, spread almost all over the Earth. Water composition in terms of nature and concentration of electrolytes is crucial for their survival.[38]

Trehalose and other osmolytes are necessary for brine shrimps. Their concentration in fact may exceed 1 m. Trehalose serves as water replacement, stabilizing the membrane structure through physical intercalation between the phospholipids headgroups.

In *Artemia*, the large amounts of the organic osmoregulating solutes (trehalose, glycerol and methylamines) may be involved in preventing macromolecular denaturation in hypersaline conditions or in the presence of "toxic" or foreign ions.

In this study we assessed the effect of different sodium salts on the vitality of *Artemia salina*. In our experiments the cysts were hydrated with artificial regenerated seawater. The breeding was thermostatted at 23° C for 17 days, then at 27° C for 11 days to promote the growth. Finally the tank was kept at 25°C. Continuous aeration was supplied.

15 individual nauplii were transferred into the salt solution, and their vitality was assessed by visual inspection upon the addition of sodium sulphate, carbonate, fluoride, chloride, bromide, iodide, thiocyanate, nitrate or perchlorate, at the same ionic strength as seawater. The number of alive nauplii was recorded as a function of time every 10 mins. The results show that the effect of the anion is:

$$Cl^-, Br^- > NO_3^- > SO_4^{2-} > CO_3^{2-} > I^- > ClO_4^- > F^-, SCN^-$$

meaning that in the presence of fluoride or thiocyanate the vitality of the brine shrimps was seriously reduced after a short period of time (about 90 mins), while the nauplii survived up to 30 hours in the presence of chloride or bromide.

The results indicate that *Artemia salina* prefers salty waters that contain weakly kosmotropic/chaotropic species (such as chloride and bromide), while the presence of strong kosmotropes (fluoride) and especially strong chaotropes (thiocyanate, perchlorate and iodide) are severely toxic. We recall here that in nature these animals develop in water environments that mainly contain chloride and sulphate ions.

The effect of fluoride as a toxic component has already been demonstrated in previous studies[39,40] and is presumably related to the inhibition of different enzymes (including phosphatases, hexokinase, enolase, succinic dehydrogenase, pyruvic oxidase) and to the interruption of metabolic processes such as glycolysis and synthesis of proteins.

The toxicity of thiocyanate ions in fish and crustaceans seems to be related to the substitution of chloride ions in the gill and to the perturbation of the ionic balance, resulting in respiratory problems.[41,42]

One possible explanation for this phenomenon relies on the mechanism through which brine shrimps breath dissolved oxygen from water. And could be related to the bubble-bubble coalescence effect of electrolytes in water. In fact sodium chloride, bromide, iodide, nitrate and sulphate are known to inhibit the coalescence of air bubble in water, while salts like sodium perchlorate and thiocyanate strongly do not affect the spontaneous coalescence of air bubbles, rendering salt water similar to fresh water in terms of size and stability of gas bubbles.[43]

The different oxygen solubility imparted by the different salts does not seem to be related to the present results, in fact sodium perchlorate and thiocyanate increases the solubility of O_2 in water, respect to NaCl and NaBr. And therefore should in principle increase the availability of dissolved oxygen in water for the crustaceans' breathing.

Further studies on the effect of different cations and on the oxygen consumption of brine shrimps in the presence of electrolytes are currently in progress.

Conclusions

In these studies we observed the behavior of living organisms, two bacterial strains and a population of a marine invertebrate, in the presence of high concentrations of electrolytes in water. Although the results are significantly different because of the different biological features of these species, however some interesting conclusions can be drawn.

The presence of salts does not hamper the vitality of these species at small concentrations, when the solution can be described in terms of electrostatic interactions and ion specificity does not emerge. Instead, the effect of the different ionic species becomes critically relevant when their concentration reaches a moderately large value, electrostatic interactions are screened and replaced by non-electrostatic (dispersion) forces.

Certainly the biochemical mechanisms that regulate the vitality of salt-loving species (*S. aureus* and *A. salina*) have adapted to an environment that is characterized by the presence of ions that reside more or less in the middle of the Hofmeister series, e.g. chloride and bromide. Strongly kosmotropes or chaotropes can be severely toxic as they affect the structures and functionalities of biological macromolecules (lipids, proteins, enzymes, etc.) and assembled structures (outer and cytoplasmic membranes). We can argue that this toxic effect involves the onset of quantum mechanical dispersion forces that are responsible for the perturbation of the structural features and functionalities of biomacromolecules and the self-assembled structures in which they are organized and through which they regulate life.

References

1. Ninham, B. W.; Lo Nostro, P. *Molecular Forces and Self Assembly. In Colloid, Nano Sciences and Biology*, Cambridge University Press, Cambridge, 2010.
2. Lo Nostro, P.; Ninham, B. W. *Chem. Rev.* **2012**, *112*, 2286-2322.
3. Boas, F. *Biochem. Z.* **1926**, *176*, 349-402.
4. Gellhorn, E. *Protoplasma* **1933**, *18*, 411-419.
5. Pearson, R. G. *J. Am. Chem. Soc.* **1963**, *85*, 3533-3539.
6. Jones, G.; Dole, M. *J. Am. Chem. Soc.* **1929**, *51*, 2950-2964.
7. Jenkins, H. D. B.; Marcus, Y. *Chem. Rev.* **1995**, *95*, 2695-2724.
8. Marcus, Y. *Chem. Rev.* **2009**, *109*, 1346-1370.
9. Marcus, Y.; Hefter, G. *Chem. Rev.* **2006**, *106*, 4585-4621.
10. Beck, T. L. *J. Phys. Chem. B* **2011**, *115*, 9776-9795.
11. Mancinelli, R.; Botti, A.; Bruni, F.; Ricci, M. A.; Soper, A. K. *J. Phys. Chem. B* **2007**, *111*, 13570-13577.
12. Ninham, B. W.; Duignan, T. T.; Parsons, D. F. *Curr. Op. Coll. Interface Sci.* **2011**, *16*, 612-617.
13. Parsons, D. F.; Ninham, B. W. *Langmuir* **2010**, *26*, 1816-1823.
14. Parsons, D. F.; Deniz, V.; Ninham, B. W. *Coll. Surf. A* **2009**, *343*, 57-63.
15. Parsons, D. F.; Ninham, B. W. *Langmuir* **2010**, *26*, 6430-6436.
16. Gutierrez, C.; Abee, T.; Booth, I. R. *Int. J. Food Microbiol.* **1995**, *28*, 233-244.

17. Wilson, G. S.; Topley, W. W. C.; Collier, L. H.; Parker, M. T. *Topley and Wilson's Principles of Bacteriology, Virology and Immunity*, Edward Arnold, London, 1990.
18. Gram-negative bacteria are not stained dark blue or violet by Gram staining. They possess a layer of lipopolysaccharide. The wall of Gram-negative bacteria is similar to a cytoplasmic membrane, only a few layers thick, and generally thinner than Gram-positive types. Gram-positive bacteria are stained dark blue or violet by Gram staining. The stain is caused by a higher amount of peptidoglycan in the cell wall, which typically lacks the secondary membrane and lipopolysaccharide layer found in other bacteria strains.
19. Suzuki, I.; Lee, D.; Mackay, B.; Harahuc, L.; Oh, J. K. *Appl. Env. Microbiol.* **1999**, *65*, 5163-5168.
20. Lo Nostro, P.; Ninham, B.W.; Lo Nostro, A.; Pesavento, G.; Fratoni, L.; Baglioni, P. *Physical Biology* **2005**, *2*, 1-7.
21. Kim, H. K.; Tuite, E.; Nordén, B.; Ninham, B. W. *Eur. Phys. J. E* **2001**, *4*, 411-417.
22. Pinna, M. C.; Salis, A.; Monduzzi, M.; Ninham, B. W. *J. Phys. Chem. B* **2005**, *109*, 5406-5408.
23. Pinna, M. C.; Bauduin, P.; Touraud, D.; Monduzzi, M.; Ninham, B. W.; Kunz, W. *J. Phys. Chem. B* **2005**, *109,* 16511-16514.
24. Lee, Y.; Cesario, T.; Owens, J.; Shanbrom, E.; Thrupp, L. D. *Nutrition* **2002**, *18*, 665-666.
25. Hallsworth, J. E.; Prior, B. A.; Nomura, Y.; Masayoshi, I.; Timmis, K. N. *Appl. Environ. Microbiol.* **2003**, *69*, 7032-7034.
26. Wood, J. M.; Bremer, E.; Csonka, L. N.; Kraemer, R.; Poolman, B.; van der Heide, T.; Smith, L. T. *Comp. Biochem. Physiol. A* **2001**, *130*, 437-460.
27. Collins, K. D. *Biophys. J.* **1997**, *72*, 65-76.
28. Herberhold, H.; Royer, C. A.; Winter, R. *Biochemistry* **2004**, *43*, 3336-3345.
29. Hurst, A.; Hughes, A.; Pontefract, R. *J. Microbiol.* **1980**, *26*, 511-517.
30. Müller, V.; Oren, A. *Extromophiles* **2003**, *7*, 261-266.
31. Jones, R. P. *J. Appl. Bacteriol.* **1987**, *63*, 153-164.

32. Robinson, R. A.; Stokes, R. H. *Electroyte Solutions*, Butterworths Scientific Publications, London, 1959.
33. Ru, M. T.; Hirokane, S. Y.; Lo, A. S.; Dordick, J. S.; Reimer, J. A.; Clark, D. S. *J. Am. Chem. Soc.* **2000**, *122*, 1565-1571.
34. Lo Nostro, P.; Fratoni, L.; Ninham, B. W.; Baglioni, P. *Biomacromolecules* **2002**, *3*, 1217-1224.
35. Pyper, N. C.; Pike, C. G.; Edwards, P. P. *Mol. Phys.* **1992**, *76*, 353-372.
36. Vittadini, E.; Chinachoti, P. *Int. J. Food Sci. Technol.* **2003**, *38*, 841-847.
37. van Stappen, G. *Introduction, biology and ecology of Artemia. In Manual on the production and use of live food for aquaculture*, P. Lavens and P. Sorgeloos, FAO, Rome, 1996.
38. Camargo, W. N.; Durán, G. C.; Rada, O. C.; Hernández, L. C.; Linero, J.-C. G.; Muelle, I. M.; Sorgeloos, P. *Saline Systems* **2005**, *1*, 9.
39. Pankhurst, N. V.; Boyden, C. R.; Wilson, J. B. *Environ. Pollut.* **1980**, *23*, 299.
40. Camargo, J. A. *Chemosphere* **2003**, *50*, 251-264.
41. Bhunia, F.; Saha, N. C.; Kaviraj, A. *Bull. Environ. Contam. Toxicol.* **2000**, *64*, 197-204.
42. Epstein, F. H.; Maetz, J.; Renzis, G. *Am. J. Physiol.* **1973**, *224*, 1295-1299.
43. Henry, C. L.; Craig, V. S. J. *Langmuir* **2010**, *26*, 6478-6483.

VI
Aqua Reticulata: topology of liquid water networks

Stephen T. Hyde

Department of Applied Mathematics,
Research School of Physical Sciences,
Australian National University, Canberra, A.C.T. 0200, Australia.
stephen.hyde@anu.edu.au

Introduction

Almost 80 years have elapsed since Bernal and Fowler explored the notion of structure at the atomic scale in liquid water.[1] While not the earliest attempt to explain the physical anomalies of liquid water, the work remains a firm starting point to rationalise the structure of liquid water as a function of temperature. At last count, these number more than sixty distinct anomalies,[2] e.g. melting and boiling temperature, density, compressibility and specific heat variations with temperature.[3] Two related principles, both deduced from relatively primitive X-ray scattering studies of water, were articulated in Bernal and Fowler's paper. First, water has a tendency to form networks with tetrahedral symmetry. The net edges are hydrogen-bonded O · · · H − O units, branched at oxygen vertices (the net nodes), with four edges at each vertex. Second, liquid water is a mixture of two distinct structural forms of low and high density, whose proportions vary on heating or cooling. The lower density form was related to silica polymorphs, namely tridymite (structurally equivalent to the Ice I_h, the most common form of ice at ambient pressure) and quartz. (Later, following the discovery of a new intermediate silicate phase (keatite), Bernal suggested this silicate phase as a better water analogue.[4]) He proposed that on heating, liquid water approaches the structure of more common ideal liquids (such as metallic melts), and adopted a cubic close-packed model for the high-density form. Eventually, Bernal rejected that "two-state" model as being too crystallographic, and focussed instead on structural features of random networks, or "heaps",[4] whose local coordination was approximately tetrahedral, initiating the field of statistical geometry. In this model, he assumed that liquid water is a single homogeneous "continuum" material rather than a mixture, similar to more recent descriptions of packed granular materials such as sand piles. Like conventional covalently bonded crystals, heaps can be described as nets, albeit geometrically disordered, given some criterion for the presence of edges linking pairs of grains (e.g. minimum separation).

Since Bernal's time, the rise of numerical simulations, and the development of synchrotron and neutron scattering as well as other experimental techniques, has seeded newer models. Variants of the two-state model have been repeatedly proposed, characterised by states of relatively lower and higher density, (see, for example,[2,5]) and the putative presence of smaller polymolecular water clusters, floating in a sea of more or less disconnected molecular species. Further, the formation of water "strings" in the liquid state has been proposed on the basis of absorption spectroscopy data,[6] since disputed.[7] It is somewhat sobering to realise that despite the central importance of water structure, our picture of the structure of liquid water remains unresolved. Evidence supporting these structurally distinct models: multi-state, string and continuum are still inconclusive.

A part of that apparent confusion surely lies in the various techniques used to probe "water structure". Distinct techniques from X-ray and neutron scattering, to X-ray Raman (XRS), x-ray absorption (XAS) and x-ray emission spectroscopy (XES) sample widely disparate time and length scales. It is agreed that fluctuations in local structure are very rapid. For example, protons freely shuttle back and forth between adjacent oxygen atoms, flipping hydrogen bonds from O $\cdots$ H − O to O − H $\cdots$ O, probably occurring at the picosecond time scale. In the analysis below, we ignore those small-scale temporal and spatial fluctuations, to construct a "zeroth-order" model, that averages over macroscopic volumes and very short time scales, corresponding to the effective structure sensed by scattering studies (that probe a macroscopic water sample at attosecond time scales). The model is strictly topological in scope, without appeal to physics or chemistry. I am interested in exploring what can be concluded about liquid water in the absence of any underlying hypotheses of geometric, chemical or physical nature. We shall see that this bold mission is in fact impossible. To arrive at some structural predictions then, I introduce some (and as little as possible) metric data, namely O $\cdots$ H − O distance data and surface to volume ratios of tessellating surfaces.

Here I revisit Bernal's original hypothesis in the light of more recent data. My primary goal is to construct as simple a model as possible that is consistent with current data. I emphasise that this analysis cannot clarify the "higher order" models guiding debates around local spatial and temporal fluctuations in bulk water. Hopefully, however, it can afford a simple foundation on which more refined decorations must rest.

How does this analysis relate to the topic of the celebrated debate between Galileo Galilei and his adversary Ludovico delle Colombe? I will not address the central issue of the debate: why these densities adopt their relative values in crystalline and amorphous states. Rather, the analysis outlined in this article relies on accurate density data as an input to deducing the network structure of liquid water. In my view, it is crucial to deduce a firm picture of the structure of liquid water before engaging in the original debate topic. Therefore, this contribution wrestles with that (difficult) preamble to the debate only.

2 Current best data on local structure of water

Scattering studies of liquid water are useful probes of bulk water liquid structure. The technique differs little from the earliest efforts to understand liquid water via X-ray diffraction by Bernal, a pioneering crystallographer without peer. Scattering methods cannot tell us the actual geometry of the arrangement of water molecules in the liquid state, due to the absence of long-range geometric order. However, they allow reconstruction of the atomic two-point correlation function (or radial density function, "rdf") between pairs of hydrogen and oxygen atoms in the sample – a coarse, but somewhat helpful structural measure.

Most recently, Skinner and colleagues have collected and analysed an impressive suite of scattering data collected from synchrotron radiation.[8] Reconstruction of the water rdf is significantly better than previous efforts, due to instrumental advances. Data was analysed from a range of different water samples (both confined and unconfined) from -38°-100°C.[9] In order to obtain $g(r)_{O-O}$ between oxygen atoms alone, the H – H and O – H contributions must be subtracted. Here, additional data is required, as direct probes of O and H contributions to the synchrotron scattering are impossible. Skinner *et al.* use neutron scattering data, arriving at a concise estimate of $g(r)_{O-O}$. The analyses reveal a linear dependence of the (spatially average) $O \cdots H - O$ distance in liquid water, between 2.795Å at 0° and 2.827Å at 100°, viz.:[9]

$$d(O - O) = 0.00032T + 2.7954 \qquad (1)$$

where $d(O - O)$ is the inter-oxygen spacing in Å and T is the temperature (°C).

Integration of $g(r)_{O-O}$, to give the number of O atoms coordinated to a central O, revealed an isosbestic crossover at $d(O - O) = 3.32$Å at all temperatures, with a coordination number of 4.3.[9] They propose this distance then as characteristic of the first coordination shell of

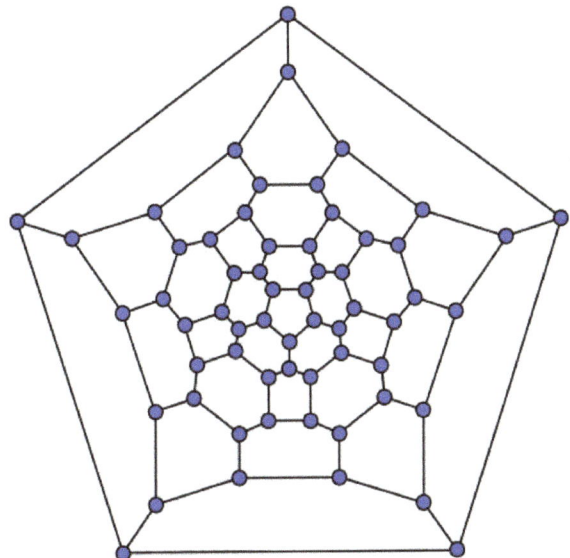

Figure VI.1. The C_{60} net (**tic** in Ref. 11) drawn as a Schlegel diagram in the two-dimensional euclidean plane, via (for example) stereographic projection from the net on the sphere to the complex plane. The two-dimensional rings of C_{60} are 5- and 6-rings only.

O atoms in the average water liquid structure, whose dimensions are fixed between melting and freezing. This assumption is arguable, given that the van der Waals radius of O atoms is 1.55Å, implying that O atoms that are not H-bonded are also counted within the first coordination shell (also going some way to explain the fact that the coordination number exceeds 4, that of an ideal tetrahedral liquid). It is however a useful working assumption for further analysis. Indeed, from additional identification of the second "coordination shell", the authors deduce O−O−O angles that are very close to the ideal tetrahedral angle (109.5°), varying between 108 − 112°.[9]

This data offers firm support then for Bernal's original structural hypothesis: that liquid water is a tetrahedral network, at least averaged over macroscopic volumes. In our topological analysis that follows, we adopt the simpler assumption that liquid (and crystalline) water is a network with (on average) four H-bonded neighbours around each O atom. From a topological perspective, a network is a spatial embedding (in our case in three euclidean dimensions) of a graph, a suite of connected edges and vertices with neither placement in space nor metric extent. Further, the location of the H atom between H-bonded O atoms (i.e. O $\cdots$ H $-$ O) assigns a direction to each edge, so the water network is a directed graph at any instant, with, at least on average, two incoming and two outgoing edges linked to each O vertex within the graph. The edge directions are relevant to liquid water physics, but we ignore them here, and consider only the undirected graph, which we assume is a degree-four graph (averaged over all vertices), corresponding to the four edges emanating from each vertex. Here I use the term "graph" to denote the topological structure of a network and "net" the spatial embedding of the graph (adopting, loosely the terminology in Ref. 10, with the exception that "nets" here can be geometrically ordered and crystalline, or disordered).

2.1 Two and three-dimensional rings in a net embedded in three-dimensional space

The graph topology is captured, in part, by its ring structure. In an infinite network (e.g. the ideal network of Ice I), cycles with arbitrarily large numbers of vertices are formed. There is, however, a minimal set of smallest rings whose union defines the entire net. "Strong rings" have been defined for periodic graphs embedded in three spatial dimensions via tilings, by O'Keeffe and colleagues.[10] However, this set of cycles (which are the three-dimensional rings) includes strong rings whose edges are also contained in more than one other ring, giving closed, polyhedral three-dimensional "tiles". We prefer to exclude those redundant rings, giving a minimal set of rings, such that each edge is common to only two rings. One way to detect those rings is to find an oriented 2-manifold (a two-dimensional surface) able to support a reticulation whose topological structure is that of the graph. This construction is clear in the case of simpler finite nets that form reticulations of the 2-sphere. Think, for example, of the degree-3 net of sp^2 carbon atoms in the fullerene C_{60}. Cycles can be found in the net with up to 20 vertices, but the net sits in the two-dimensional sphere such that all cycles can be made up of sums of edges of the elemental 5- and 6-rings in the net (see Figure VI.1). We intuit a 2-sphere embedding of the buckminsterfullerene net, since the smallest cycles in the net form a reticulation of that 2-manifold. An underlying manifold for more complex nets is often less evident, and two-dimensional geometry, which classifies 2-manifolds according to its topological genus and underlying Gaussian curvature, offers a useful guide.

For example, the graphene net (**pbz**) is often described as planar, since it reticulates the usual two-dimensional euclidean plane. (For convenience, I label nets and their underlying topological graphs by a three-letter code in **bold script** from the RCSR catalogue)[11] Since this net has three hexagons around each vertex, we describe it by the two-dimensional vertex symbol (6.6.6). The two-dimensional vertex symbol encodes in cyclic order the ring-sizes of all the faces that share a vertex in the net. In contrast, the fullerene C_{60} net (**tic**), with vertex symbol (5.6.6), reticulates the two-dimensional sphere. If 8-rings are inserted into the graphite network, giving a net with vertex symbol (6.8.8), the resulting net reticulates hyperbolic 2-manifolds, with negative Gaussian curvature (in place of the zero and positive Gaussian curvature characteristic of the plane and two-dimensional sphere respectively). Examples, reticulating the G, D and P triply periodic minimal surfaces, can be found on-line (at *epinet.anu.edu.au/sqc12886*, *epinet.anu.edu.au/sqc9271* and *epinet.anu.edu.au/sqc9265* respectively.

All three crystalline nets are derived from a single underlying hyperbolic (6.8.8) net (*epinet.anu.edu.au/UQC19*). This hyperbolic, two-dimensional picture of crystalline nets is at first encounter an odd way to describe these structures, that are, after all, three-dimensional crystals. For example, the **pbz** net (see *rcsr.anu.edu.au/nets/pbz*), a novel form of graphite, is more conventionally described as a cubic array of twisted 6-rings, arranged in a three-dimensional lattice. But this net topology is exactly reproduced by the (6.8.8) reticulation of the D-surface, mentioned above.

Like the plane and sphere, the D-surface is a two-dimensional space, or a 2-manifold. König's Theorem guarantees that we can find a 2-manifold for any net. More accurately, the theorem asserts that any connected graph may be embedded in an orientable surface to form the edges and vertices of a map.[12] A map is a set of simply-connected regions (faces on the surface), bounded by the edges and vertices of the graph. Note that a face may have a single edge that appears more than once on its boundary, in which case the face winds surrounds a channel of the surface. Given that possibility, the theorem is almost trivial, since for any graph we can form a map simply by inflating all edges to tubes, and merging the tubes smoothly, to give a sponge-like (and almost inevitably hyperbolic) surface, with one tube per graph edge. That embedding has just one face, with ring-size equal to the total number of edges on the graph. In general, however, this is a redundant embedding, with far more tubes, and far larger faces, than necessary. Among the various possible two-dimensional embeddings for the net, we choose the minimal embedding,[13] with the simplest topology among all oriented surfaces. (Note that here the term "minimal" refers to the surface genus, and is unrelated to "minimal surfaces", discussed below.) This topological constraint is a sensible one, since it implies (via Euler's formula, below) that we find the set of smallest faces, or two-dimensional rings. That choice is equivalent to fitting an oriented 2-manifold through the three-dimensional net such that it passes through all net nodes and edges, and maximises the two-dimensional density of nodes in the manifold (the number of nodes per unit area). (This two-dimensional density is distinct from the three-dimensional density.) For example, a graph of edges of the tetrahedron can be embedded on a tubular surface with three channels emerging from all four distinct nodes, and meeting up to form a tubular surface whose tubes lie on edges of the tetrahedron. That embedding (which is maximal) is on a genus-two surface, with one face of ring-size equal to the number of edges, six (and two-dimensional vertex symbol (6)). In contrast, the minimal embedding is on the genus zero sphere, with four triangular faces (with two-dimensional ring-size equal to three, and two-dimensional vertex symbol (3.3.3)). In general, the detection of the (minimal, oriented) 2-manifolds for an arbitrary infinite net, crystalline or disordered, is a difficult problem. (The inverse problem, to enumerate infinite crystalline nets as reticulations of 2-manifolds, forms the basis of the *Epinet* project.[14]) The net topology software *Topos* can, in many cases, detect 2-manifolds that support infinite crystalline nets and offers a useful route,[15] though the full power of this numerical approach requires further exploration. In general, the minimal manifold is not unique. However, in some cases, we do know the 2-manifold minimal embedding. A fuller description of some known examples can be found elsewhere.[16] In general, the two-dimensional rings are a subset of the rings described by the extended three-dimensional vertex symbol.[17] They are also a subset of the rings that describe faces of the three-dimensional tiling, introduced by O'Keeffe (Ref. 10 and listed on-line at *rcsr.anu.edu.au/nets*).

The relationship between two-and three-dimensional rings is clear for simpler infinite nets. For example, in the **sod** net (the alumino-silicate skeleton of the zeolite sodalite, *rcsr.anu.edu.au/nets/sod*) a subset of the three-dimensional rings lie in the 2-manifold, forming the two-dimensional rings, while the remaining three-dimensional rings form "collars" around channels of the manifold. The three-dimensional ring symbol (the vertex symbol) for **sod** is (4.4.6.6.6.6).[17] If a pair of 4-rings sharing a common vertex (and no edges) is selected as collar rings, the P surface emerges as the support surface, and the two-dimensional rings are the remaining rings, giving two-dimensional symbol (6.6.6.6) (as illustrated at *epinet*.anu.edu.au/*UQC3*); selection of a pair of 6-rings as the collars results in the minimal embedding on the D surface, with two-dimensional symbol (4.6.4.6) (illustrated at *epinet.anu.edu.au/UQC4*). (Here just one symbol describes all vertices, since they are topologically equivalent in the **sod** net.). In fact, a number of embeddings of hyperbolic nets on minimal surfaces give the **sod** net, as shown at *epinet.anu.edu.au/sqc970*. The minimal embedding for sod is the (4.6.4.6) reticulation of the D-surface, since this has the smallest average two-dimensional ring-size. A more detailed discussion of this issue can be found at Ref. 18.

It is useful to introduce an average two-dimensional ring-size,

$$\langle n^{(2)} \rangle := \sum_i \frac{z_i}{\sum_j \frac{1}{n^{(2)}_j}}$$

where $\sum_i$ denotes the sum over all net vertices, and $\sum_j$ the sum over all z_i two-dimensional rings that share vertex i. So, if the two-dimensional vertex symbol for a degree-four net is (n_1, n_2, n_3, n_4),

$$\langle n^{(2)} \rangle := \frac{4}{n_1^{-1} + n_2^{-1} + n_3^{-1} + n_4^{-1}}$$

(so that, for example, $\langle n^{(2)} \rangle = 24/5$ for the (4.6.4.6) minimal embedding of **sod** discussed above. Values of $\langle n^{(2)} \rangle$ for various silicates can be taken from earlier work.[16] Note that the estimated values of $\langle n^{(2)} \rangle$ for keatite (**lon**) and coesite (**coe**) are unknown, however, they can be bounded above and below from the face-sizes in the tilings of these nets.[11] Keatite (**kea**, described at *rcsr.anu.edu.au/nets/kea*) has three-dimensional rings (5.5.5.7.8.8) (4 x multiplicity per unit cell) and (5.7.5.7.5.7) (8x multiplicity). Likely two-dimensional ring symbols are therefore ((5.5.5.7) (4x) and (5.7.5.7) (8x), where the first vertex type is assumed to have larger 8-rings surrounding channels (collar rings). The two-dimensional ring-size is then:

$$\left\langle n^{(2)} \right\rangle = \left(\frac{4.8}{5^{-1}+5^{-1}+5^{-1}+7^{-1}} + \frac{4.8}{5^{-1}+7^{-1}+5^{-1}+7^{-1}} \right) \cdot (12)^{-1} = 5.68$$

Similarly, the three-dimensional rings of the rings of the **coe** net (*rcsr. anu.edu.au/nets/coe*) have symbol (4.8.4.9.6.8) and (4.6.4.6.8.9). The two-dimensional ring-sizes are therefore bounded below by (4.8.4.9) and (4.6.4.6) and above by (4.9.6.8) and (4.6.8.9), corresponding to an average two-dimensional ring-size, $\left\langle n^{(2)} \right\rangle$, within the interval [5.12, 6.13]. (I have used the notation $\left\langle n^{(2)} \right\rangle$ to emphasise that this refers to the two-dimensional measure of ring-size, which excludes all rings except those that bound faces that tile the underlying 2-manifold. This is distinct from the usual three-dimensional ring-size, which includes collar-rings and other cycles whose homotopies are more complex than the (null homotopic) cycles contributing to $\left\langle n^{(2)} \right\rangle$).

3 Silicate network polymorphism

Water and silica are the dominant chemical species on the earth's surface. They share structural features: for example, both typically form tetrahedral networks. It is often stated that – like framework silicates – water exhibits an unusual degree of structural flexibility, evidenced by the wealth of ice phases formed at low temperature under pressure. However, its polymorphism pales in comparison to silicates. While less than twenty phases of ice are known, over a hundred distinct crystalline (alumino-)silicate frameworks are recognised and catalogued.[19]

Silicates can form zero-, one-, two-, or three-dimensional nets, according to the number of independent translations vectors that define their structure. Three-dimensional "framework" (or "tecto") silicates are conveniently sorted into three classes, according to their three-dimensional framework density: "dense", "intermediate" and "rare" (Table 1).

Table 1: Tetrahedral silicate networks, listed according to their idealised density, from dense silicates to low-density zeolite frameworks (all assumed to have stoichiometry SiO_2). Density and ring-size parameters are defined in the main text.

class	examples	net ID	density, ρ_g	topological density, TD_{10}	2d ring size, $\langle n^{(2)} \rangle$
dense	coesite	**coe**	869	1318	5.12-6.13
	α-quartz	**α-qtz**	790	1231	6
	keatite	**kea**	746	1225	5.68
	β quartz	**β-qtz**	686	1231	6
	tridymite	**lon**	666	1027	6
	amorphous silica		660		
intermediate (clathrasils)	melanophlogite	**mep**	580	1058	5
rare (zeolites)	sodalite	**sod**	530	791	4.80
	analcime	**ana**	570	933	4.80
	gmelinite	**gme**	450	694	4.57
	zeolite ZK-5	**kfi**	450	681	4.57
	Linde Type-A	**lta**	430	641	4.80
	zeolite rho	**rho**	430	641	4.36
	faujasite	**fau**	380	579	4.36

The physical density of framework silicates is a measure that often depends on the presence of interstitial or intra-framework species and partial substitution of silicon atoms by aluminium or other cations in the framework, as well as variable degrees of framework collapse on dehydration. In order to compare silica with water polymorphs,

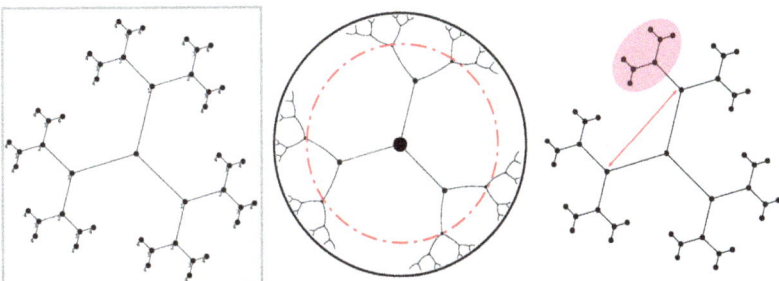

Figure VI.2. (a) A fragment of a degree-three tree. Vertices of the tree are labelled with integers corresponding to the number of edges between the central (origin) node of the tree and those vertices. Shells within the tree fragment are formed by the set of all vertices with equal indices. (b) Embedding of the degree-three tree in the (Poincaré model of the) hyperbolic plane, such that all edges are of equal length, and the tree is regular with symmetrically identical vertices and edges. Vertices in equivalent shells now lie on the perimeters of hyperbolic circles. (One shell is indicated by the dashed circle.) (c) Formation of a three-ring in the tree by fusing the arrowed tree vertices. Closure of an *n*-sided ring is accompanied by pruning of all vertices in one branch beyond one vertex in shell]*n*/2[(where] [denotes the integer part of the fraction).

consider an idealised silicate net, with silicon atoms at net nodes, and oxygen anions at mid-edges. We therefore consider the hypothetical net, containing only tetrahedral atoms at nodes, in its fully expanded form, realised by the most symmetric, barycentric embedding of the net topology.[20] A dimensionless measure of the geometric net density,

$\rho_g := 10^3 \frac{vl^3}{V}$ where v is the number of tetrahedral nodes in the

volume V, and l is the length of net edges (which are generally all equal in the maximally symmetric and least dense form).

Table 1 shows a general trend of increasing (three-dimensional) density with increasing ring-size. This correlation, at first glance counter-intuitive, nevertheless emerges naturally within a two-dimensional perspective by the following argument (adapted from the original discussion of Bukowinski et al.[21]). First, we define a measure of three-dimensional density that is independent of the geometry of the net: the topological density. This number is formed from a decomposition of the net into topological "shells", formed as follows. Each vertex is labelled with an index, v_i, equal to the number of edges in a shortest path traversed from that vertex back to a vertex (v_0) assigned as the origin. The net is thus layered into concentric shells containing the set of vertices with $v_i = 0,1,2,3,...$, and the number of vertices in each successive shell, N_i, forms an integer coordination sequence.[22] Figure VI.2(a) illustrates this construction for a degree-three tree. We define the topological density as:

$$TD_{10} := \sum_{i=0}^{10} N_i$$

The maximum possible topological density results from a (degree-z) net that is a tree, i.e. with no closed cycles, since then each node in the net acts a source for $z - 1$ nodes in the next shell, giving a topological density,

$$TD_S = \frac{S(z-2) + ((z-1)^S - 1)z}{z-2}$$

which grows exponentially fast with the maximum shell number S.

Exponential growth is consistent with the growth of shell perimeter with radius, itself exponential if the shells form a series of concentric circles of linearly increasing radius in the hyperbolic plane (H^2). This conflation of a topological feature, namely the topological density, with metric dimensions is a loose one, but qualitatively exact if one embeds the net such that all edges have equal length, in which case a tree indeed reticulates H^2 (Figure VI.2(b)).[23]

The formation of a cycle in the tree involves the pruning of all vertices from the tree due to a root vertex in the cycle, since that root vertex is "grafted" onto another vertex to form the closed cycle (Figure VI.2(c)). The smaller the ring, the closer to the origin of the net the root vertex lies, so the larger the number of vertices pruned from the tree to form the net. Conversely, large rings lose fewer vertices during this pruning and grafting operation. This argument explains qualitatively the paradoxical direct correlation of topological density with ring size in a net.

In order to derive a quantitative correlation, consider the embedding of the net in a 2-manifold, such that each vertex occupies an (assumed constant for simplicity) area of Ω in the surface. From differential geometry, we can express the geometric density, ρ_g, defined above, in terms of the net topology, via the degree of the vertices (the bond valency), z and the two-dimensional ring-size of the net, $\langle n^{(2)} \rangle$. The geometric density $\rho_g = \dfrac{A}{\Omega V}$ where A is the area of the 2-manifold within volume V. The Gauss-Bonnet Theorem allows us to express A in terms of the surface-average of the Gaussian curvature of the manifold over area A contained within the volume V, $\langle K \rangle := \dfrac{\iint_A K \, da}{A}$ namely $A = \dfrac{2\pi\chi}{\langle K \rangle}$, where χ is the Euler-Poincaré characteristic of the portion of the surface contained within volume V. The geometric density can then be rewritten in terms of a dimensionless surface-to-area measure, the "homogeneity index", $h := \dfrac{A^{3/2}}{\left(2\pi|\chi|\right)^{1/2} V}$

The geometric density is then: $\rho_g = \dfrac{h}{\Omega^{3/2}} \left(\dfrac{2\pi|\chi|}{V} \right)^{1/2}$.

The final form of the density relation follows from Euler's formula, which relates the characteristic χ to the net topology as follows. The usual form of Euler's formula is $\frac{\chi}{v} = 1 - \frac{e}{v} + \frac{f}{v}$ where v, e and f denote the number of vertices, edges and faces (bounded by the two-dimensional rings with size $\langle n^{(2)} \rangle$) of the net in the volume V. But $\frac{e}{v}$, the number of edges emanating from a single vertex, is equal to $\frac{z}{2}$, since each edge is associated with a pair of vertices. Likewise $\frac{f}{v}$ is equal to $\frac{z}{\langle n^{(2)} \rangle}$ since each vertex contains $\frac{1}{\langle n^{(2)} \rangle}$ of each of the z rings around that vertex. Euler's formula can then be rewritten in terms of the net topology only, viz.:[16]

$$\rho_g = \frac{h}{\Omega^{3/2}} \left(\frac{\pi\left((z-2)\langle n^{(2)} \rangle - z\right)^{1/2}}{\langle n^{(2)} \rangle} \right)$$

In keeping with the paper's focus on topology to the exclusion of geometry (as far as possible), we prefer to discuss topological density (TD_{10}) in preference to the geometric density measure ρ_g. This measure of density correlates well with geometric density as shown in Figure VI.3. Indeed, the correlation holds from the least dense framework silicate (faujasite, **fau**) to the densest tetrahedral silicate polymorph, coesite (**coe**), with an approximate relation:

$TD_{10} = k\rho_g$ and $k \approx 1.58$ for the silicate data of Figure VI.3.

It makes sense, then, to recast the issue of geometric structure, and derivative quantities, including geometric density, in terms of topological measures alone. We can write a "normalised" topological density as:

$$TD_{10} = k \frac{h}{\Omega^{3/2}} \left(\frac{\pi\left((z-2)\langle n^{(2)} \rangle - z\right)^{1/2}}{\langle n^{(2)} \rangle} \right) \qquad (2)$$

This equation describes surprisingly well the variation of topological density with two-dimensional ring-sizes, with a single parameter, $c := k\dfrac{h}{\Omega^{3/2}} = 524$, as shown in Figure VI.4. Though possibly treasonable, it is nevertheless not unreasonable to claim that the very simple picture sketched here, of three-dimensional chemical networks as two-dimensional reticulations of hyperbolic manifolds is a valid first-order approximation.

Three parameters influence the magnitude of the topological density: the scaling between topological and geometric densities, k, normalised surface-to-volume ratio, h, and the area in the 2-manifold occupied by each vertex (i.e. SiO_2 group), Ω. The silicate density data suggests that k is constant. The fraction $\dfrac{h}{\Omega^{3/2}}$ is therefore also

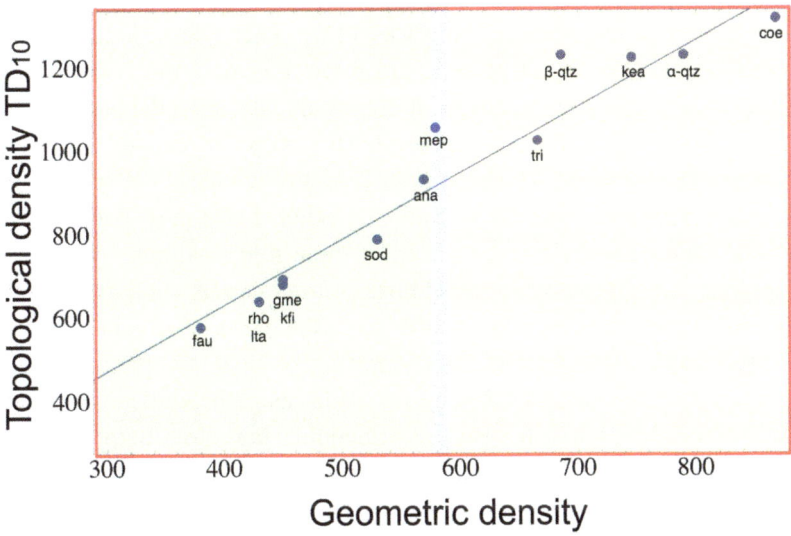

Figure VI.3. Topological *vs.* geometric density for a number of degree-four (tetrahedral) silicate nets. Polymorphs are labelled by their three-letter net code from Ref. 11 and split into three regions: zeolites, clathrates and dense silicates, listed in Table 1.

approximately constant for the range of silicates, from rare zeolites to the dense frameworks. The value of h is dependent on the geometry of the 2-manifold and its embedding in three-space. The simplest hypothesis is that these 2-manifolds, like those for many zeolites, are in fact 3-periodic minimal surfaces, or, simpler still, embeddings of the canonical hyperbolic manifold, one of uniform negative Gaussian curvature in three-space, with zero mean curvature. Differential geometric arguments then imply that $h = \frac{3}{4}$.[24] If we assume that this holds for the tetrahedral silicates, the area per silica node, Ω, must also be fixed, regardless of three-dimensional density. In other words, although the three-dimensional density of tetrahedral tecto-silicates varies from very porous zeolites to dense coesite, their two-dimensional density remains fixed. That must follow if h is indeed fixed for the 2-manifolds in which the tetrahedral nets are embedded, and close to its value known for some zeolites from their reticulation on three-periodic minimal surfaces (3/4). This question cannot be resolved definitively at present, since we not know the 2-manifolds that give minimal embeddings of denser silicates than the zeolites. In fact, it is certain that melanophlogite (the **mep** net) cannot reticulate an intersection-free hyperbolic surface via a subset of its strong rings.[25] This may also be a feature of other dense nets, particularly those with edges lying along axes of three-fold rotational symmetry in their most symmetric embeddings in three-space, such as **dia** and **lon** nets. In those cases, eq. 3 fails, since the embedding is in surfaces with degree-three branch lines, forming cellular complexes rather than 2-manifolds. Alternatively, we could insist the embedding be in a 2-manifold, in which case the embedding may be a map, but one whose faces have edges appearing more than once in their bounding two-dimensional rings (analogous to the maximal embedding of the tetrahedral graph mentioned on p. 150). For example, the **dia** net embeds in the D-surface as a tree, with a single face and unbounded two-dimensional ring-size, so that formally $\langle n^{(2)} \rangle = \infty$.[24,26] In that case, Euler's formula fails, and so eq. 3 no longer holds.

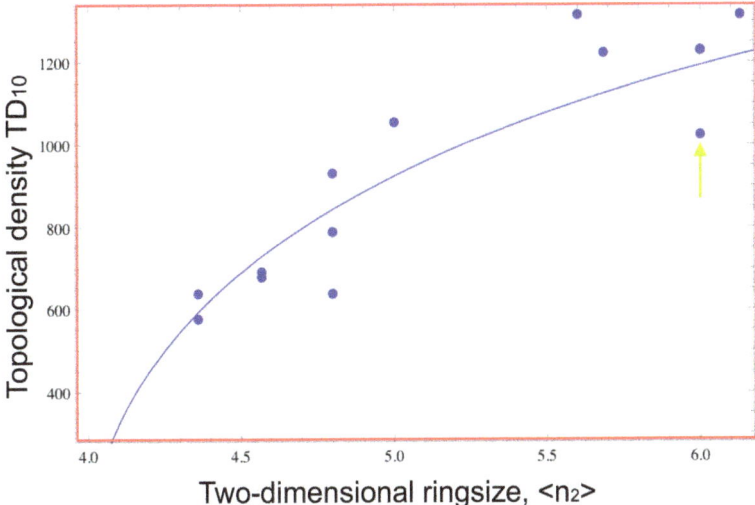

Figure VI.4. Topological density as a function of two-dimensional ring-size for a number of degree-four (tetrahedral) silicate nets. The arrow marks tridymite data. The curve is a best fit to the data via equation 3.

4 Amorphous silica

We can use these fits to infer the simplest topological feature of amorphous or fused silica as follows. Though it is clearly simplistic to suppose that "amorphous silica" is a well-defined polymorph of silica, like the zeolites, clathrates and dense silicates, this assumption is a reasonable one in silicate science, given the characteristic physical properties of amorphous silica. Since it has no geometrically defined crystalline structure, this may seem naïve. However, lack of crystalline (and hence geometric) order, need not imply lack of topological order, to some degree. And the topological analysis developed above can in principle allow us to probe the topological order of fused silica, knowing only its density. Recall, that the analysis is predicated on the notion of a reticulation of a 2-manifold. That picture explains *a priori* the counter-intuitive relation of increasing three-dimensional density with increasing ring-size inferred from the full range of crystalline tetrahedral polymorphs of silica, from low-density zeolites to dense tectosilicates. Limited simulations of fused silica, inspired by studies of shock-induced densification of fused silica, are consistent with this two-dimensional view. In common with the relation between ring-size and density explained here in terms of 2-manifolds, those simulations also show increasing ring-size on densification.[27] Those studies report a (three-dimensional!) average ring-size distribution between 3- and 10-rings, and a significant change in the network topology on densification. In the absence of a full discussion of their ring-size detection algorithm, quantitative comparisons between $\langle n^{(2)} \rangle$ and the ring distributions reported from these simulations are uncertain. Kubota *et al.* write that laser-induced densification of amorphous silica causes a

> persistent increase in the 3 and 4-membered rings compared with the unshocked configuration. In addition, the 7- and larger-membered rings also show persistent increases in their contributions to the ring distribution. These persistent changes in the ring coordination indicate the formation of a stable phase of permanently densified fused silica. The large increases observed in the larger rings also suggest the disruption of the fused silica network, and may indicate the existence of cracks, and microvoids. The increases in the smaller strained rings are a plausible connection to the observed permanent densification. The role of the larger rings upon densification is not clear. However the ability for larger rings to exist in compact non-planar configurations suggests that their impact on density is small.

Within a two-dimensional perspective, their report of an increase in large rings (and hence, we suspect, of $\langle n^{(2)} \rangle$) on densification is very reasonable, and consistent with data for crystalline silicates. Indeed, those larger rings are responsible for the densification! The quote above attempts to write off the influence of large rings, ascribing densification to a very minor change in the number of three-rings, consistent with intuition from a three-dimensional picture. Their results are sensible, their discussion less so. The ring statistics found in simulations form a distribution between 3- and 10-sided rings, peaked between 5- and 6-rings (Figure VI.5).

Their work supports the working hypothesis of amorphous silica as a two-dimensional hyperbolic pattern. Now the density of (relaxed and unshocked) amorphous silica (fused silica glass) is equal to 2.203 g cm^{-3}. It is reasonable to assume that the Si-Si distance in amorphous silica is the same as that of its crystalline counterparts, namely 3.1Å, so that the geometric density $\rho_g = 658$. We can then use the linear fit relating geometric and topological density for silicates

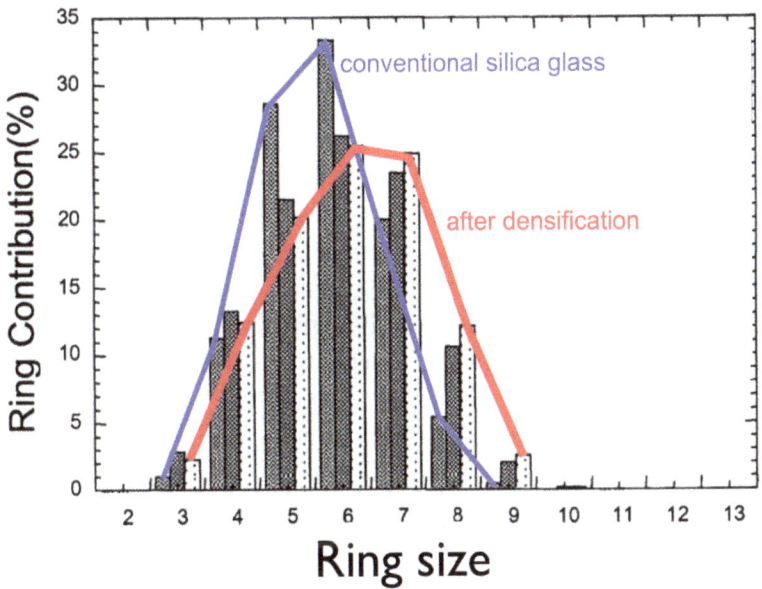

Figure VI.5. Distribution of ring-sizes before, during and after laser-induced densification of amorphous silica, deduced from numerical simulations, adapted from data reported in Ref. 27.

(Figure VI.3) to deduce the topological density, namely $TD_{10} = 1050$. Lastly, eq. 3 (with constant c from the fit shown in Figure VI.4) leads to an estimate of the average two-dimensional ring-size, $\langle n^{(2)} \rangle \approx 5.4$, illustrated in Figure VI.6. Though direct comparison with the ring statistics reported in Ref. 27 is uncertain, this value is close to the average reported by Kubota *et al.* (see Figure VI.5).

These authors also reported a densification of 20% on laser-shock treatment, which (assuming no change in Si-Si distances) implies that the denser phase has topological density $FD_{10} \approx 1150$, realised by a hyperbolic network with $\langle n^{(2)} \rangle \approx 5.8$. This increase too is consistent with the ring statistics reported in Ref. 27 (see Figure VI.5).

5 Ices

The close parallels between silicate and ice frameworks mean that the foregoing analysis that allows an estimate of the two-dimensional ring-size of the amorphous silica framework is easily modified to predict $\langle n^{(2)} \rangle$ for liquid water.

Unfortunately, the wealth of tetrahedral framework silicates is not matched by "pseudo-tetrahedral" ice phases. (I use this term to denote ice/water nets with four edges per node. Typically, the edges radiate from the centre of each tetrahedron to their four vertices. But since the analysis is topological rather than geometric, the edge geometry is irrelevant, and any degree-four netwotk is admissable.) Six distinct pseudo-tetrahedral ice crystalline polymorphs are known. Their structural data are tabulated in Table 2.

Table 2: Tetrahedral ice networks. Density and ring-size parameters are defined in the main text (*cf.* Table 1).

class	examples	net ID	density, ρ_g	topological density, TD_{10}	2d ring size, $\langle n^2 \rangle$
dense	ice II	**ict**	833	1333	6
	ice III	**kea**	833	1225	5.68
	ice Ih	**lon**	637	1027	6
	ice Ic	**dia**	649	981	6
intermediate	clathrate I	**mep**	549	1058	5
(*clathrates*)	clathrate II	**mtn**	545	1049	5

Like the silicate data, the topological and geometric densities for the ice frameworks are related by $TD_{10} = k'\rho_g$, where a best fit

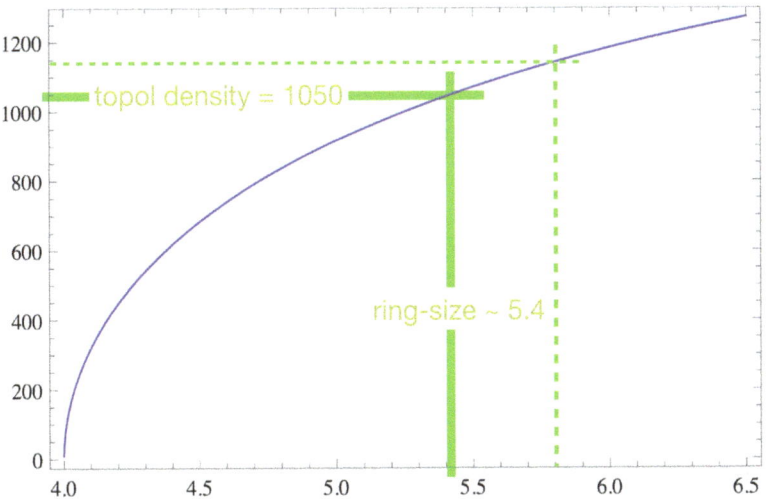

Figure VI.6. Estimate of the two-dimensional ring-sizes of normal and densified amorphous silica from their framework densities (see main text). The data suggests that amorphous silica has a two-dimensional ring-size, $\langle n^{(2)} \rangle \approx 5.4$ (full lines). The dotted lines show the corresponding data for laser-shocked densified silica.

gives $k' = 1.62$ (Figure VI.7). This is slightly larger than the constant k found for silicates, though the difference is insignificant given the scatter of the data.

The limited data also is also consistent with eq. 2, where the best fit is found for $c' = \dfrac{k'h}{\Omega^{3/2}} = 832$, as shown in Figure VI.8. Again, the paucity of data (compared with silicates) does not allow definitive estimate of the constancy or otherwise of k'. The data are, however, consistent with the functional form of eq. 2 assuming that c' is constant.

6 Liquid water

The density of liquid water at atmospheric pressure has been measured carefully between 0 and 100°C.[28] Combining that data with the recently measured O···H − O distances in liquid water (eq. 1) leads to the conclusion that the topological density of liquid water changes only

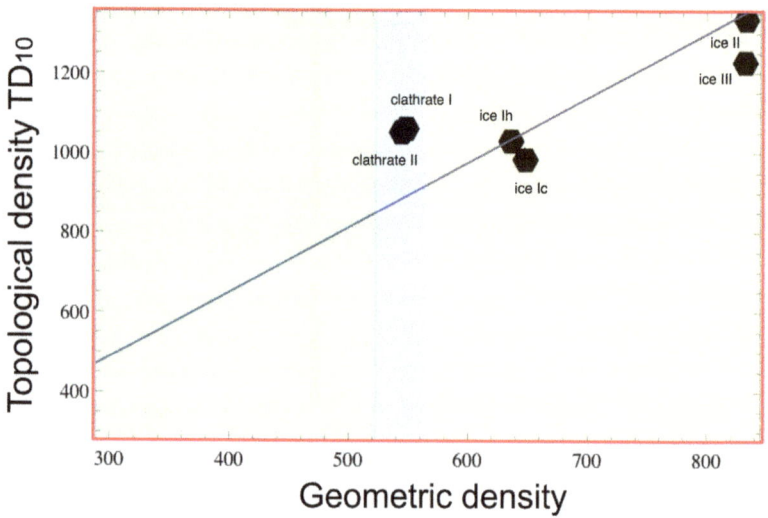

Figure VI.7. Topological vs. geometric density of crystalline pseudo-tetrahedral ice polymorphs (*cf.* Table 2).

Aqua Reticulata: topology of liquid water networks 169

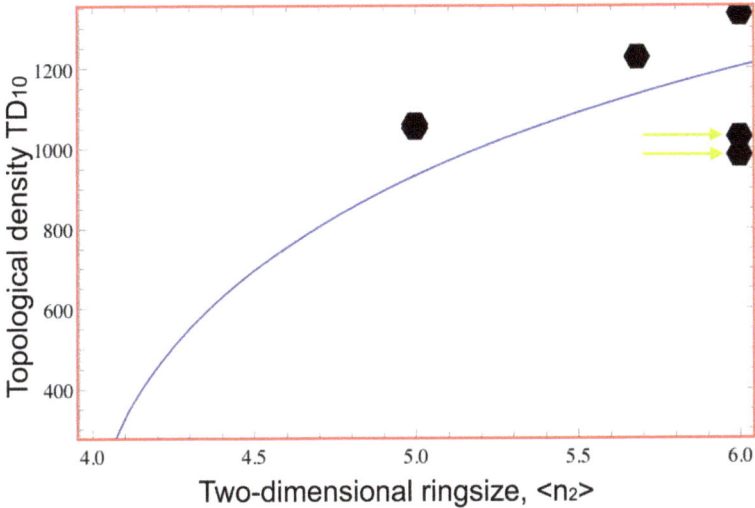

Figure VI.8. Topological density as a function of two-dimensional ring-size for ice polymorphs (*cf.* Figure VI.4.) The arrows mark data for hexagonal and cubic forms of ice I (lower and upper respectively).

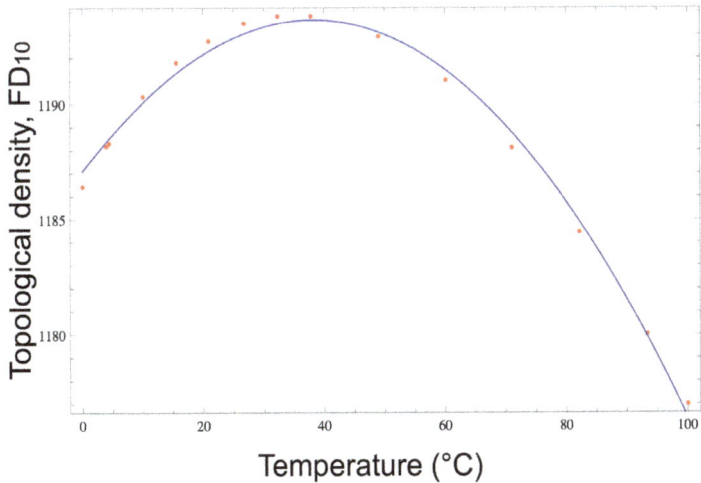

Figure VI.9. Topological density of water, TD_{10}, as a function of temperature at atmospheric pressure.)

marginally (between 1194 and 1176, peaking in density very close to body temperature) over the entire range from 0-100°C, as shown in Figure VI.9.

According to the fit deduced from the ice polymorphs (Figure VI.8), this variation in topological density is accounted for by a variation of the two-dimensional ring-size, $\langle n^{(2)} \rangle$ between [5.83 − 5.88], i.e. varying only very slightly over the complete temperature range up to 100°C, as plotted in Figure VI.10.

7 Closing and an auto-critique

What can be gleaned from this analysis? First, the concept of water as a tetrahedral framework, due to Bernal, remains reasonable. However, his and Fowler's original concept of liquid water as a continuum of states, from relatively open tetrahedral nets at lower temperatures, to denser patterns closer to its boiling point – reminiscent of close-packed structures – is not consistent with the most recent scattering analyses. These data suggest that the (geometric) density variations with temperature are not accompanied by significant changes in topological density, so that the network topology of liquid water is rather constant with temperature. This conclusion also argues against more recent "two-state" models of water, all of which imply measureable changes in water network topology with temperature. In addition, the supposed presence of water "strings" is not supported by this analysis.[6] Strings have a degree of just two, and the presence of strings would be expected to lower the (average) degree of the bonding net in liquid water, z, below four. In fact, as mentioned in the **Introduction**, scattering studies suggest an average first-shell coordination slightly larger than four.[8] Further, our analysis suggests that density is expected to increase with increasing z and $\langle n^{(2)} \rangle$ (cf. eq. 2). Since melting of ice is accompanied by an increase in density, were water strings to form (thereby lowering z) a compensating increase in $\langle n^{(2)} \rangle$ (beyond 6)

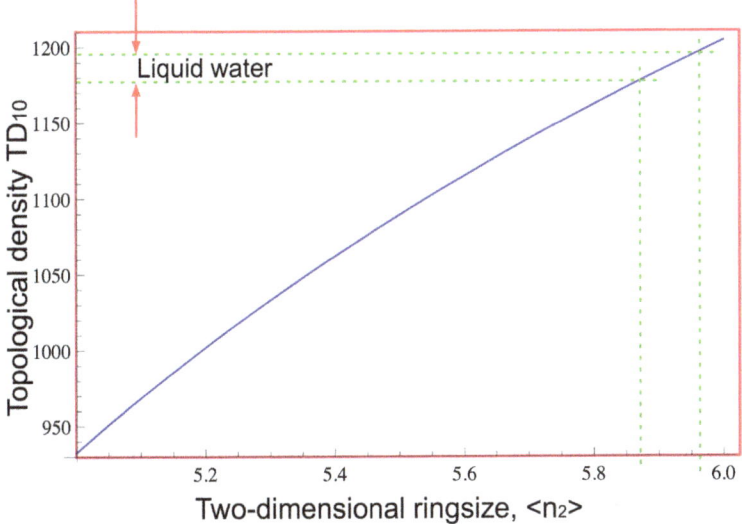

Figure VI.10. Estimate of the two-dimensional ring-sizes of liquid water between 0-100 °C (*cf.* Figure VI.6).

would be required. For example, if $\langle n^{(2)} \rangle = 6$ and $z = 4$ (corresponding to ice I), the topological density for ice/water (assuming $c' = 832$) is $T\Delta_{10} = 1183$. If z drops to (e.g.) 3.8 to accommodate strings, $\langle n^{(2)} \rangle$ must increase to larger than 6.6 to achieve a denser melt than ice, a severe topological change, for which there is no supporting evidence. Other water models have argued for clathrate-like clusters, forming finite domains (see, for example, Ref. 5). Evidently, finite cluster have $z < 4$, due to the boundary. So here too, large rings are essential to achieve the increased density within the cluster (and still larger to achieve the required bulk density, due to low density interstitial regions between clusters). Such models therefore appear unlikely candidates for liquid water structures, within this two-dimensional view.

This estimate of the two-dimensional ring-size of liquid water is dependent on the assumption that the "constants" linking topological and geometric density (k') and topological density and ring-size (c')

have fixed values for all pseudo-tetrahedral ice polymorphs as well as liquid water. Unfortunately, minimal 2-manifold embeddings for the ice polymorphs are unknown, in common with dense silicate networks. Given that difficulty, I can only estimate the crucial parameters governing the topology of these nets $\langle n^{(2)} \rangle$, h and hence Ω. It is therefore likely that the single topological measure extracted from this analysis, $\langle n^{(2)} \rangle$ for amorphous silica and liquid water, is an approximation only. To go further, more detailed explorations of minimal embeddings of (particularly dense) nets are required.

Reflection on the central issue that drove the original debate between Galileo and delle Colombe reveals the uncertainty in my simple analysis. I have argued that the (topological) density grows with increasing two-dimensional ring-size. How can it be then, that conventional ice (Ic), with $\langle v^2 \rangle \rangle \langle n^{(2)} \rangle$ floats on water $\langle v^2 \rangle$ ($\langle n^{(2)} \rangle$ Inspection of the density-ring-size data (Figs. VI.4 and VI.9) reveals that the topological density for ice I (cubic and hexagonal forms) is substantially lower than that predicted from the best fit for ice polymorphs (see arrowed data points in Figure VI.8). Recall that the topological density depends on ring-size ($\langle n^{(2)} \rangle$), network degree (z), and the constants h, Ω and k (or k'). It is notable that the tridymite polymorph of silica, which also has $\langle n^{(2)} \rangle$ = 6 (and is structurally equivalent to ice Ih) is similarly anomalous compared with other silicates, with an unexpectedly low density for its ring-size (arrowed data in Figure VI.4). In the absence of known minimal 2-manifold embeddings for **lon** (tridymite, Ice Ih) and **dia** (ice Ic), this anomaly, common to both amorphous silica and liquid water, remains unresolved. It is likely a central issue surrounding the issue of relative densities of liquid and frozen water. We can, however, conclude that both amorphous silica and liquid water have similar network topologies, namely degree-four nets with average ring-sizes likely between 5.5-6. Further, the density and accompanying ring-size variations between crystalline and amorphous states of water and silica are similar, suggesting that these relative densities may be a feature of tetrahedral frameworks in general, rather than a peculiarity of H_2O nets.

Perhaps the most valuable aspect of this analysis lies in the topological approach, rather than the quantitative details. Bernal's concept of structural similarities between liquid water and (possibly multi-phase) crystalline silicate networks initiated rigorous discussions of the concept of water structure. While useful, direct analogies with crystalline networks are overly simplistic and misleading. Indeed, the current controversy over two-state models and related crystalline phases appears to have overlooked the potent arguments of Fletcher that the phenomenon of super-cooling of liquid-state water is incompatible with a network that resembles crystalline frameworks. As he wrote in 1971 *"It is clear, from considerations based on the possibility of supercooling water at various pressures, that the liquid does not contain any significantly large fraction of cluster with structure identical with that of one of the ices."*[29] Quantitative investigations and comparisons of ordered and disordered structures require approaches that jettison the usual geometric concepts underlying crystalline patterns. The ideas discussed here are a simplistic attempt to explore structural frameworks at this broader level. They offer a starting point for further investigations.

Acknowledgements

I thank Lawrie Skinner and John Parise for generously sending me their water diffraction data prior to publication.

References

1. Bernal, J. D; Fowler., R. H. *J. Chem. Phys.* **1933**, *1*, 515.
2. Huang, C.; Wikfeldt, K. T.; Tokushima, T.; Nordlund, D.; Harada, Y.; Bergmann, U.; Niebuhr, M.; Weiss, T. M.; Horikawa, Y.; Leetmaa, M.; Ljungberg, M. P.; Takahashi, O.; Lenz, A.; Ojamae, L.; Lyubartsev, A.

P.; Shin, S.; Pettersson, L. G. M.; Nilsson, A. *Proc. Natl. Acad. Sci. USA* **2009**, *106*, 15241.

3. Chaplin, M. *Anomalous properties of water.* http://www.lsbu.ac.uk/water/anmlies.html

4. Bernal, J. D. *Proc. Roy. Soc. Lond. A* **1964**, *280*, 299-322.

5. Martin, C. *The icosahedral water clusters.* http://www1.lsbu.ac.uk/water/clusters.html

6. Wernet, Ph.; Nordlund, D.; Bergmann, U.; Cavalleri, M.; Odelius, M.; Ogasawara, H.; Näslund, L..; Hirsch, T. K.; Ojamäe, L.; Glatzel, P.; Pettersson, L. G. M.; Nilsson, A. *Science* **2004**, *304*, 995-999.

7. Head-Gordon, T; Johnson, M. E. *Proc. Natl. Acad. Sci. USA* **2004**, *103*, 7973-7977.

8. Skinner, L. B.; Huang, C.; Schlesinger, D.; Pettersson, L. G. M.; Nilsson, A.; Benmore, C. J. *J. Chem. Phys.* **2013**, *138*, 074506.

9. Skinner, L. B.; Benmore, C. J.; Neuefeind, J. C.; Parise, J. B. A structural description of the compressibility minimum in water. *preprint*, 2013.

10. Delgado-Friedrichs, O.; O'Keeffe, M. *J. Solid State Chem.* **2005**, *178*, 2480-2485.

11. O'Keeffe, M.; Peskov, M. A.; Ramsden, S. J.; Yaghi, O. *Acc. Chem. Res.* **2008**, *41*, 1782-1789.

12. Lindsay, J. H. *Am. Math. Monthly* **1959**, *16*, 117-118.

13. Firby, P. A; Gardiner, C. F. *Surface Topology*. Ellis Horwood, New York, 2001.

14. Hyde, S. T.; Robins, V.; Ramsden, S.. *Epinet*. http://epinet.anu.edu.au, **2008**.

15. Blatov, V. A. *Struct. Chem.* **2012**, *23*, 955-963.

16. Hyde, S. T. *Acta Cryst. A* **1993**, *50*, 753-759.

17. O'Keeffe, M.; Hyde, B. G. *Crystal Structures. 1. Patterns and Symmetry*, Mineralogical Society of America, Washington D.C., 1996.

18. Hyde, S. T. in *Defects and Processes in the Solid State. Some Examples in Earth Sciences*, Elsevier, Amsterdam, **1993**.

19. Baerlocher, Ch.; McCusker, L. B. *Database of zeolite structures.* http://www.iza-structure.org/databases/

20. Delgado-Friedrichs, O.; O'Keeffe, M. *Acta Cryst. A* **2003**, *59*, 351-360.
21. Stixrude, L.; Bukowinski, M. S. T. *Science* **1990**, *250*, 541-543.
22. Brunner, G. O.; Laves, F. *Zeit. Techn. Univ. Dresden* **1971**, *20*, 387-390.
23. Hyde, S. T.; Ramsden, S. in *DIMACS Series in Discrete Mathematics and Theoretical Computer Science* **2000**, *51*, 203-224,.
24. Hyde, S. T.; Ramsden, S. in *Chemical Topology: Applications and Techniques*. Gordon and Breach Science, Amsterdam, **2000**.
25. Corkery, R. *Artificial biomineralisation and metallic soaps.* Ph.D. Thesis, Australian National University, **1998**. Available at *http://hdl.handle.net/1885/46251*
26. Hyde, S. T.; Oguey, C. *Eur. Phys. J. B*, **2000**, *16*, 613-630.
27. Kubota, A.; Caturla, M.-J.; Davila, L.; Stolken, J.; Sadigh, B.; Quong, A.; Rubenchik, A.; Fe, M. D. *Proc. SPIE 4679* **2001**, *108*, 387-390.
28. Kell, G. S. *J. Chem. Eng. Data* **1975**, *20*, 97-105.
29. Fletcher, N. H. *Rep. Prog. Phys.* **1971**, *34*, 913-994.

VII
Supercooled water: two liquids?

Francesco Sciortino

Dipartimento di Fisica, Università di Roma "La Sapienza",
Piazzale A. Moro 2, I-00185, Rome, Italy.
francesco.sciortino@uniroma1.it

In this brief article, I will describe my personal view of the evolution of an idea, germinated while interpreting results of one of the early simulation study of supercooled water, which has permeated a large part of the water community. The idea of a liquid-liquid critical point in a one-component system, despite can not be possibly experimentally proved in water, has been extremely fruitful to understand the behavior of several different compounds.

The thermodynamic behavior of water at low temperatures, and even more in supercooled states, is unconventional. Several response functions are characterized by non monotonic temperature or pressure dependence. Well known examples are the isobaric density, the isothermal compressibility, the constant-pressure specific heat.[1] The anomalous behavior of these quantities has attracted the attention of several researchers and several propositions have been put forward to explain it. Back in the eighties, two main ideas were actively discussed: the so-called re-entrant spinodal scenario, proposed by Speedy and Angell[2] and the percolation of four-coordinated molecules, by Stanley and Teixeira.[3] In the re-entrant spinodal scenario, the response functions were predicted to diverge at some unreachable temperature due to the encounter of a line of thermodynamic instability, which was thought to be the continuation of the gas-liquid spinodal at low T. The spinodal line, that for common liquids is always negatively sloped in the P−T plane, was predicted to turn around toward positive pressures on encountering the locus of density maxima. On cooling along a constant pressure path, the spinodal line would be encountered. According to Stanley and Teixeira instead,[3] the anomalies were the result of the increasing number of hydrogen bonds on cooling. Water indeed can be considered a fully connected network, which is constantly restructuring on a pico-second time scale. Special emphasis was put on the four-bonded nodes of the network, which were assumed to be characterized by low potential energy, large local volume and high orientational order. The authors argued that the clustering process of these four-coordinated particle and the associated random percolation process could be responsible of the water anomalies.

In the eighties, computer simulation of classical model of water become accessible. One of the early model, the ST2 potential,[4] represented water as a rigid five-site molecule. The central site modeled the oxygen atom (as a Lennard-Jones center) while the other four sites were oriented in a tetrahedral geometry, modeling the two hydrogen via a positive partial charge and the two lone pairs, via a negative partial charge. This model was able to reproduce the density maximum and the compressibility minimum, although at displaced temperatures compared to water. Naturally, tests of the aforementioned theories started to be conducted, despite the limited number of water molecules (a few hundreds) and the limited time interval (hundreds of picoseconds) accessible via simulations. The hydrogen bond network picture of water was confirmed as well as the coupling between local energy and local volume.[5]

The numerical investigation of the behavior of the response function in supercooled states was naturally the next goal. In Boston, where I was doing my post-doc in the group of H. E. Stanley, such a project was started. Indeed, the two theories were predicting a different behavior for the temperature dependence of the compressibility. A divergence in the re-entrant spinodal and a maximum in the percolation model. The numerical study was quite challenging due to the slowing down of the dynamics on cooling and the need to calculated the equation of state in a wide region of temperatures and pressures. The outcome of this investigation was quite rewarding. Indeed divergence of the compressibility was not observed at ambient pressure and the spinodal was shown to be not re-entrant. But also the percolation theory was not able to describe the numerical data. A novel thermodynamic scenario, incompatible with the random percolation assumption, was indeed disclosed.[6] At relatively high pressures, the numerical estimate of the ST2 equation of state were compatible with a van der Waals loop, suggestive of a unexpected liquid-liquid critical point, despite the one-component nature of the system. If the results of the ST2 model would apply to water, they would indicate that two different liquid forms of water, differing in their density and their local structure, should exist below a critical temperature. A line of first-order transition, emanating from the critical point, would separate these two liquid states. The anomalous behavior of all thermodynamic quantities was related to the presence of the liquid-liquid critical point. Equally interesting was the connection between the predicted existence of two liquid forms and the previously experimentally discovered existence of two disordered form of amorphous water, the so-called low and high density amorphous ices.[7]

In the following years, several numerical investigations were performed to test the liquid-liquid critical point hypothesis, both with different water models as well as with more refined evaluations of the equation of state for the ST2 model itself. Figure VII.1 shows the very preliminary data reported in my laboratory notebook of three investigated isotherms and the very first speculation (drawn with a pencil !) of a van der Waals loop. It was quite brave to hypothesize the presence of a maximum and a minimum in the equation of state,

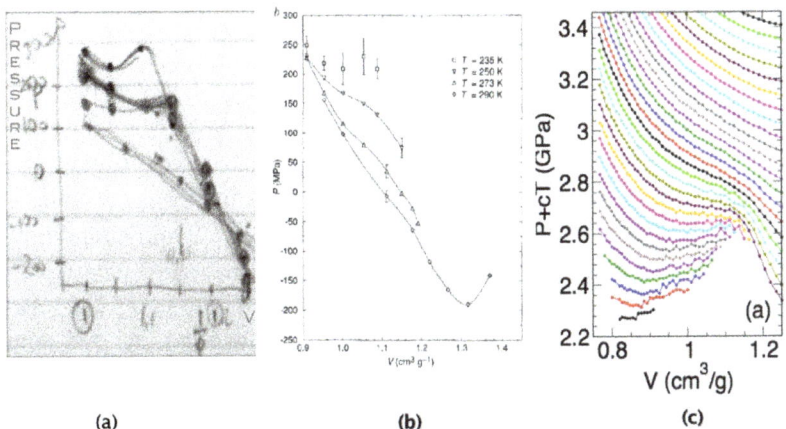

(a) (b) (c)

Figure VII.1. Equation of state of ST2 water, including supercooled states: (a) reproduction of the very first results reported in my laboratory notebook, suggestive of a van der Waals loop around $P = 250$ MPa and $T = 235$K. (b) Pressure-volume data reported in the original publication (1992).[6] (c) Equation of state for the same model re-calculated in 2005.[8] Here different temperatures are shifted by a constant value.

based on very few points and with still a large numerical uncertainty. The more refined data, still based on very few isotherms, confirmed the initial observation and were published in the original Nature article. Figure VII.1 shows them together with the evaluation of the same equation of state performed after fifteen years on a canadian cluster of processors,[8] each of them running its own state point. The presence of the van der Waals loop at low temperature was always recovered, strengthening the initial observation and the associated interpretation in term of novel critical phenomena. In the last years powerful techniques, whose application has been made possible by the increased computer power, have definitively confirmed the original results for the ST2 potential.[9-12]

Numerous experimental studies have also attempted to provide evidence of the second critical point. Unfortunately its location is well within the region where water spontaneously crystallizes and the evidence of the liquid-liquid critical point and of the onset of the associated fluctuations is unfortunately indirect. The condition to observe critical fluctuations requires indeed that the system survives in its metastable state longer than the time requested to equilibrate and that the experimental observation time is smaller than the spontaneous nucleation time. These conditions are meet in the simulation study, in which heterogeneous nucleation is absent by construction and homogeneous nucleation is slowed down by the small size of the investigated sample. In addition, simulation studies do not extend in the ms region, where crystallization would possibly take over even numerically. Despite the fact that experiments that access the critical point can not be possibly realized in water, more and more evidence is accumulating about the absence of re-entrant of the spinodal line at negative pressure and the presence of a so-called Widom-line,[13] a line of extrema of response functions which is expected theoretically to converge at the critical point. Also, the amorphous ice structures have been found to be related to the structure in the liquid at low and high pressure, respectively. Current finite size investigations as well as experiments on aqueous solutions can provide further support to the liquid-liquid critical point idea. I also like to recall that the available thermodynamic data for water can be properly modeled with a theoretical expression that assumes the existence of the liquid-liquid critical point.[14]

Despite the difficulties in providing definitive experimental evidence of the liquid-liquid critical point in water the idea of a transition between two different local structures does not limit its applicability to water, but it is possibly general in a wider class of tetrahedral network former atoms. Silica, silicon, carbon, among others, have been discussed as possible candidates of liquid-liquid transitions and research is progressing along these lines. The structural changes associated to the liquid-liquid transition possibly have also an important role in determining a cross-over in the dynamical properties (fragile-strong cross-over).

I believe the experimental observation of a liquid-liquid critical point in a one-component network former will arise from novel studies in soft-matter systems. Soft-matter is currently undergoing a astonishing development, based on the synthesis and production in bulk quantities of on-demand particles. The ideas of a bottom-up self-assembly are permeating the field and different types of tetrahedral colloids are appearing.[15] As an example I like to recall the recently reported experimental study of a network of tetrahedral DNA-stars, DNA-made particles with four arms.[16] For this system, a numerical study does show the possibility of liquid-liquid phase-separation.[17] Patchy particles with controlled valence and patch width are also extremely promising candidates.[18] Differently from the atomic and molecular world, soft-matter scientists have the possibility to tune the interactions, playing with the numerous parameters controlling the formation of the network (softness, bonding-strength, bond flexibility). All these parameters may control in different ways the location of the expected liquid-liquid critical point compared to crystal nucleation, offering perhaps a system where the idea proposed more than twenty years ago can finally find a clean experimental detection.

Acknowledgment

I wish to thank the large group of colleagues which have contributed to the ideas briefly reviewed in this article.

References

1. Debenedetti, P. G.; Stanley, H. E. *Physics Today* **2003**, 56, 40.
2. Speedy, R. J.; Angell, C. A. *J. Chem. Phys.* **1976**, *65*, 851–858.
3. Stanley, H. E.; Teixeira, J. *J. Chem. Phys.* **1980**, *73*, 3404–3422.
4. Stillinger, F. H.; Rahman, A. *J. Chem. Phys.* **1974**, *61*, 4973–4980.

5. Geiger, A.; Stillinger, F. H.; Rahman, A. *J. Chem. Phys.* **1979**, *70*, 4185–4193.
6. Poole, P. H.; Sciortino, F.; Essmann, U.; Stanley, H. E. *Nature* **1992**, 360, 324–328.
7. Mishima, O. *J. Chem. Phys.* **1994**, *100*, 5910–5912.
8. Poole, P. H.; Saika-Voivod, I.; Sciortino, F. *J. Phys.: Condens. Matter* **2005**, *17*, L431–L437.
9. Liu, Y.; Panagiotopoulos, A. Z.; Debenedetti, P. G. *J. Chem. Phys.* **2009**, *131*, 104508–1/7.
10. Sciortino, F.; Saika-Voivod, I.; Poole, P. H. *Phys. Chem. Chem. Phys.* **2011**, *13*, 19759–19764.
11. Poole, P. H.; Bowles, R. K.; Saika-Voivod, I.; Sciortino, F. *J. Chem. Phys.* **2013**, *138*, 034505–1/7.
12. Liu, Y.; Palmer, J. C.; Panagiotopoulos, A. Z.; Debenedetti, P. G. *J. Chem. Phys.* **2012**, *137*, 214505.
13. Xu, L.; Kumar, P.; Buldyrev, S.; Chen, S.; Poole, P. H.; Sciortino, F.; Stanley, H. E. *Proc. Natl. Acad. Sci. USA* **2005**, *102*, 16558–16562.
14. Fuentevilla, D. A.; Anisimov, M. A. *Phys. Rev. Lett.* **2006**, *97*, 195702.
15. Wang, Y.; Breed, D.; Manoharan, V.; Feng, L.; Hollingsworth, A.; Weck, M.; Pine, D. *Nature* **2012**, 491, 51–56.
16. Biffi, S.; Cerbino, R.; Bomboi, F.; Paraboschi, E. M.; Asselta, R.; Sciortino, F.; Bellini, T. *Proc. Natl. Acad. Sci. USA* **2013**, 15633–15637.
17. Hsu, C. W.; Largo, J.; Sciortino, F.; Starr, F. W. *Proc. Natl. Acad. Sci. USA* **2008**, *105*, 13711–13715.
18. Smallenburg, F.; Sciortino, F. *Nat. Phys.* **2013**, *9*, 554–558.

VIII
Is water chiralic? (Experimental evidence of chiral preference of water)

Yosef Scolnik

IYAR, Israel Institute for Advanced Research, Rehovot, Israel.
yosefsc@walla.com

Abstract

Mirror-image asymmetric molecules, i.e., chiral isomers or enantiomers, are classically considered as chemically identical. It is known, however, that parity violation by the nuclear weak force induces a tiny energy difference between chiral isomers. Upon combination with a massive amplification process, expansion of this difference to a detectable macroscopic level may be achieved. Yet, experimental tests of this possibility, where one enantiomer is compared to the other in solution, are hampered by the possible presence of undetectable impurities. In this study we have overcome this problem by comparing structural and dynamic features of synthetic D- and L-polyglutamic acid and polylysine molecules each of 24 identical residues. In these water-soluble polypeptides helix formation is an intramolecular autocatalytic process amplified by each turn, which is actually unaffected by low level of putative impurities in the solvent. The helix and random coil configurations and their transition were determined in this study by circular dichroism (CD) and isothermal titration calorimetry (ITC) in water and deuterium oxide. Distinct differences in structure and transition energies between the enantiomeric polypeptides were detected by both CD and ITC when dissolved in water. Intriguingly, these differences were by and large abolished in deuterium oxide. Our findings suggest that deviation from physical invariance between the D- and L-polyamino acids is induced in part by different hydration in water which is eliminated in deuterium oxide. Based on findings by Tikhonov and Volkov we suggest that ortho-H_2O, which constitutes 75% of bulk H_2O, has a preferential affinity to L-enantiomers.[1] In other words, this and other studies to be quoted, **prove that indeed water has chiral preferences, i.e. are chiralic**. Accordingly, Heavy Water has no chiralic preference. Differential hydration of enantiomers may have played a role in the selection of L-amino acids by early forms of life.

Introduction

Space symmetry of physical laws asserts that in any macroscopic chemical or physical reaction, where achiral molecules are converted to chiral products, the system as a whole remains absolutely racemic. However, when an autocatalytic arm is associated with such a process, a slight excess of one enantiomer can be expanded considerably.[2-4] In the autocatalytic process of crystallization, such an enantiomeric enhancement can be induced by mechanical agitation,[5] stirring,[6] or even β-irradiation.[7] Furthermore, it is now widely accepted that chiral isomers are inherently at a slightly different energy state due to the parity violation of the electro-weak nuclear force (parity violation energy difference, PVED). The tiny excess of one enantiomer in a racemic mixture due to PVED can, in principle, be amplified by an external autocatalytic process to a level of detectable macroscopic difference. Whether such a process could explain the initial selection which led to the homochirality of amino acids, saccharides and nucleic acids in the biological realm, remains uncertain In an earlier study,[8] similar trends were observed. Micelles of N-palmitoyl or N-stearoyl D- or L-serine in water displayed intense CD bands of opposite directions which are abolished by disruption of the micelles with ethanol. These specific CD bands were attributed to chiral surfaces where the serine head groups are arranged in a unique set of spines which are integrated to a chiral micellar surface. However, the absolute level of the CD band in the D-serine micelles was about 50% stronger than that of the enantiomeric L-serine micelles. This difference indicated a tighter correspondence between the residues on the surface of the D-serine micelles due to a lesser interaction with the surrounding aqueous layer. Experimental verification of the intriguing assertion that enantiomers, like natural amino acids, are not fully identical, is hampered by the possibility that when comparing bulk systems the presence of impurities, even in undetectable concentrations, can tip the apparent macroscopic balance to an erroneous level. In the following study we have overruled this notorious problem by comparing structures and their transitions of two separate systems of water soluble enantiomeric polypeptides: poly L- and poly D-glutamic acid and poly L- and poly D-lysine, each of 24 identical residues. In these polypeptides -helical and random coil configurations and their transitions involve intramolecular autocatalytic processes where low level of impurities in the solvent can be considered as negligible. Polyglutamic acid and polylysine are water soluble polypeptides which undergo structural changes related to the degree of ionization of their side chains.[9-11] When in the ionized state, these polypeptides are at an equilibrium among fluctuating unstructured conformations, generally termed as "random coil". In the neutral state they assume a well characterized α-helical structure which has a distinctive circular dichroism (CD) spectrum.[12] These structures and their transitions were studied in our report.[13]

Results and discussion

Circular dichroism

The CD spectrum of the right handed α-helix of poly-L-glutamic acid or poly-L-lysine is expected to be a mirror image of the spectrum of the left handed helices of their D-polypeptides under the same conditions.[12] Furthermore, the energetics associated with helix formation or breaking is assumed to be identical for any pair of polypeptides of identical size composed of the D or the L enantiomers.[14] However, parity violation energy differences (PVED) between chiral enantiomers which are extremely small ($\sim 10^{-17}$eV), can in principle be increased to a detectable level when associated with an amplifying mechanism. Helix formation in such polypeptides is a typical autocatalytic process, where each turn enhances propagation to the next turn and beyond as the helix builds up. The energetics associated with helix formation in enantiomeric polypeptides, such as poly-L- or poly-D-glutamic acid, may be thus slightly different, a possibility which has far reaching implications Poly-L-glutamic acid, poly-D-glutamic acid, poly-L-lysine and poly-D-lysine, each blocked at the carboxylic terminus as amide, and each of precisely 24 monomers [poly $(L-Glu)_{24}$, poly$(D-Glu)_{24}$, poly$(L-Lys)_{24}$, and poly$(D-Lys)_{24}$, respectively], were synthesized by solid phase stepwise addition of monomers, then isolated by preparative HPLC and analyzed Their purity was the highest we could achieve. The precise molecular weight recorded for the polypeptides and the absence of opposite enantiomeric residues in the starting materials,[15] led us to assess a purity of >99%. The CD spectra of poly $(L-Glu)_{24}$ and poly $(D-Glu)_{24}$ at their α helix (measured at pH 2.5) and random-coil (measured at pH10.5) configurations are presented in Figure VIII.1 and VIII.2. As shown the CD spectra in Figure 1 are mirror images typical of right handed and left handed α-helices, respectively. Similar CD mirror image spectra were also recorded with poly $(L-lys)_{24}$ and poly $(D-Lys)_{24}$ in their α-helix region (recorded in 0.1 M NaOH, not shown). The identity of the absolute CD bands further supported the assessment of high purity of the studied polypeptides since any presence of a contaminant in the polypeptide matrix is expected to induce a marked effect on the helix formation reflected in the CD band. In the random coil region of

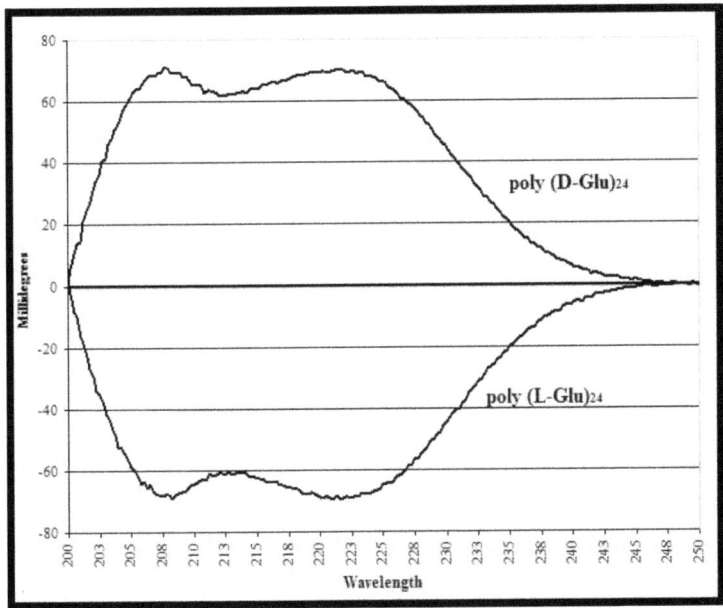

Figure VIII.1. CD spectra of poly (D-Glu)$_{24}$ and poly (L-Glu)$_{24}$ in water at pH 2.5. The spectra were recorded in a 1 mm path-length, at a concentration of 3 mM per residue.

both sets of polypeptides, the CD spectra of the enantiomeric couples were not identical mirror images, as presented for the polyglutamic acid in Figure VIII.2, indicating a net difference in the equilibrium state of their random coil conformations. As shown in Figure VIII.2, D$_2$O markedly affected the CD spectrum of poly (L-Glu)$_{24}$ but had a significantly smaller effect on the spectrum of poly (D-Glu)$_{24}$ It is reasonable to assume that in this region, small differences in energy of the fluctuating conformations, which determine the equilibrium, could account for the observed deviation from mirror image spectra in the random coil region of poly (D-Glu)$_{24}$ and poly (L-Glu)$_{24}$. In the α-helix region, on the other hand, the energies associated with the hydration and the intramolecular hydrogen bonding are presumably

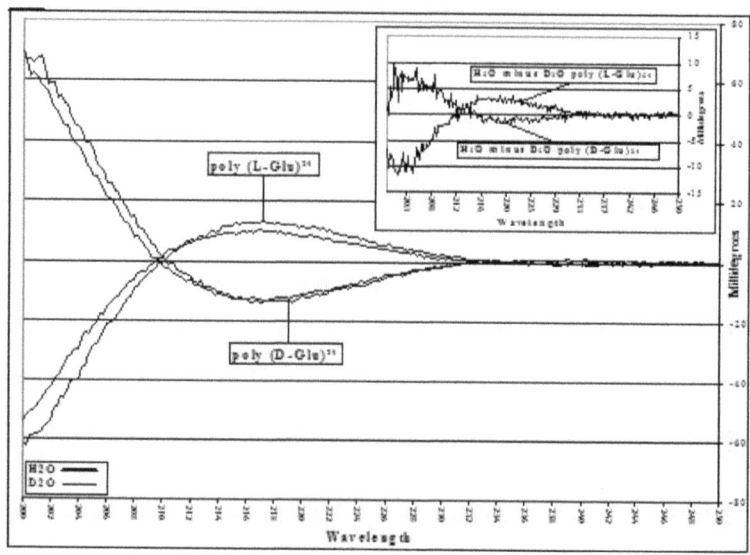

Figure VIII.2. CD spectra of poly (D-Glu)$_{24}$ and poly (L-Glu)$_{24}$ at pH 10.5 in H$_2$O and D$_2$O. The insert represents the spectral difference between H$_2$O and D$_2$O. The spectra were recorded using a 1 mm path length, at a concentration of 3 mM per residue.

much larger and could mask small energy differences between these peptides which appear to be identical in their CD spectra.

Isothermal titration calorimetry

The above suggestion could be tested by comparing the relative transition energies between α-helical and random coil configurations. We therefore conducted a detailed energetic determination of the helix-to-coil transition in these sets of poly peptides under isothermal conditions. Isothermal titration calorimetry (ITC) profiles at increments of decreasing pH were determined at 30 C, either in H$_2$O or in a 4:1 (v/v) mixture of D$_2$O and H$_2$O. Typical profiles are displayed in Figure VIII.3. The ITC profiles in Figure VIII.3 can be divided into three distinct regions related to the degree of ionization of the glutamic acid side chains: pH > 6, where the polypeptides retain equilibrium among random coiled structures, pH ~ 6–3, the range of transition to α-helix, and pH 0-3, where the polypeptides are at their α-helix conformation. This classification was verified by CD scanning in a series of buffers of pH 2.5–10 which indicated α-helix-to-coil transition at pH ~ 6 for both polypeptides (not shown). In the α-helix region, the ITC profiles indicated marginal enthalpy change for both polypeptides. Similar ITC profiles of low enthalpy change were observed for acetic acid and propionic acid in water in the pH range of 2–8 (not shown), which suggests that the carboxylic side chains of poly(L-Glu)24 and poly(D-Glu)24 behave as isolated monomers at their α-helix region. On the other hand, in the random coil region (pH>6) complex ITC profiles were recorded for both poly peptides, which indicated correspondence in enthalpy changes between the polypeptide backbone and the carboxylic head groups. The ITC profiles of the two enantiomeric polypeptides in this pH region were similar in shape but not identical (see Figure VIII.3), in agreement with the structural differences recorded by CD in this region (see Figure VIII.2). The most pronounced differences between the ITC profiles of poly(L-Glu)$_{24}$ and poly(D-Glu)$_{24}$ in H$_2$O were recorded in the transition region. The onset of the transition of poly(D-Glu)$_{24}$ was 0.2–0.3 pH units above that of poly(L-Glu)$_{24}$, namely, at a lower level of protonated side chains. This transition point is independent on concentration. In other words, the transition to α-helix of poly(D-Glu)$_{24}$ started at a point of higher proportion of ionized side chains than in poly(L-Glu)$_{24}$ (at pH 6.2 compared to 5.8, respectively), indicating a stronger tendency of poly-D-glutamic acid to adopt an α-helix structure. Both the degree of cooperativity and the associated change in enthalpy of the transition to α-helix of poly(D-Glu)$_{24}$ were considerably higher than that of poly (L-Glu)$_{24}$ .It is important to stress, that the abolition of the differences described above between the enantiomeric polypeptides in water by D$_2$O, overrules the possibility of an undetectable flaw in their synthesis.

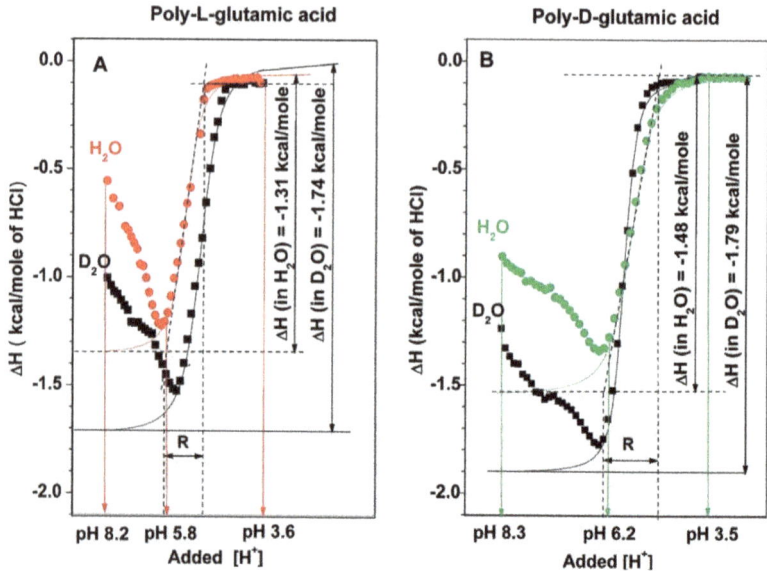

Figure VIII.3. ITC profiles of poly(L-Glu)$_{24}$ and poly(D-Glu)$_{24}$ in H$_2$O and D$_2$O–H$_2$O (4 : 1 v/v) at 30 C.

Under such a hypothetical case the results in both solvents would be identical. The most plausible interpretation for the above differences is that poly(DGlu)$_{24}$ has a higher α-helix stability than poly(L-Glu)$_{24}$. According to this interpretation, each residue of D-glutamic acid in the turns which build the α-helix conformation, contributes approximately 0.1 kcal /mol more than that of L-glutamic acid in their enantiomeric helical turns. Taking 3.6 as the number of residues in each turn of α-helix, the excess energy of 0.4 kcal/mol turn in poly(D-Glu)$_{24}$, is approximately equivalent to that of 10% of a hydrogen bond. In practice, ten α-helical turns of poly-D-glutamic acid would have an excess energy of an additional hydrogen bond compare to its poly-L-glutamic acid enantiomer. It is of great interest that in D$_2$O–H$_2$O 4:1, the above differences were greatly diminished due to an almost selective effect on the ITC profile of poly(L-Glu)$_{24}$ (see Figure VIII.3).

This unexpectedly selective sensitivity of poly(L-Glu)$_{24}$ to replacement of H_2O by D_2O, displayed in the CD spectra (Figure VIII.2) and the ITC profiles, provide the key issue in our suggested hypothesis which is presented below. ITC profiles of poly(L-Lys)$_{24}$ and poly(D-Lys)$_{24}$ were measured and analyzed analogously to those presented above. The recorded ITC profiles of these polypeptides were qualitatively of a similar pattern to those shown in Figure VIII.3.

General discussion

The application of polymers for the detection of chiral deviations has two main advantages. The first relates to the notorious effects of impurities in tested solutions, which in polymers is reduced to defects in the polymer matrix which can be eliminated by building it from highly purified monomers. The other advantage corresponds to intramolecular autocatalytic amplification of a slight chiral deviation which propagates along the polymer backbone. An example of such amplification was found in polyisocyanates, where a single stereospecific deuterium substitution in the side chains was found to induce a strong CD signal corresponding to a unidirectional helix formation.[16] The energy difference between the left- and right-handed helices in these polymers is extremely small and the specific deuterium substitution sufficed for tipping the perfect balance between the enantiomeric helices. Analogously, the energy difference in enantiomeric polypeptides, like those used in this study, is undoubtedly very small. As stated above, even such small energy difference could be amplified to a detectable level.

The mechanism we suggest for the chiral amplification process in polypeptides is presented below. In the absence of an external discriminatory factor such as circularly polarized light, PVED remains the only physical effect which can lead to a selective and repetitive chiral enhancement.[17-21]

As PVED between isolated chiral isomers is extremely small, any expansion to the macroscopic realm must be associated with additional processes which lead to PVED amplification. In our poly-amino acid systems, amplification of PVED could, in principle, operate in two independent, yet cooperating, processes. The first is autocatalysis of helix formation or breaking, as clearly indicated in the ITCprofiles of helix formation of our polypeptides. In the α-helix conformation of poly-amino acids, 3.6 residues constitute a complete turn. Since each turn contributes to the catalysis of helix formation, the 6–7 turns in poly(Glu)$_{24}$ or poly(Lys)$_{24}$ presumably correspond to a significant autocatalytic amplification of the inherent PVED of the monomers and could thus account in part for the observed differences in these sets of enantiomeric polypeptides. **The additional putative amplification arm is exclusive for aqueous solutions and corresponds to PVED induction of differences in the hydration layer between poly L-amino acids and their enantiomeric poly D-amino acids.** In 2002 Tikhonov and Volkov 24 reported an unexpected effect concerning proton exchange between the two spin isomers of H_2O, namely ortho-H_2O, where the proton spins are parallel, and para-H_2O, where the proton spins are anti-parallel. They have shown that the exchange rates between these isomers are much slower than expected (half life of 30–50 min.). Further evidence for this unexpected slow exchange has been recently reported.[22] One reasonable implication of this finding is that bulk water can be practically viewed as a mixture of ortho-H_2O and para-H_2O in a 3 : 1 ratio (due to the three degenerate states of ortho-H_2O). We have previously proposed that, since ortho-H_2O bears a magnetic field, it has a slight preference to react with L-enantiomers due to their PVED induced magnetic component.[23,24] As a result, in aqueous solutions of racemic mixtures, solvation preference of L-enantiomers may take place, which in the extreme cases may lead to chiral enhancement and further to a selective separation.[23,24] In line with this hypothesis, a polypeptide of L-amino acids in water might be solvated slightly more than its mirror image poly-D amino acids, so that the latter adopts an apparently more hydrophobic nature. In poly(L-Glu)$_{24}$, poly(D-Glu)$_{24}$, poly (L-Lys)$_{24}$ and poly(D-Lys)$_{24}$, where the asymmetrical conformations are determined almost exclusively by solvation energy and intramolecular hydrogen bonding, the structures and their transition energies in water could thus become significantly different. If this hypothesis is correct, then the spin isomers of H_2O and their putative selective effect on chiral isomers should be greatly diminished in D_2O or D_2O-H_2O mixtures. Indeed, this is what we have found. In our ITC experiments we have used a mixture of D_2O-H_2O 4 : 1 (v/v) as a system analogous to H_2O but with "scrambled" spin isomers, namely, devoided of ortho and para water. In the steady state, this mixture is constituted of 4% H_2O, 32% HDO and 64% D_2O and from all physical and chemical aspects resembles pure H_2O. As shown in Figure VIII.3, the pronounced differences in ITC profiles in H_2O between poly(D-Glu)$_{24}$ and poly(L-Glu)$_{24}$ are by and large abolished in 80% D_2O. Furthermore, the main effect of this mixture is on (L-Glu)$_{24}$ and to a considerably smaller extent on poly(D-Glu)$_{24}$. These observations, in addition to the differences presented in Figure VIII.2, comply with the above hypothesis that the preferred interaction of ortho-H_2O with the L-enantiomer can account for a significant part of the structural and dynamic differences between enantiomeric polypeptides in water which were described in this study. The slight difference in solvation between the enantiomeric polypeptides described above may have a considerable contribution to a difference in the entropy of transition, alongside the observed difference in the enthalpy of transition, as implied by the hierarchic model.[25] The slightly less solvated poly-D glutamic acid, may restrict the conformation balance in the random coil region, as compared with the poly-L-glutamic acid enantiomer. Therefore, the steering effects arising from hydrogen bonds and intramolecular interactions between residues in poly-D-glutamic acid may, by virtue of the relatively higher hydrophobicity of the D-isomer, be more pronounced. These will cause a higher restriction of the number of conformations, and create in the random coil a favorable backbone conformation of a higher tendency towards α-helix formation in poly-D-glutamic acid. Results indicating chiral discrimination between D- and L-alanine at their crystalline or aggregated states have already been provided by Wang *et al.*[26-28] The results described here were confirmed and strengthen by further work done by us, not published yet, with other systems:

-The pioneering work of Meir Shinitzki already mentioned,[8] on micelles formation differences between L and D stearoyl serine was repeated, including the crucial comparison with D_2O. The results show again the same differences between the water interaction with L and D racemates already described.

- Diffusion coefficients of *R* and S2 amino butanol were found to be different in water, again in accordance with the results described above.

- L alanine and D alanine form aggregates in water differently, again with accordance with the pattern described above – the racemates behave exactly the same in D_2O, only L alanine is affected by changing from H_2O to D_2O.

Others groups work support our findings, albeit without the crucial comparison to D_2O: Kodona *et al.* got similar results using Polyglutamic acid L and D;[29] H. Liang *et al.* found differences akin to ours in the enthalpy of dilution of R and S amino butanols.[30]

The results reported here imply that polypeptides comprised of D and L amino acids may have slightly different structures and transition energies than their mirror image enantiomeric polypeptides. This difference could be reflected in a difference in catalytic activity between such pairs of polypeptides, which may provide a clue for the selection of L-amino acids in the origin of life. In future work, this possibility could be tested with enantiomeric peptides which bear a catalytic site. This may clarify the thermodynamic advantage in the presumed chiral selection proposed in this study.[31]

As shown by the reported results –water reacts differently with L and R racemates – in other words – **water is indeed chiralic**.

References

1. Tikhonov, V. I.; Volkov, A. A. *Science* **2002**, *296*, 2363.
2. Morovitz, H. J. *J. Theor. Biol.* **1969**, *25*, 491.
3. Thiemann, W.; Darge, W. *Origins Life* **1974**, *5*, 263.
4. Soai, K.; Shibata, T.; Morioka, H.; Choji, K. *Nature* **1995**, 378.
5. Buhse, T.; Durand, D.; Kondepudi, D. K.; Laudadio, J.; Silker, S. *Phys. Rev. Lett.* **2000**, *84*, 4405.
6. Kondepudi, D. K.; Kaufman, R. J.; Singhi, N. *Science* **1990**, 250.
7. Mahurin, S.; McGinnis, M.; Bogard, J. S.; Hulett, L. D.; Pagni, R.; Compton, R. N. *Chirality* **2001**, *13*, 636.
8. Shinitzky, M.; Haimovitz, R. *J. Am. Chem. Soc.* **1993**, *115*, 12545.
9. Imahori, K.; Tanaka, J. *J. Mol. Biol.* **1959**, *1*, 359.
10. Rosenheck, K.; Doty, P. *Proc. Natl. Acad. Sci. USA*, **1961**, *46*, 1775.
11. Makovec, T. *Biochem. Mol. Biol. Ed.* **2000**, *28*, 244.
12. Greenfield, N.; Fasman, G. *Biochemistry* **1969**, *8*, 4108.
13. Scolnik, Y.; Portnaya, I.; Cogan, U.; Tal, S.; Haimovitz, R.; Fridkin, M.; Elitzur, A. C.; Deamer, D. W.; Shinitzky, M. *Phys. Chem. Chem. Phys.* **2006**, *8*, 333.
14. Doty, P.; Yang, J. T. *J. Am. Chem. Soc.* **1956**, *78*, 498.
15. The optical purity examined by Nova Biochem was 0.5% while enantiomeric impurities were undetectable and were therefore assessed as 0.1%. The overall purity of the reacting compounds was assessed by HPLC to be higher than 99.5%.
16. Selinger, J. V. *Angew. Chem. Int. Ed.* **1999**, *38*, 3138.
17. Hegstrom, R. A.; Rein, D. W.; Sandars, P. G. H. *Chem. Phys.* **1980**, *73*, 2329.
18. Mason, S. F.; Tranter, G. E. *Mol. Phys.* **1984**, *53*, 1091.

19. Salam, A. *J. Mol. Evol.* **1991**, *33*, 105.
20. Bonner, W. A. *Orig. Life Evol. Biosphere* **1991**, *21*, 59.
21. Shinitzky, M.; Nudelman, F.; Bard, Y.; Haimoviz, R.; Cohen, E.; Deamer, D. W. *Orig. Life Evol. Biosphere* **2002**, *32*, 285.
22. Miani, A.; Tennyson, J. *J. Chem. Phys.* **2004**, *120*, 2732.
23. Shinitzky, M.; Elitzur, A. C.; Deamer, D. W. *Progress in Biological Chirality*, ed. G. Playi, C. Zucchi and L. Caglioti, Elsevier, New York, 2004.
24. Cogan, U.; Shpigelman, A.; Portnaya, I.; Rutenberg, A.; Scolnik, M.; *Chirality* **2012**, *24*, 500-505.
25. Baldwin, R. L.; Zimm, B. H. *Proc. Natl. Acad. Sci. USA* **2000**, *97*, 12391.
26. Wang, W. Q.; Sheng, X. G.; Jin, H. F.; Wu, J.; Yin, B.; Li, J.; Zhao, Z. X.; H. S. Yang, H. S.; F. M. Lou, F. M.; Z. Z. Zhuang, Z. Z. *J. Biol. Phys.* **1996**, *22*, 65.
27. Wang, W. Q.; Yi, F.; Ni, Y.; Zhao, Z.; Jin, Y.; Tang, Y. *J. Biol. Phys.* **2000**, *26*, 51.
28. Wang, W. Q. *Biophys. Chem.* **2003**, *103*, 289.
29. Kodona, E. K. *J. Coll. Interface Sci.* **2008**, *72*, 312.
30. Liang, H. *Chirality* **2012**, *24*, 374.
31. Elitzur, A. C. *J. Theor. Biol.* **1995**, *176*, 349.

IX
The topological and quantum structure of zoemorphic water

Marc Henry

UMR 7140, Chimie Moleculaire du Solide,
Institut Le Bel, Strasbourg, France.
henry@unistra.fr

1. Introduction

The status of water in the cell is intimately linked to the outstanding challenge for a scientific explanation of the so-called "water anomalies" without which no life will be possible on Earth. Accordingly, the mere fact that ice floats on water, has protected life on Earth even during long glaciation periods. The surprising fact that liquid water (LW) expands at 4°C either on heating or cooling was already the subject of a memorial debate in Florence between Galileo Galilei and Ludovico delle Colombe during the summer of 1611.[1] Since, many physical explanations for water's density anomaly have been advocated and more than 400 years after, water remains "aqua incognita" for anybody using concepts derived from classical physics (CP). The ruling out of CP should be evident as expansion on heating or cooling simply means that each H_2O molecule tends to increase its volume, i.e. holds more vacuum for the same amount of matter. But in CP, vacuum existence is forbidden leading to an UV catastrophe for a black body emitting light and to average lifetime of about 10^{-10} s for atoms. CP reasoning being clearly useless to understand LW properties, one may be surprised that molecular dynamics (MD) based on Newton's laws of motion coupled to purely electrostatic interactions could be of some help for understanding water structure. The main reason for the apparent success of such simulations may in fact reside in *ad-hoc* hypotheses and adjustable parameters that allow performing clever mathematical curve fitting.

One of the principal *ad-hoc* hypotheses is surely the occurrence of a highly directional short-range force, the so-called "hydrogen bond (HB)", introduced by Linus Pauling for explaining the increase in vacuum on crystallisation and justifying the high boiling point of LW. Associated to HBs, two rather mysterious long-range forces, hydrophilic and hydrophobic interactions also haunt biology. These two interactions being related to the H-bonded nature of LW, could they be useful tricks with no real physical content, being invoked just to classify and organise experimental facts? The controversy around the HB's covalence and about its bonding or anti-bonding character illustrates the fuzzy nature of this concept. For instance, periodic intensity variations observed in Compton's profile anisotropies of ice I_h at distances of 1.72 Å (HB-length) and 2.85 Å (nearest-neighbour O-O distance) were first interpreted as a proof for HB's partial covalent character and for the failure of a purely electrostatic (classical) bonding model.[2] However, a quantum mechanical (QM) study at the Hartree-Fock level of the H-bonded water dimer using a basis set that virtually eliminates the basis set superposition error (BSSE) showed that such oscillations were just the result of anti-symmetrising the product of monomer's wave functions.[3] Worse, at the O-O distance in ice, calculations indicate a net anti-bonding contribution from overlap effects to the total energy.[3,4] It follows that HB stabilisation energy mainly involves donation and back-donation of charge between the oxygen lone pair and the O-H anti-bonding orbitals on neighbouring molecules.[5] However, in order to get a strong attractive electrostatic interaction, an internal (s-p)-hybridisation is needed in order to minimise the repulsive charge overlap between filled orbitals on bonded atoms. It was also argued that charge induction from the surrounding medium fails to properly describe the internal charge redistribution upon HB formation. The trouble is that s-p hybridisation takes place only in orbitals at the time of bond formation. Real electrons are never involved during hybridisation that is just a mathematical trick transforming delocalised MO into localised ones providing better overlapping than pure s, p and d-orbitals. Orbitals having no physical reality by themselves, hybridisation concepts are thus of no help for explaining the real nature of HBs.

Another problem is how the water molecule in its electronic ground-state (GS) configuration of C_{2v} symmetry,

$$(1a_1)^2(2a_1)^2(1b_2)^2(3a_1)^2(1b_1)^2(4a_1)^0(2b_2)^0$$

allows the emergence of force fields giving rise to the highly directional HBs. Leaving aside the $(1a_1)$ molecular orbital (MO) corresponding to 1s-orbital (O-atom's core), the next two MOs $(2a_1, 1b_2)$ may describe the two covalent O-H σ-bonds, leaving two outer non-equivalent "lone-pairs" $(3a_1, 1b_1)$ available for HB with other water molecules. Now, the $3a_1$ (HOMO-1) and the $1b_1$ (HOMO) have two very different topologies and energies. X-ray emission spectroscopy (XES) clearly showed that HB in LW involved mainly the $3a_1$ (HOMO-1) valence-orbital.[5,6] As shown in Figure IX.1, the $3a_1$ MO lies in the molecular plane, whereas the HOMO $1b_1$ lies in a plane perpendicular to σ-bonds. Partial HB's covalence is thus possible only when the in plane fully occupied $3a_1$-(HOMO-1) overlaps with an empty $4a_1$-LUMO of a neighbouring molecule, the overlap being obviously zero using the other out-of-plane $1b_1$-HOMO. Strong HBs are then expected by interacting with the $3a_1$-level while van der Waals (VdW) interactions are expected for the $1b_1$-level. Are, the changes in the $3a_1$-orbital revealed by XES be experimental evidence for electron sharing in HBs?[6] No, as in frontier's orbitals theory the assumed covalence would primarily affect the HOMO outmost $1b_1$-orbital and definitively not the $3a_1$-HOMO-1. Covalence in a QM sense, i.e. HOMO-LUMO interaction, is thus not supported by experimental data. In other words, a valid CP or QM picture of HB seems impossible.

Owing to these conceptual difficulties, *in-silico* LW simulations

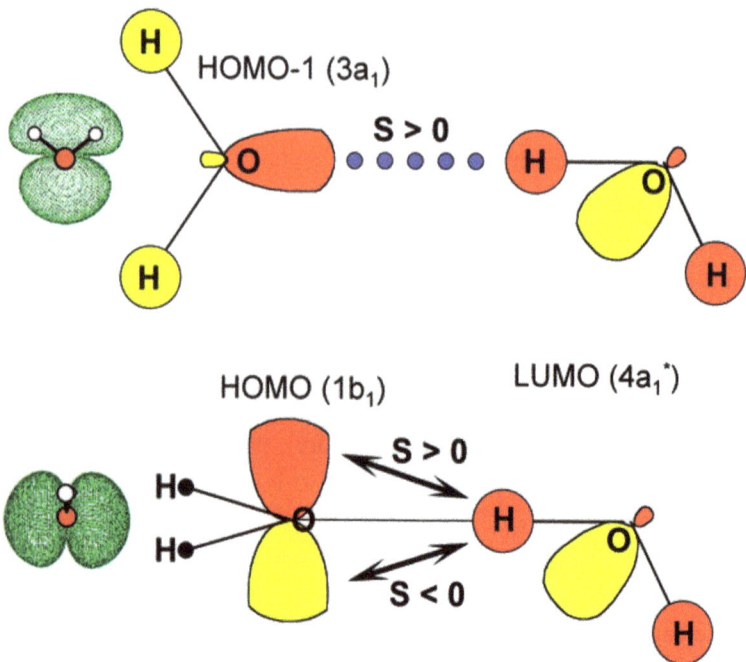

Figure IX.1. HOMO-LUMO interactions between two water molecules.

using Monte-Carlo methods are obliged to describe HBs through a two-body potential whose action is maximum at well-defined distances and angles. Strong deviations from the *ad-hoc* geometry lead to "broken HBs", allowing molecules to approach at closer distances. In order to fit the experimental data,[7] another *ad-hoc* dynamical scenario is invoked where each pair of molecules has a probability p_{HB} to make an HB, with a lifetime τ_{HB} function of temperature comprised between 1ps (T=300K) and 20ps (T=250K).[8] Consequently, the kind of dynamical network generated by simulations depends on the definition adopted for the HB. Owing to the lack of HB observable in QM, *ab-initio* MD-simulations[9] are characterised by complete arbitrariness in the choice of what is actually measured.[10] This leads to a quite strange situation where some scientists claim that LW should be a distorted, partly broken and fluctuating tetrahedral network,[11] whereas other scientists claim that the average coordination in LW should be 2 and not 4 .[12] Inclusion of nuclear quantum effects in MD-simulations is of no help yielding to some degree of over-structuring in the quantum simulation, indicating a residual error in the description of the potential energy surface.[13] The fact that no model potential is able to reproduce in every detail the properties of LW despite forty years of active research thus leaves a taste of incompletion.[14]

The big question is then how much space is left in the quest for a force field sufficiently accurate to be really helpful to understand the real nature of LW? Can the HB-paradigm introduced more than eighty years ago by Linus Pauling be theoretically justified? A first clue is that corpuscular QM (CQM) cannot be a realistic theory of the quantum world.[15] The second clue is that according to Bader's topological analysis electronic density, the HB-interaction involving closed-shell molecules cannot be distinguished from VdW bonding.[4] As topology ignores all metrical details focussing on fundamental properties that are scale-invariant, it follows that the HB concept may be in fact a nice empty shell. Accordingly, both VdW bonds and HBs are cooperative operating either at a microscopic or at a macroscopic scale. If both kinds of interaction are of similar nature, quantum electrodynamics (QED), i.e. the QM description of electromagnetic (e.m.) fields (EMFs) should be the right framework for understanding HB. In other words, is LW is a macroscopic quantum super-fluid, displaying full QM behaviour as in other collective phenomena such as ferromagnetism, superconductivity and super-fluidity? CQM being the "first quantisation" of a system of isolated particles displaying a finite number of degrees of freedom, one should then move to "second quantisation" or quantum field theory (QFT), i.e. to quantisation of fields having an infinite number of degrees of freedom to get new insights about the LW's structure.

First quantisation or Corpuscular Quantum Mechanics

In medieval times, following Aristotle's philosophy, there was no distinction between kinematics (from the Greek work κινήματος meaning "motion") and dynamics (from the Greek work δυνάμεως meaning "force" or "power"). Motion was nothing but the process by which bodies, once perturbed, reach their "natural place". Galileo Galilei was the first scientist to point out that instantaneous motion should not be viewed as a process but rather as a "state" of a material system, dynamics being the quantitative description of physical causes that change one state into another as time flows. CP has really started with Descartes's view of empty space as a 3D-manifold whose points may be put in a one-to-one correspondence with the 3D-continuum of real numbers, whereas time forms a 1D-continuum of real numbers, which depends on the observer. A physical system is then a space-region (x,y,z,m) that differs from what is perceived as empty space (x,y,z) by the presence of masses m. This has allowed distinguishing clearly between kinematics, the mathematical description of the correlation between the space domains spanned by a generic physical system in time, and dynamics, the quantitative description of the time evolution of a physical systems under the action of external forces (Newton's laws). In order to specify the correlations between two configurations at different times one needs also to assign either velocities (v_x, v_y, v_z) from Lagrange's viewpoint, or momenta (p_x, p_y, p_z) from Hamilton's viewpoint, defining a 6D-(x,y,z,p_x,p_y,p_z) classical phase-space (PS) for each mass. Finally, by gluing together with appropriate internal forces N material points, physical objects of any complexity may be generated leading to a 6D-PS. In CP there is no basic distinction between the state of the system at time t and physical observables $u(q_i, p_i, t)$ assuming a well defined real numerical value, once the totality of $q_i(t)$ and $p_i(t)$ have been determined. The most general and fruitful formulation of CP involves the use of Poisson's brackets (PBs) $\{u, H\}$ with the Hamiltonian $H(q_i, p_i, t)$, sum of kinetic (KE) and potential energies (PE):

$$\{u,v\} = \sum_i \frac{\partial u}{\partial q_i}\cdot\frac{\partial v}{\partial p_i} - \frac{\partial u}{\partial p_i}\cdot\frac{\partial v}{\partial q_i} \Rightarrow \dot{u} = \frac{du}{dt} = \{u,H\} + \frac{\partial u}{\partial t}$$

CP has also physical entities $\varphi(x,y,z,t)$, named "fields", that are defined for everywhere in space at all times and irreducible to any "mechanistic" description by a PS. EMFs are characterised by a scalar potential φ associated to volume charge densities ρ and a magnetic vector potential (A_x, A_y, A_z) associated to surface density currents (j_x, j_y, j_z) propagating at the speed of light c:

$$\begin{cases} x^\alpha = (ct, x, y, z) \Rightarrow \partial^\alpha = \frac{\partial}{\partial x_\alpha} = \left(\frac{1}{c}\cdot\frac{\partial}{\partial t}, -\frac{\partial}{\partial x}, -\frac{\partial}{\partial y}, -\frac{\partial}{\partial z}\right) \\ A^\mu = \left(\frac{\varphi}{c}, -A^x, -A^y, -A^z\right) \Rightarrow \partial_\alpha A^\alpha = \partial^\alpha A_\alpha = 0 \\ J^\alpha = (c\rho, j^x, j^y, j^z) \Rightarrow \partial_\alpha J^\alpha = 0 \end{cases}$$

These potentials obey Maxwell's field equations through the electromagnetic tensor $F_{\mu\nu}$, the metric tensor $g^{\mu\nu}$ of $(+,-,-,-)$ signature and Levi-Civita tensor $\varepsilon^{\alpha\beta\gamma\delta}$ (0 for equal indices, ±1 for even or odd permutations):

$$F_{\mu\nu} = \partial_\mu A_\nu - \partial_\nu A_\mu \Rightarrow \begin{cases} \partial^\mu \partial_\mu A^\nu = g^{\mu\alpha}\partial_\alpha\partial_\mu A^\nu = \mu^\circ J^\nu \\ \partial_\alpha\left(\frac{1}{2}\varepsilon^{\alpha\beta\gamma\delta}F_{\gamma\delta}\right) = 0 \end{cases}$$

Surprisingly, in CP the non-existent vacuum has both an electric permittivity ε_0 as well as a magnetic permeability μ_0 such that $\varepsilon_0\mu_0 c^2=1$. Moreover, according to Maxwell's equations (Langevin's theory) LW should have a relative dielectric constant given by:

$$\varepsilon_r = 1 + \frac{N_A\cdot\rho\cdot p_0^2}{3k_B T\cdot\varepsilon_0} = 1 + \frac{18500\cdot[\rho/g\cdot cm^{-3}]\cdot[p_0/D]^2}{[T/K]\cdot[M/g\cdot mol^{-1}]}$$

With p_0=1.85498 D, ρ=1g·cm^{-3} and M=18 g·mol^{-1}, one predicts

$\varepsilon_r(300K)=13$ instead of $\varepsilon_r=80$. In order to realise the electrical rigidity of water molecules (polarisability $\alpha=1.47 Å^3$) it is worth computing their classical induction energy at their closest possible distance ($R=2.4Å$):

$$U_{ind} = -\frac{2\alpha \cdot p_0^2}{(4\pi\varepsilon_0)^2 R^6}$$

$$\Rightarrow (U_{ind}/eV) = -1.2483\frac{(\alpha/Å^3) \cdot (p_0/D)^2}{(R/Å)^6} \approx 0.03 eV$$

With an ionisation potential, $IP \approx 12.6 eV$, one may also consider the dispersion energy at the same distance:

$$U_{disp} = -\frac{3\alpha^2 \cdot I}{4(4\pi\varepsilon_0)^2 R^6}$$

$$\Rightarrow (U_{disp}/eV) = -0.75\frac{(\alpha/Å^3)^2}{(R/Å)^6} \cdot (I/eV) \approx 0.12 eV$$

Finally, for the dipole-dipole interaction energy at $T=300K$, one gets:

$$U_{dd} = -\frac{2p_0^4}{3(4\pi\varepsilon_0)^2 \cdot k_B T \cdot R^6}$$

$$\Rightarrow (U_{dd}/eV) = -3013.81\frac{(p_0/D)^4}{(T/K) \cdot (R/Å)^6} \approx 0.05 eV$$

The HB-strength being $E_{HB} \approx 0.22$ eV, the HB-interaction is the deep enigma of CP.

The discovery at the dawn of the XXth century that Poisson's brackets and Maxwell's equations were unable to explain the atomic structure of matter and light emission from atoms has led to the development of QM through joined efforts of Werner Heisenberg (Germany), Erwin Schrödinger (Austria) and Paul Dirac (UK). In its first quantisation version, the real PS is discarded and replaced by an abstract complex Hilbert vector space (HVS) where each system's state is represented by a "ket" $|\psi>$ holding all the measurable information. This HVS is also equipped with a scalar product $<\varphi|\psi>$ defining a dual HVS inhabited by conjugated "bra" $<\varphi|$ describing the complex probability amplitude of being in a state $|\varphi>$ if the system is a given superposed state $|\psi>$. This HVS also have Hermitian operators $\hat{O}$ allowing switching between states: $\hat{O}|\psi>=|\varphi> \Leftrightarrow <\varphi|=<\psi|\hat{O}^{\dagger}$ and self-adjoint operators characterised by $\hat{O}=\hat{O}^{\dagger}$ being called "observables" owing to the fact that their eigenvectors, such that $\hat{O}|\psi>=\lambda|\psi>$, have only real eigenvalues λ. First quantisation is then a process involving the correspondence: $i\hbar\{O,A\} \rightarrow [\hat{O},\hat{A}]=\hat{O}\hat{A}-\hat{A}\hat{O}$ between classical PBs and quantum commutators where $\hbar=h/2\pi$ (h is the Planck's quantum of action introduced by Max Planck), leading to canonical commutation relationships (CCRs) of QM:

$$\{q_i,p_i\}=1 \Rightarrow [\hat{q}_i,\hat{p}_i]=i\hbar$$

This non-vanishing commutator between two conjugated dynamical variables means that the probability amplitude $<q|p>$ for a quantum particle (QP) to be at position q with a well-defined momentum p is:

$$\hat{p}|p\rangle = p|p\rangle \Rightarrow \langle q|p \rangle = \frac{1}{\sqrt{2\pi\hbar}}\exp\left(i\frac{p \cdot q}{\hbar}\right)$$

This corresponds to a wave delocalised over the entire space, i.e. the QP is found everywhere at the same time. On the other hand, any measurement of the position q of this same QP will show that it is at a definite space location (Dirac's peak). Consequently, in QM, it is impossible to measure both positions (x,y,z) and associated momenta (p_x,p_y,p_z). This irreducible contradiction between a corpuscular and a wave-like aspect for QPs has led to the Copenhagen interpretation of QM by Niels Bohr, where experimental reality is modelled by two mutually exclusive and complementary aspects engraved into the celebrated Heisenberg's uncertainty principle (HUP): $\Delta q \cdot \Delta p \geq \hbar/2$.

Owing to this intrinsic fuzziness of QPs, the only way to characterise their motion is to introduce quantum numbers for representing quantum states $|\psi\rangle = |n,l,m,\sigma\rangle$. In Schrödinger's representation of CQM, a complex time-dependent wave-function $\psi(x,y,z,t)=\langle x,y,z,t|\psi\rangle$ is introduced solution of the following equations involving time-independent differential operators:

$$\hat{H} = \frac{\hat{p}^2}{2m} + V(x,y,z) \Rightarrow i\hbar \frac{\partial}{\partial t}\psi(x,y,z,t) = \hat{H}\psi(x,y,z,t)$$

$$\psi(x,y,z,t) = \varphi(x,y,z) \cdot \exp\left(-i\frac{E \cdot t}{\hbar}\right) \Rightarrow \hat{H}\varphi(x,y,z) = E\varphi(x,y,z)$$

In another representation due to Heisenberg, time-dependent operators act on time-independent quantum states according to the following equation:

$$O(q_i, p_i, t) \Rightarrow i\hbar \frac{d\hat{O}}{dt} = [\hat{O}, \hat{H}] + i\hbar \frac{\partial \hat{O}}{\partial t}$$

This formulation allows establishing Ehrenfest's theorem ruling the time-evolution of statistical distributions $\langle \hat{O} \rangle = \langle \psi | \hat{O} | \psi \rangle$ for operators $\hat{O}$ that have no explicit time-dependence ($\partial \hat{O}/\partial t = 0$):

$$i\hbar \frac{d\hat{O}}{dt} = \left[\hat{O}, \hat{H}\right]$$

$$\Rightarrow \frac{d\langle \hat{O} \rangle}{dt} = \frac{d\langle \Psi | \hat{O} | \Psi \rangle}{dt} = \langle \Psi | \frac{d\hat{O}}{dt} | \Psi \rangle = \frac{1}{i\hbar}\left\langle \left[\hat{O}, \hat{H}\right] \right\rangle$$

The time $\Delta \tau$ needed for a displacement of the centre of $\langle \hat{O} \rangle$ equal to its dispersion ΔO is then such that:

$$\Delta O \cdot \Delta H \geq \frac{\hbar}{2}\left|\frac{d\langle \hat{O} \rangle}{dt}\right| \Leftrightarrow \frac{\Delta O}{\left|d\langle \hat{O} \rangle/dt\right|} \cdot \Delta H = \Delta \tau \cdot \Delta H \geq \frac{\hbar}{2}$$

The uncertainty in energy being $\Delta E=2\cdot\Delta H$, the lifetime $\Delta\tau$ of a quantum state is linked to its energy spreading according to: $\Delta\tau\cdot\Delta E\geq\hbar$. The trouble with these two representations of QM is that both use operators associated to an abstract HVS that tends to obscure the physical meaning of the calculations. Fortunately, US physicist Richard Feynman has introduced a quite general correspondence between quantum amplitudes $<q_f,t_f|q_i,t_i>=e^{i\varphi}$ and classical PS variables (q_t,p_t). In this "Path-Integral" (PI) representation of CQM, a quantum phase $\varphi=S/\hbar$ related to the classical action integral (AI), $S=\int L(t)\cdot dt=\int(KE-PE)\cdot dt$, is associated to each PS path starting at point q_i at time t_i and ending at point q_f at time t_f. Knowing the Lagrangian $L(t)$, complex amplitudes are then just summed over all possible paths. For a macroscopic body such that $S>>\hbar$, the only paths that do not interfere together are those for which $\delta S\approx 0$, leading directly to the Lagrange-Hamilton-Poisson equations of CP. On the other hand, when $S\approx\hbar$, one gets a bundle of possible trajectories surrounding the classical one, in full agreement with the HUP.

Second quantisation or Quantum Field Theory (QFT)

CQM is haunted by two concepts, localisation and separability, leading to interminable philosophical debates about what is exactly meant by physical reality. In CQM, the QP is quite distinct from the quantum introduced by Einstein and Planck to explain the black body spectrum. It is rather a metaphysical object, much like the Newtonian mass-point, but with the fundamental and puzzling difference that the very physical means to define it, by following its trajectory, is in principle unavailable. In this dualistic view of nature, the microscopic world is ruled by the Schrödinger equation whereas the macroscopic world, to which belongs all physical observers with their measuring devices, has only access to statistical distributions $<\hat{O}(t)>=<\Psi(t)|\hat{O}|\Psi(t)>$ that follow the laws of CP. According to J.A. Wheeler, the Schrödinger wave function bears to the unknowable physical reality the same relationship that a weather forecast bears to the weather.[15] Consequently, the Copenhagen physicist adopting the wave-particle complementarity viewpoint has given up his ambition to describe the physical world as it is, being satisfied by accounting for what he may say about the world, whose reality remains fundamentally inaccessible. Another deep criticism that may addressed to CQM is the asymmetrical role played by spatial positions that have associated Hermitian operators and time that is a mere label with no associated QM operator. In fact, thanks to the PI-representation, it is possible to build a realistic quantum theory that do not rely on the troublesome wave-particle duality where both space and time are treated on an equal foot, a prerequisite for having a relativistic compliant quantum theory. The key concept in QFT is the field $\Psi(x,y,z,t)$ characterised by its ground state $|0>$ and where each point in space holds an infinite collection of harmonic oscillators (HOs) being all in their state of zero-point energy ZPE=$\frac{1}{2}\hbar\omega$. From this field zero-point ground-state (FZPGS) describing a state of vacuum or empty space, the field may be excited to produce any amount of quanta or particles associated to the field. Being produced by the same field, all quanta are exactly the same with no possibility of distinguishing a given quantum from another one. In theories involving several different kinds of fields there may be several different species of particles. But again, there are states with all possible numbers of each kind of particle, and all the members of a given species are exactly the same. As particles just reflect the state of excitation of the field, they have no physical reality by themselves or dynamical autonomy. It is thus impossible for them to hold a proper HVS as in CQM. All the reality is thus lying in the field, a non-local and non-separable entity by nature. Quanta are observed only upon interactions with other fields, the role of the field being to describe how a given space region differs from the FZPGS $|0>$. This new way of thinking means obviously that at any time bosons (integer spin) or fermions (half-integer spin) may spontaneously emerge from the FZPGS, popping up apparently from nowhere. Based on the lifetime-energy inequality one have to distinguish between real QPs such that $\Delta\tau\cdot\Delta E \geq \hbar$ (RQPs) and virtual ones when $\Delta\tau\cdot\Delta E < \hbar$ *(VQPs)*. Obviously VQPs cannot be observed, their role being to justify the existence of classical field's lines introduced by Michael Faraday. Accordingly, these fictive lines become an experimental reality only after putting somewhere a test mass (gravitational field), a test charge (electric field) or a magnet (magnetic field). The immaterial aspect of classical fields as well as their infinite range of action is thus clearly explained on going from CP to QFT. For instance, let a charge q_A emits virtual photons over a distance Δx reaching charge q_B after a time $\Delta t=\Delta x/c$. As virtual photons cannot be detected, the maximum energy that could be exchanged is $\Delta E \cdot \Delta t \approx \hbar$, that is, $\Delta E = \alpha \cdot \hbar c (q_A \cdot q_B)/\Delta x$, a nice QFT-explanation of the $1/r$-variation of Coulomb's potential. Consequently, at a QFT-level the same mechanism explains covalent bonding (short Δx, large ΔE), HB (medium Δx, moderate ΔE) or just VdW-bonding (large Δx, weak ΔE). Taking into account the fact that VQPs originate from the FZPGS, the proportionality constant may be written $\alpha = e^2/4\pi\varepsilon_0 \hbar c$ in SI units. Now, if the FZPGS is not able to reabsorb a previously emitted VQP after the time $\Delta t \approx \Delta E/\hbar$, this VQP becomes a RQP and classical observers will speak of spontaneous emission or decay, seeing a real particle leaving the system.

The field and its cohorts of VQPs are thus real things that cannot be observed but allow to know if a spatial region is in the FZPGS, where by definition any observation should give a null result. Localisation and separation, two concepts that haunt CQM cause no trouble in QFT as the definition of space or time belongs to observers through their measuring devices (rulers and clocks) and not to the field that describe the physical state of the observed space-time region. The nice point is that there is no need to use classical devices ruled by deterministic CP laws as in CQM. All is needed is an exchange of energy (quantum or classical) between the field and the measuring apparatus. Any measurement thus corresponds to a field's transition between two states whose energy difference corresponds to a small space-time region defined by the measuring device itself.

Mathematically speaking, the lack of time operator in CQM is solved by transforming spatial coordinates into mere labels with no associated Hermitian operators, recovering full symmetry between space and time coordinates. Let's now consider for simplicity a 1D-field (Figure IX.2) and associate to each point in space two field operators, one corresponding to the classical amplitude of the field $\varphi(q,t)$ at position q and the other one corresponding to a momentum $d\varphi/dt=\pi(q,t)$. The move from classical fields to quantum fields implies giving CCRs involving the couple (field amplitude, field momentum) instead of the couple (position, linear momentum) in CQM:

$$\left[\hat{\Psi}(q,t),\hat{\Pi}(q',t)\right]=i\delta(q-q')$$

$$\Rightarrow\left[\hat{\Psi}(q,t),\hat{\Psi}^{\dagger}(q',t)\right]=\delta(q-q')$$

From these new CCRs the eigenvalue problem for energy and momentum in the interval $-L/2<q<+L/2$ is easily solved after Fourier decomposition of the field operators treating each independent wave-mode p of the field as a 1D-HO, each mode having associated annihilation $\hat{a}_p|n>=|n-1>$ and creation $\hat{a}_p^{\dagger}|n>=|n+1>$ operators with new CCRs:

$$\begin{cases} p=n\frac{2\pi}{L} \quad \hat{\Psi}(q,t)=\frac{1}{\sqrt{L}}\sum_p \hat{a}_p(t)\cdot\exp(ip\cdot q) \\ \left[\hat{\Psi}(q,t),\hat{\Psi}^{\dagger}(q',t)\right]=\delta(q-q') \Rightarrow \left[\hat{a}_p(t),\hat{a}_{p'}^{\dagger}(t)\right]=\delta_{pp'} \end{cases}$$

This implies the existence of a unique GS $|0>$ as for all p-modes, one has $\hat{a}_p|0>=0$ and that the number of particles for a given p-mode becomes an observable $\hat{a}_p^{\dagger}\hat{a}_p$ having various possible outcomes and such that:

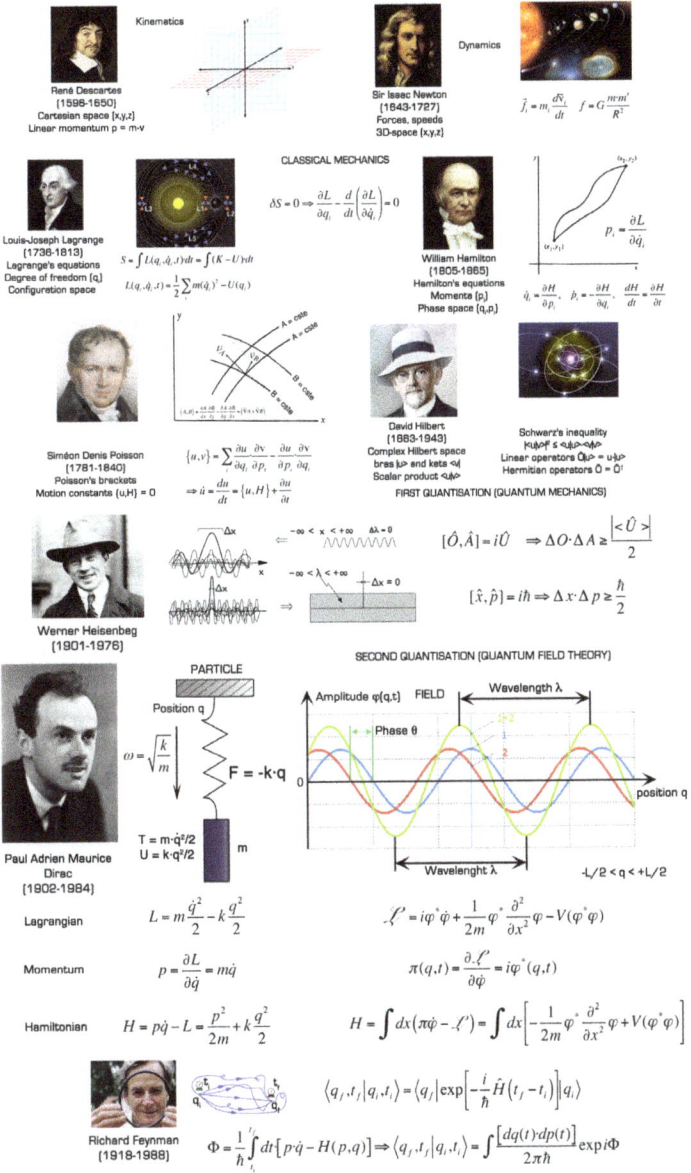

Figure IX.2. From CP to QFT, 400 years of science for explaining Nature.

$$\hat{N}_p = \hat{a}_p^\dagger \hat{a}_p \Rightarrow \hat{H} = \sum_p \left(\frac{p^2}{2m}\right)\left(\hat{N}_p + \frac{1}{2}\right) \quad \hat{P} = \sum_p p\left(\hat{N}_p + \frac{1}{2}\right)$$

The remarkable feature is that both Hamiltonian and total momentum are now simultaneously diagonal in the $|\{n_p\}\rangle$-basis, i.e. in the complete set of eigenvectors of the number operator. The generic state $|\{n_p\}\rangle$ thus corresponds to a field configuration where the wave-mode p is populated by n_p quanta:

$$\hat{N}_p = \hat{a}_p^\dagger \hat{a}_p \Rightarrow \hat{H} = \sum_p \left(\frac{p^2}{2m}\right)\left(\hat{N}_p + \frac{1}{2}\right)$$

$$\hat{P} = \sum_p p\left(\hat{N}_p + \frac{1}{2}\right)$$

The kinematical independence of the HOs pertaining to different p-values implies that the corresponding space of states, the so-called "Fock space" (FS), is nothing but the infinite-dimensional tensor product of the infinite-dimensional HVS for each wave-mode p. Moreover, due to the cancellation of opposite momenta, the FZPGS characterised by $n_p=0$ has, as expected, zero momentum. However, owing to a sum extending over an infinite number of wave-modes p, the total energy content E_{FZPGS} of the FZPGS is infinite, explaining why QFT has been regarded with high suspicion at its very beginning. Todays, this causes no more troubles as it follows from general relativity that any mass m cannot have a radius R smaller than its Schwarzschild radius R_S: $R \geq R_S = 2G \cdot m/c^2$, where G is Newton's gravitational constant. But according to QFT, any empty region of size $\Delta R = c \cdot \Delta t$ fluctuates with an energy $\Delta E = \Delta m \cdot c^2$ such that $\Delta E \cdot \Delta t \geq \hbar$, leading to $\Delta m \geq \hbar/(c \cdot \Delta R)$. Combining the quantum structure of empty space with its maximum possible curvature leads to: $\hbar/(c\Delta R) \leq \Delta m \leq c^2 \cdot \Delta R/G$. Consequently, when $\hbar/(c\Delta R) = c^2 \cdot \Delta R/G$, i.e. $\Delta R = L_p = [\hbar G/c^3]^{1/2} \approx 10^{-35}$ m, it becomes impossible to make a clear distinction between empty space (vacuum) and matter (mass). Empty space and matter are thus two equivalent aspect of a same quantum reality: the field. At this quantum gravity limit, space should no more be considered as a continuum, establishing a natural cut-off momentum $P_p \approx h \cdot c / L_p \approx 10^{20}$ GeV, implying a finite value for E_{FZPGS}. It can thus be safely subtracted, leading to finite energy differences between the FZPGS and the energy of any state of Fock's space: $\Delta E(\{n_p\}) = E(\{n_p\}) - E_{FZPGS}$.

Coherent states representation

Having recognised the crucial importance of the number operator of a p wave-mode, one may introduce its canonically conjugated phase operator:

$$\hat{a}_p^\dagger \hat{a}_p = \hat{N}_p \Rightarrow \hat{a}_p = e^{i\hat{\theta}_p} \hat{N}_p^{1/2} \Rightarrow \left[e^{i\hat{\theta}_p}, \hat{N}_p \right]$$

$$= i \cdot \left[\hat{\theta}_p, \hat{N}_p \right] \cdot e^{i\hat{\theta}_p} = e^{i\hat{\theta}_p} \Rightarrow \left[\hat{\theta}_p, \hat{N}_p \right] = -i$$

Owing to this non-vanishing commutator and to the topological differences existing between the straight line ($-\infty < x < +\infty$) and the circle ($-\pi \leq \theta_p \leq +\pi$) one has considering the function $Y(\theta_p) = \theta_p \, mod(2\pi)$:[17]

$$\left[1 - \sum_{n=-\infty}^{n=+\infty} \delta\{\hat{\theta}_p - (2n+1)\pi\} \right] \Rightarrow \Delta\theta_p \cdot \Delta N_p \geq \frac{1}{2} \left| 1 - 3 \left(\frac{\Delta\theta_p}{\pi} \right)^2 \right|$$

For a fixed number of quanta as in CQM, $\Delta N_p = 0$, showing that the quantum phase is completely undetermined: $\Delta\theta_p = \pi/\sqrt{3}$. Conversely, large fluctuations of N_p mean less dispersed quantum phases with $\Delta\theta_p \cdot \Delta N_p \geq 1/2$ for small phase dispersion. Now, the eigenvectors |n>

of a given number operator $\hat{a}^\dagger\hat{a}$ and those $|\alpha\rangle$ of the corresponding annihilation operator $\hat{a}$ are known to be such that:[15]

$$\begin{cases} \hat{N} = \hat{a}^\dagger\hat{a}, \quad \hat{N}|n\rangle = n|n\rangle \Rightarrow |n\rangle = \frac{1}{\sqrt{n!}}(\hat{a}^\dagger)^n|0\rangle \\ \hat{a}|\alpha\rangle = \alpha|\alpha\rangle \Rightarrow |\alpha\rangle = \exp\left(-\frac{\alpha\cdot\alpha^*}{2}\right)\sum_{n=0}^{+\infty}\frac{(\alpha)^n}{\sqrt{n!}}|n\rangle \\ \qquad = \exp\left(-\frac{\alpha\cdot\alpha^*}{2}\right)\cdot\exp(\alpha\hat{a}^\dagger)|0\rangle \end{cases}$$

It then follows that such $|\alpha\rangle$ states are non-orthogonal with a well-defined quantum phase θ, implying large fluctuations, $\Delta N = |\alpha| \to \infty$ as $\Delta\theta \to 0$:

$$\begin{cases} \langle\alpha|\alpha'\rangle = \exp\left(\alpha^*\alpha' - \frac{\alpha\alpha^*}{2} - \frac{\alpha'\alpha'^*}{2}\right) \neq \delta(\alpha - \alpha') \\ \alpha = |\alpha|e^{i\theta} \Rightarrow \langle\alpha|e^{i\hat{\Phi}}|\alpha\rangle = \langle\alpha|\hat{a}\hat{N}^{-1/2}|\alpha\rangle \approx e^{i\theta} \end{cases}$$

With their well-defined quantum phase for a given Hamiltonian $H(\alpha^*, \alpha)$ these "coherent states" may be used to define a new QFT-based coherent state representation (CSR) $\psi(\alpha)$ of macroscopic quantum states using the PI-formulation of QM:

$$L(t) = \frac{m}{2}(\dot{q}^2 - \omega^2 q^2)$$

$$\Rightarrow \langle\alpha_f|\alpha_i\rangle = \int\frac{da(t)da^*(t)}{2\pi i\hbar}\exp\left[\frac{i}{\hbar}\int_{t_i}^{t_f}dt\cdot L(t)\right]$$

$$q = \sqrt{\frac{\hbar}{2m\omega}}\cdot[a(t) + a^*(t)]$$

$$a(t) = \alpha\cdot e^{-i\omega t} \Rightarrow \frac{L_{em}}{\hbar} \approx \left[\frac{\dot{\alpha}\dot{\alpha}^*}{2\omega} + \frac{i}{2}(\dot{\alpha}\alpha^* - \alpha\dot{\alpha}^*)\right]$$

By varying the AI against $\alpha^*(t)$ for an HO such that $H(\alpha^*,\alpha)=\omega\cdot\alpha^*(t)\cdot\alpha(t)$, an oscillating wave solution is immediately retrieved, showing that the CSR represent the best quantum approximation of a classical behaviour.

$$\delta\left(\int_{t_i}^{t_f} dt\left[\alpha^*(t)i\dot\alpha(t)-\omega\alpha^*(t)\alpha(t)\right]\right)=0$$

$$\Rightarrow i\dot\alpha(t)-\omega\alpha(t)=0 \Rightarrow \alpha(t)=\alpha\cdot e^{-i\omega t}$$

The same procedure applied to a Lagrangian density corresponding to a free EMF of frequency ω leads after neglecting the rapidly oscillating $exp(\pm 2\omega it)$ terms to:

$$L(t)=\frac{m}{2}(\dot q^2-\omega^2 q^2) \Rightarrow \langle\alpha_f|\alpha_i\rangle=\int\frac{da(t)da^*(t)}{2\pi i\hbar}\exp\left[\frac{i}{\hbar}\int_{t_i}^{t_f} dt\cdot L(t)\right]$$

$$q=\sqrt{\frac{\hbar}{2m\omega}}\left[a(t)+a^*(t)\right] \quad a(t)=\alpha\cdot e^{-i\omega t} \Rightarrow \frac{L_{em}}{\hbar}\approx\left[\frac{\dot\alpha\dot\alpha^*}{2\omega}+\frac{i}{2}(\dot\alpha\alpha^*-\alpha\dot\alpha^*)\right]$$

For matter fields, let $|\psi_n\rangle$ be the eigenvector and E_n the energy of a single molecule: $\hat H|\psi_n\rangle=E_n|\psi_n\rangle$ and let $\hat a_n$ be the corresponding annihilation operator obeying the (anti-)commutation relations: $[\hat a_n,\hat a_m^\dagger]_\pm=\delta_{nm}$, according to whether our identical systems have Fermi or Bose character. As usual, the operator $\tilde N_n=\hat a_n^\dagger\hat a_n$ allows specifying the occupation number of the energy level $|\psi_n\rangle$. Then, using the Natural Unit System ($\hbar=c=k_B=1$), the CSR allows introducing a coherent matter amplitude $a_n(t)$:

$$\hat\Pi_n=i\hat\Psi_n^\dagger \Rightarrow L_n(t)=\int_V\left(\hat\Pi_n\dot{\hat\Psi}_n-\hat H_n\right)dV=ia_n^*(t)\cdot\dot a_n(t)-E_n a_n^*(t)a_n(t)$$

Now, taking into account coupling between matter-fields and EMFs means that:

$$H\propto\frac{(\vec p-e\vec A)^2}{2m}=\frac{p^2}{2m}-\frac{e}{m}\vec p\cdot\vec A+\frac{e^2}{2m}\vec A\cdot\vec A$$

Considering a two-level system with transitions from a GS having a coherent amplitude $a_0(t)$ towards an excited level having a coherent amplitude $a_q(t)$ and using the PI-CSR leads to:

$$\langle \alpha_f | \alpha_i \rangle = \int \left[\frac{da_0(t)da_0^*(t)}{2\pi i} \cdot \frac{da_q(t)da_q^*(t)}{2\pi i} \cdot \frac{da(t)da^*(t)}{2\pi i} \right] \exp\left[i \int_{t_i}^{t_f} dt \cdot L(t) \right]$$

$$L(t) = \overbrace{ia_0^*\dot{a}_0 - E_0 a_0^* a_0 + ia_q^*\dot{a}_q - E_q a_q^* a_q}^{\text{matter fields}} + \underbrace{\frac{\dot{\alpha}\dot{\alpha}^*}{2\omega_q} + i\frac{\dot{\alpha}\alpha^*}{2} - i\frac{\alpha\dot{\alpha}^*}{2}}_{\text{e.m. field}}$$

$$\underbrace{- g_q \left(a_q^* a_0 \alpha e^{-i\omega_q t} + a_0^* a_q \alpha^* e^{i\omega_q t} \right)}_{\text{dipolar transitions}} - \underbrace{\mu_q \cdot \alpha \alpha^*}_{\text{dispersion}}$$

Here, fields have been averaged over a volume V holding N particles while g_q and μ_q are coupling constants ruled, in the dipolar approximation, by oscillator strengths f_{mn} or $f_0(\omega)$:

$$f_{mn} = \frac{2m_e}{3} |E_n - E_m| |\langle n | \vec{\varepsilon} \cdot \vec{r} | m \rangle|^2 \Rightarrow g_q = e \sqrt{\frac{2\pi f_{0q}}{m_e V}}$$

$$\lambda_q = -\frac{3\omega_q}{2} \left[\sum_{n \neq q} \frac{f_{nq}}{\omega_n^2 - \omega_q^2} + \int_{\omega_{cont}}^{+\infty} \frac{f_0(\omega)d\omega}{\omega_n^2 - \omega_q^2} \right] \Rightarrow \mu_q = \frac{e^2 \lambda_q}{m_e} \left(\frac{N}{V} \right)$$

Performing a $N^{1/2}$ scaling of the fields then leads to:

$$L(t) = N \left[\overbrace{i\chi_0^*\dot{\chi}_0 + i\chi_q^*\dot{\chi}_q}^{\text{matter fields}} + \underbrace{\frac{\dot{A}\dot{A}^*}{2\omega_q} + \frac{i}{2}(\dot{A}A^* - A\dot{A}^*)}_{\text{e.m. field}} \right.$$

$$\left. \underbrace{- g_q \left(\chi_q^* \chi_0 A + \chi_0^* \chi_q A^* \right)}_{\text{dipolar transitions}} - \underbrace{\mu_q \cdot AA^*}_{\text{dispersion}} \right]$$

Finally, variation of the AI against the three interacting fields allows writing three coherence equations (CEs), two describing absorption and emission by matter of a virtual photon from or towards the FZPGS and the last one describing the status of the FZPGS holding both matter and EMFs:

$$\frac{\delta}{\delta\chi_0^*}\left(N\int_{t_i}^{t_f}dt\cdot L(t)\right)=0 \Rightarrow i\dot{\chi}_0 - g_q\chi_q A^* = 0$$

$$\Leftrightarrow i\dot{\chi}_0(\tau) = \frac{g_q}{\omega_q}\chi_q(\tau)\cdot A^*(\tau)$$

$$\frac{\delta}{\delta\chi_q^*}\left(N\int_{t_i}^{t_f}dt\cdot L(t)\right)=0 \Rightarrow i\dot{\chi}_q - g_q\chi_0 A = 0$$

$$\Leftrightarrow i\dot{\chi}_q(\tau) = \frac{g_q}{\omega_q}\chi_0(\tau)\cdot A(\tau)$$

$$\frac{\delta}{\delta A^*}\left(N\int_{t_i}^{t_f}dt\cdot L(t)\right)=0 \Rightarrow -\frac{\ddot{A}}{2\omega_q}+i\dot{A}-g_q\chi_0^*\chi_q-\mu_q\cdot A = 0$$

$$\Leftrightarrow -\frac{\ddot{A}(\tau)}{2}+i\dot{A}(\tau)-\frac{\mu_q}{\omega_q}A(\tau)=\frac{g_q}{\omega_q}\chi_0^*(\tau)\chi_q(\tau)$$

These three CEs refer to any piece of macroscopic matter of volume $V \approx L^3$ inside which any space-dependence of the fields is neglected and that should be in equilibrium with a FZPGS holding a collection of virtual oscillators having any frequency ω. Any A-EMF such that $\omega \neq \omega_q = (E_q - E_0)$, will remain in a perturbative ground state (PGS), meaning that any fluctuation arising from the FZPGS is doomed to disappear with a lifetime given by $\Delta t \cdot \Delta E \approx \hbar$. When $\omega \approx \omega_q$, the matter-fields χ_0 and χ_q becomes strongly coupled with the A-EMF. If the matter density (N/V) is large enough, the system may condensate into an array of coherence domains (CDs) having a size L of the order of the wavelength of the selected EMF-mode: $L \approx \lambda \approx 2\pi/\omega_q$. Within each CD the dynamical behaviour of matter and EMF is coherent and homogeneous. All fields are normalised to be dimensionless in such a way that all conserved quantities (total number of particles N, momentum Q and hamiltonian H) refer to a single molecule that may be found anywhere within the CD:

$$\omega_p = e\sqrt{\frac{N}{m_e V}} \Rightarrow g = \frac{g_q}{\omega_q} = \frac{\omega_p}{\omega_q}\sqrt{2\pi f_{0q}}$$

$$\mu = \frac{\mu_q}{\omega_q} = -\frac{3\omega_p^2}{2}\left[\sum_{n \neq q}\frac{f_{nq}}{\omega_n^2 - \omega_q^2} + \int_{\omega_{cont}}^{+\infty}\frac{f_0(\omega)d\omega}{\omega_n^2 - \omega_q^2}\right]$$

$$\chi_0^*\chi_0 + \chi_q^*\chi_q = 1$$

$$Q = A^*A + \frac{i}{2}(A^*\dot{A} - \dot{A}^*A) + \chi_0^*\chi_0$$

$$H = Q + \frac{\dot{A}^*\dot{A}}{2} + \mu A^*A + g(A^*\chi_q^*\chi_0 + A\chi_0^*\chi_q)$$

Now let's consider an initial configuration ($\tau=0$) with all molecules in their GS, the FZPGS being in its PGS:

$$\left|A(0) \simeq \frac{1}{\sqrt{N_{CD}}}, \quad \chi_q(0) \simeq \frac{1}{\sqrt{N_{CD}}}, \quad \chi_0(0) \simeq 1\right.$$

At very short time τ, the three CEs becomes:

$$\tau \approx 0 \Rightarrow \chi_0(\tau) = \chi_0^*(\tau) \approx 1$$

$$\Rightarrow -\frac{1}{2}\ddot{A}(\tau) + i\ddot{A}(\tau) - \mu\dot{A}(\tau) - g\dot{\chi}_q(\tau) = 0$$

$$i\dot{\chi}_q(\tau) = g\chi_0(\tau)A(\tau) \approx gA(\tau)$$

$$\Rightarrow -\frac{1}{2}\ddot{A}(\tau) + i\ddot{A}(\tau) - \mu\dot{A}(\tau) + ig^2 A(\tau) = 0$$

$$A(\tau) \propto \exp(ip)$$

$$\Rightarrow i\left(\frac{p^3}{2} - p^2 - \mu \cdot p + g^2\right)A(\tau) = 0$$

When $g \leq g_c$, all three roots of the third equation are real and the PGS remains stable. However when $g > g_c$, one root is real and the

The topological and quantum structure of zoemorphic water 219

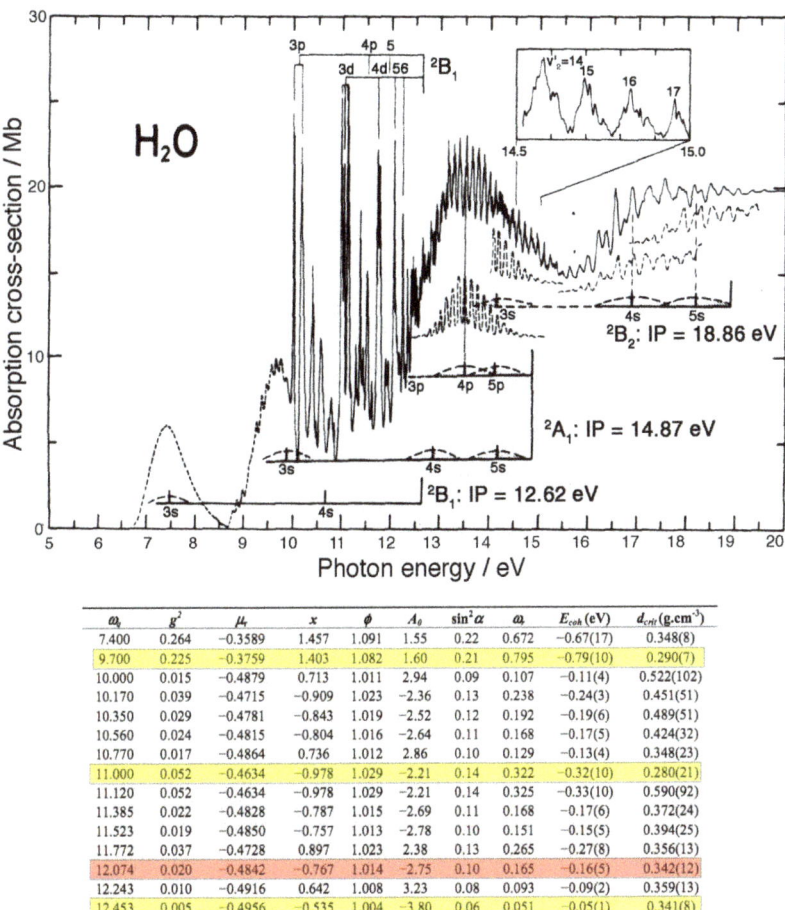

Figure IX.3. Coherent states for LW predicted from the experimental absorption spectrum of water vapour.

other two are complex conjugate leading to an exponential increase in $A(\tau)$ that overcomes rapidly its $O(N_{CD}^{-1/2})$ initial value. An avalanche process thus drives the system away from the PGS towards a new coherent ground state (CGS) releasing an energy corresponding to the latent heat of vaporization to the environment. The critical g-value g_c separating the two regimes is given by:

$$\begin{cases} \dfrac{d}{dp}\left[\dfrac{1}{2}p^3 - p^2 - \mu p + g_c^2\right] = 0 \\ \dfrac{1}{2}p^3 - p^2 - \mu p + g_c^2 = 0 \end{cases}$$

$$\Rightarrow g_c^2 = \dfrac{8}{27} + \dfrac{2}{3}\mu + \left(\dfrac{4}{9} + \dfrac{2}{3}\mu\right)^{3/2}$$

Everything depends on the values of the dispersive coupling constants μ that are ruled by the full absorption spectrum, including the continuum tale, of the considered substance. The stationary solution of the CEs corresponding to the CGS may be found by writing the fields in a phase representation:[18]

$$\begin{cases} \chi_0(\tau) = \cos\alpha \cdot \exp[i\theta_0(\tau)] \\ \chi_q(\tau) = \sin\alpha \cdot \exp[i\theta_q(\tau)] \\ A(\tau) = A_0 \exp[i\phi(\tau)] \end{cases}$$

$$\Rightarrow \begin{cases} A_0 = \dfrac{x\dot{\phi}}{2g} \quad \alpha = \dfrac{1}{2}\arctan x \quad \omega_r = \omega_q |1 - \dot{\phi}| \\ \dot{\phi} = \dot{\theta}_q - \dot{\theta}_0 = \dfrac{gA_0}{\tan\alpha} - gA_0 \tan\alpha \\ E_{coh} = \omega_q \left[A_0^2(1 + 2\mu_r) + \sin^2\alpha - \dfrac{3}{2}gA_0 \sin 2\alpha\right] \end{cases}$$

This CGS has five unknowns: the mixing angle α between the two energy levels, the amplitude A_0 of the trapped EMF and three rates of variation for each phase: $d\theta_0/d\tau$, $d\theta_q/d\tau$, $d\phi/d\tau$ linked by five relations. The most remarkable fact is the phase locking where matter oscillates coherently and collectively between two molecular levels in tune with a coherent EMF, which unlike what happens in a laser, remains trapped in the matter bouncing from molecule to molecule within a given CD. During the process of condensation (g, μ, ω_q)-values are renormalized, giving rise to a highly non-linear system. Moreover, on going from the PGS to the CGS, a coherence gap E_{coh} is opened stabilising the system against thermal fluctuations. The first energy term is positive meaning that energy should be borrowed from the FZPGS to allow EMF-fluctuations. The second one is also positive allowing rising water molecules from the GS to an excited state $|q>$ borrowing again energy from the FZPGS. The third one is negative, allowing keeping alive EMF-fluctuations without any expense of energy from the FZPGS.

Liquid water

Solving the CGS equations means knowing the coupling constants g and μ as well as the frequencies ω_q of the excited levels of the considered substance. Figure IX.3 shows the excitation spectrum of water vapour in the visible-range,[19] the UV-excitation part up to 200 eV being also available.[20] The mathematical procedure for solving the CEs for LW has been published elsewhere.[18] By searching for the level

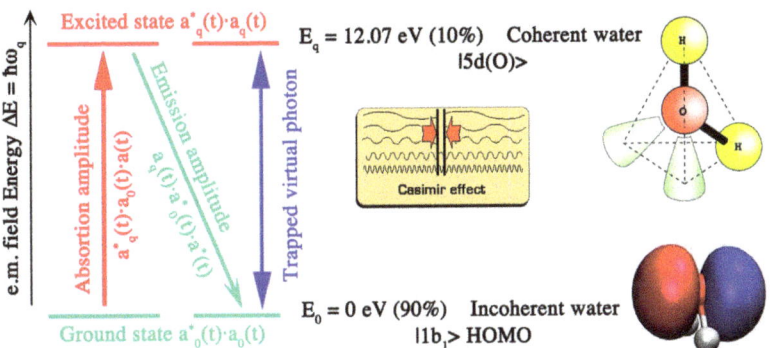

Figure IX.4. Water self-lasering between a GS and an excited state leading to emergence of coherence.

having a critical density close to the experimental value $\rho^*=0.322$ g·cm^{-3} associated to a coherent gap close to the HB-energy in crystalline ice ($E_{HB}\approx 20$ kJ·mol$^{-1}\approx 0.21$ eV, it is thus found that the best match (Figure IX.3) is provided by an excited level $|5d(O)\rangle$ located 12.07 eV above the GS[19] and characterised by $\rho^*=0.342$ g·cm^{-3} and $E_{HB}\approx 0.16$ eV. The consequences for LW are the following:

i) The coherence gap plays at a QFT-level the role of the HB-concept at the CP/CQM-level. The association energy of about 0.2eV·molecule^{-1} comes out naturally from the sole knowledge of the excitation spectrum of the water vapour. As $\Delta\varphi\cdot\Delta N \geq \frac{1}{2}$, phase coherence within a CD ($\Delta\varphi\rightarrow 0$) is achieved only at the expense of strong fluctuations in the number of associated quanta ($\Delta N\rightarrow +\infty$). At a CP/CQM-level this means that the HB-network image for LW is doomed to fluctuate, as observed in all MD-simulations. In QFT, HB or VdW-bonding have the same origin, HB meaning evolving towards a CGS with phase-locking between matter and the vacuum and VdW-bonding meaning a stable PGS with decoupling between matter and the PGS (phase incoherence).

ii) In the CGS water molecules oscillate between two levels, $|CGS\rangle = \cos\alpha\cdot|0\rangle + \sin\alpha\cdot|5d(O)\rangle$, with $\cos^2\alpha=90\%$ and $\sin^2\alpha=10\%$, leading to a larger molecular radius: 1.5Å versus 1.1Å in the GS. Among the d-manifold, two orbitals (z^2, x^2-y^2) displaying a_1-symmetry in C_{2v} symmetry may be mixed with the two MO-orbitals ($2a_1$, $3a_1$) describing O-H covalent bonding and/or oxygen lone-pairs. From this mixing involving four occupied molecular states, one may expect a more or less tetrahedral arrangement of electrons and H-atoms in order to minimise electronic repulsions (Figure IX.4).

iii) Let r_0 be the radius of the smallest spherical matter-CD in equilibrium with the EMF. Within a matter-CD ($r<r_0$), the strong water-EMF coupling shifts the exciting frequency ω_q downwards ($\omega_r<\omega_q$) causing the vacuum/water interface to act a total reflection mirror. This prevents the EMF to leak out into the vacuum outside the CD, where the field would propagate with a frequency ω_q. Outside the matter-

CD ($r>r_0$), the EMF should obey the free field equation, $\partial^2 A(r,t)/\partial t^2 - \Delta A(r,t)=0$ or $(-\omega_r^2-\Delta)A(r)=0$ after time-dependence factorisation. Owing to the spherical symmetry, the angular part of the Laplacian Δ at a border of the matter-CD ($r\approx r_0$), is approximately equal to $\omega_q^2 \cdot A(r)$. By joining the amplitude and its first radial derivative within a matter-CD to the exponentially decaying solution outside the matter-CD, one gets:[21]

$$\begin{cases} r<r_0 \Rightarrow A(r,t) = A_0 \cdot \dfrac{\sin(\omega_q r)}{\omega_q r} \cdot \exp(-i\omega_r t) \\ r>r_0 \Rightarrow \dfrac{d^2(rA)}{dr^2} - (\omega_q^2 - \omega_r^2)(rA) = 0 \end{cases}$$

$$\Rightarrow \begin{cases} r_0 \approx \dfrac{3\pi}{4\omega_q} \\ A(r) \approx \dfrac{A_0}{\sqrt{2}} \cdot \dfrac{\exp\left[-\sqrt{\omega_q^2 - \omega_r^2}\,(r-r_0)\right]}{\omega_q r} \end{cases}$$

Excitation of water at $\lambda=2\pi/\omega_q$ (field-CD) corresponds to a matter-CD diameter $L=2r_0\approx(\tfrac{3}{4})\lambda\approx 75$ nm for $\omega_q=12.07$ eV. Assuming 100% of coherence at the lowest super-cooling temperature reached so far ($T_S\approx 228$K) for a density close to that of ice ($\rho=0.92$ kg·cm^{-3}) means that $V_{CD}=\pi L^3/6=220{,}893$ nm^3 or $N_{CD}=30.8 \cdot V(\text{nm}^3)\approx 6.8$ millions water molecules. A CD is thus definitively not a water cluster but rather a coherent nanodrop. Close packing of many CDs at minimum inter-domain distance L leads to a superposition of the inner trapped fields and the evanescent tails of neighbouring domains (Figure IX.5). To a good approximation[21] within each matter-CD ($\omega_q r=\pi x$, $0\leq x\leq\tfrac{3}{4}$):

$$A(x) = A_0 \cdot F(x) = A_0 \left[\dfrac{\sin(\pi x)}{\pi x} + \dfrac{\sqrt{2}}{\pi} \cdot \dfrac{\exp[-\pi(3/4-x)]}{3-2x}\right]$$

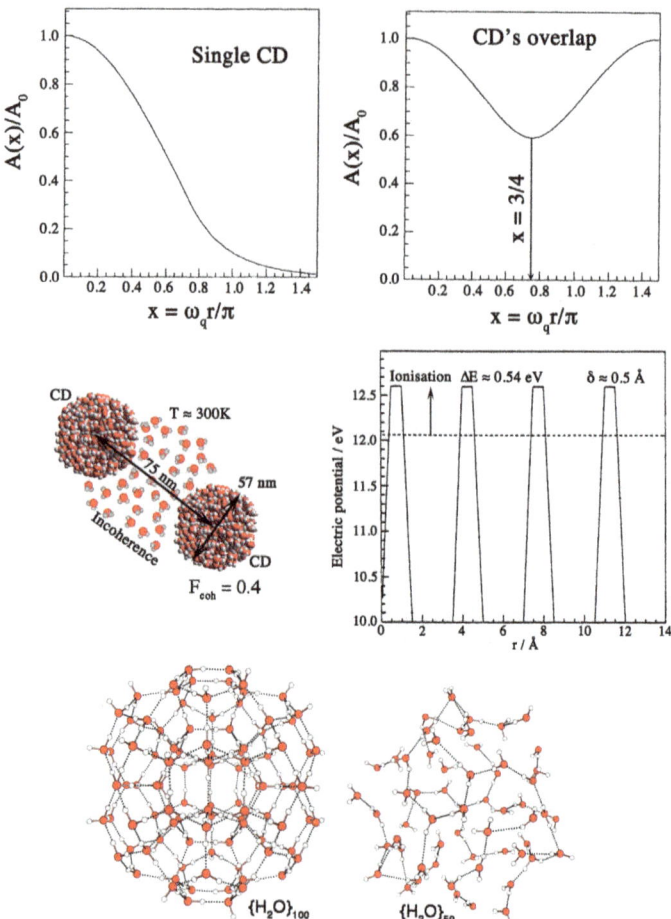

Figure IX.5. Close packing of CDs leading to EMFs overlapping (top), LW structure at T=300K (middle-left), electric potential with a CD (middle-right) and two large water clusters experimentally observed through single-crystal X-ray diffraction at the heart of polyoxomolybdate nanocapsules (bottom).

This mutual penetration of the field CD's inside the matter CD's produces an actual interaction among domains responsible for a long-range cohesion of water molecules. The inter-domain attraction being lower than the attraction between water molecules within one domain, it requires much less energy to produce a bubble than to vaporise a corresponding quantity of water. As gas bubbles should be localised in the voids generated by the stacking of the CDs in a highly unstable configuration favouring coalescence, the fact that a series of common electrolytes were found either to reduce the rate of bubbles coalescence in water or to have no effect[22] is not surprising. Each CD being strongly coupled to a trapped near-infrared EMF ($\lambda \approx 1/\omega_r \approx 1.2\ \mu m$), solute species may enter a CD if resonance between an ion's vibrational mode and the trapped EMF is possible. For such ions, bubbles coalescence is expected as in pure water. Without resonance, ions should remain outside the CDs stabilising nanobubbles above a certain critical concentration. Entering or staying outside a CD being a pure QFT-effect, applying CP/CQM-concepts to explain bubbles' behaviour is doomed to fail.

Strictly speaking, all the above conclusions apply at $T=0K$. If $T>0$, thermal fluctuations may excite out of the CGS an incoherent fraction $F_{inc}(T)$ of water molecules located at the boundaries of the CDs, starting from $x=¾$ and proceeding inwards toward $x=0$ with increasing temperature. By analogy with super-fluidity in liquid ^{4}He, Landau's two fluids model may be used.[23] In the normal incoherent fluid, water molecules may be excited either by phonons at low temperature of by quantised rotations (rotons) at higher temperature. Knowing the minimum energy δ_0 needed for exciting rotons around a characteristic momentum k_0 and assuming a smooth connection with the excitation spectrum of phonons characterised by a sound velocity v_s (Figure IX.6), the following partition function may be derived:[21]

$$Z = \left(\frac{V}{N}\right) \cdot \left(\frac{k_0^2}{2\pi}\right) \cdot \sqrt{\frac{mT}{2\pi}} \cdot \exp\left(-\frac{\delta_0}{T}\right)$$

At temperature T, the incoherent fraction at a distance x within the

226 Aqua Incognita: why ice floats on water and Galileo 400 years on

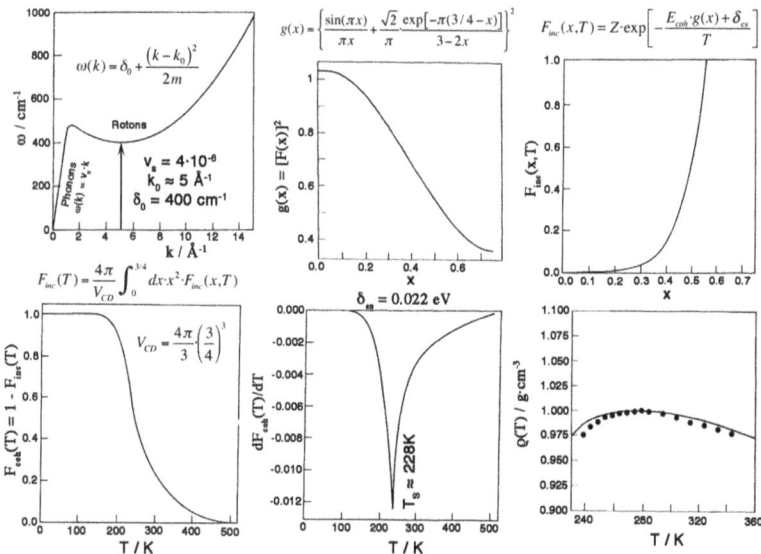

Figure IX.6. The thermodynamics of LW according to QFT (see text for details).

CD is thus: $F_{inc}(x,T)=Z \cdot exp[-\Delta(x)/T]$ with an energy gap protecting the coherent phase from thermal fluctuations that depends on the amplitude F(x) of the trapped EMF (Figure IX.5) plus a short-range contribution δ_{es}. Integration of the incoherent fraction over a whole CD allows expressing the coherent fraction as $F_{coh}(T)=1-F_{inc}(T)$ and computing its temperature derivative. The occurrence of a deep minimum in $dF_{coh}(T)/dT$ is then the QFT-signature of the existence of a super-cooled state for LW, with a melting temperature $T_S \approx 230K$ corresponding to $\delta_{es}=0.022$ eV or $\tau_{es}=0.2$ ps. Comparison with experimental rotation diffusion times $\tau_{rot}=\tau_0 \cdot exp(E_a/k_BT)$, $E_a=7,7$ $kJ \cdot mol^{-1}$ and $\tau_0=0.0485ps$,[7] show that $\tau_0 \approx \tau_{es}/5$, a not too bad guess. The observed water's density may thus be written as:

$$\rho(T) = 0.92 \cdot F_{coh}(T) + [1 - F_{coh}(T)] \cdot \rho_n(T)$$

Here $\rho_n(T)$ is a decreasing function of temperature,

$$\rho_n(T)=\rho_n(T_0)-a \cdot (T-T_0)$$

with $a>0$. Normalisation to $\rho=1 g \cdot cm^{-3}$ at $T_0=277K$ allows fixing the two constants a and $\rho_n(T_0)$:

$$\begin{cases} \rho(T_0 = 277K) = 1 \\ d\rho(T_0)/dT = 0 \end{cases} \Rightarrow \begin{cases} \rho_n(T_0) = \dfrac{1-0.92 \cdot F_{coh}(T_0)}{1-F_{coh}(T_0)} \\ a = -\dfrac{0.08}{[1-F_{coh}(T)]^2} \cdot \dfrac{dF_{coh}}{dT}(T_0) \end{cases}$$

The celebrated density anomaly of liquid water is then well reproduced (Figure IX.6), the density maximum being the point where the increase of the incoherent fraction $[1-F_{coh}(T)]$ of higher density is just compensated by the decrease of the normal phase density with temperature. Other thermodynamic properties of LW may be derived along the same lines.[21]

From Figure IX.6 as $F_{coh} \approx 0.4$ around $T \approx 300K$, it follows that LW should be viewed as an array of CDs having a radius $R_{CD}=28.5$ nm separated by a dense incoherent liquid of thickness $2\delta \approx 18$ nm. A water molecule within a CD oscillates with a frequency $\omega_r=0.165eV \approx 1330 cm^{-1} \approx 4 \cdot 10^{13}Hz$ between the GS and an excited level located 12.07 eV above the GS, spending here about 10% of its time. As the ionisation threshold of the water molecule is 12.6eV, this means that in the excited state an electron has to jump over an energy barrier $\Delta E=0.53eV$ at a distance $\Delta r=0.5 Å$ for reaching another water molecule with an attenuation amplitude (tunnel effect) given by:

$$p = exp\left(-\frac{\Delta r \cdot \sqrt{2m_e \cdot \Delta E}}{2\hbar}\right)$$
$$= exp\left[-0.2563 \cdot (\Delta r/Å) \cdot \sqrt{(\Delta E/eV)}\right]$$

With $p \approx 0.91$, this means that 0.10 electrons per water molecule are in a highly delocalised state and may produce super-currents owing to their well-defined phase patterns. The large separation $2\delta \approx 18$ nm between the super-conducting CDs obviously prevents LW to be a good macroscopic conductor. However, these freely moving electrons behave as a charged fluid performing rigid rotations with a moment of inertia and angular momentum given by:

$$I = \frac{2}{5}\left[\frac{4\pi}{3} R_{CD}^3 \cdot \left(\frac{Nm_e}{V}\right)\right] \cdot R_{CD}^2 \Rightarrow \vec{L} = I \cdot \vec{\omega}$$

This endows each CD with a magnetic moment $\mu = e \cdot L/2m_e$ a QFT-explanation of the sensitivity of LW to magnetic fields B. With $B = 50\mu T$ for Earth, the associated Larmor pulsation is $\omega_L = e \cdot B/2m_e$ corresponding to a frequency $\nu \approx 1.4 MHz$. Consequently, if EMF-waves have no effect on LW in CP or CQM, this is no more the case in QFT where LW becomes sensitive to infrared as well as radio waves.

Experimental evidence for the existence of such CDs within LW is provided by the observation of coherent librations of water molecules in deionised Milli-Q water using four-photon Rayleigh-wing spectroscopy, in the range 0-100cm^{-1} and showing frequencies coinciding with the rotation spectrum of gaseous water vapour.[24] In particular, ortho- and para-isomers of water molecule spectral lines have been clearly identified in LW. By comparison, it was observed that nuclear spin conversion was fast in water dimers and small clusters,[25] meaning that emergence of coherence cannot occur below a certain critical number of water molecules (several millions). Encapsulation of large water clusters in polyoxo-molybdate nanocapsules (Figure IX.6) have also evidenced existence of an ordered fully tetrahedral $\{H_2O\}_{100}$ water cluster having a density $\rho \approx 0.69$ g·cm^{-3} and a non-tetrahedral disordered $\{H_2O\}_{59}$ water cluster with $\rho \approx 0.36$ g·cm^{-3}.[26,27] This is a convincing experimental proof that when density is high enough, a transition towards tetrahedral behaviour is observed. Finally, coherent patterns can be observed even in classical MD-simulations, either as large vortex-like structure of dipole field [28] or through Voronoï polyhedra topological analysis showing the existence of two-states patches in liquid water with clear evidence of associated isobestic points.[29]

Zoemorphic water

The emergence of quantum coherence in LW at a macroscopic scale above a certain density threshold should have profound consequences in biology. Figure IX.7 shows the typical chemical composition of *E. Coli* bacterial cell[30] having a mass $m=0.95$ pg.[31] Using pure numbers,

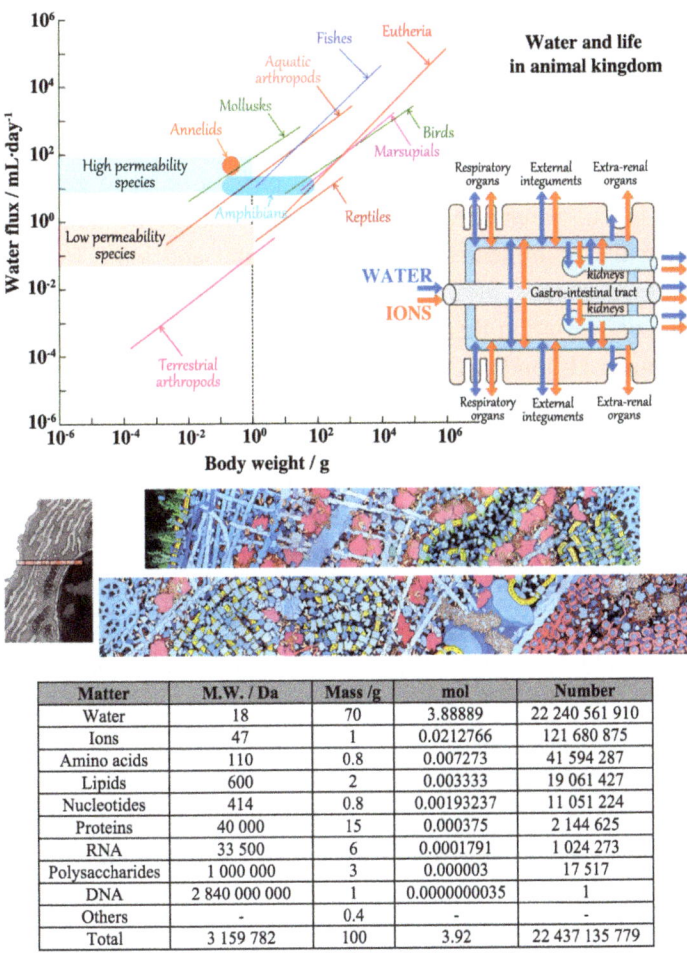

Matter	M.W. / Da	Mass /g	mol	Number
Water	18	70	3.88889	22 240 561 910
Ions	47	1	0.0212766	121 680 875
Amino acids	110	0.8	0.007273	41 594 287
Lipids	600	2	0.003333	19 061 427
Nucleotides	414	0.8	0.00193237	11 051 224
Proteins	40 000	15	0.000375	2 144 625
RNA	33 500	6	0.0001791	1 024 273
Polysaccharides	1 000 000	3	0.000003	17 517
DNA	2 840 000 000	1	0.0000000035	1
Others	-	0.4	-	-
Total	3 159 782	100	3.92	22 437 135 779

Figure IX.7. Importance of water for living organisms (top) and cells (bottom) and packing of organic matter within a cell (middle).

i.e. topological invariants, we see that the two major life-giving ingredients are water (more than 99 mol%) followed by inorganic ions. This provides a scientific basis for the well-known assertion that *"water is life"* and to the topological fact that all living organisms are based on a polar tube made of proteins and crossed by water, ions and few organic small molecules. DNA that carries the genetic information is relegated at the end of this topological classification showing the irony of calling it the *"molecule of life"*. Accordingly, if life is motion, then DNA is obviously the worst candidate for moving. Moreover, the average lifetime of a DNA double helix in the vacuum without its hydration shell is about 50 ps.[32] In order to get a stable double helix, water and magnesium ions should be added, again stressing the prominent role played by water and ions. This means that storing genetic information is more an affair of solvation and charge compensation than an affair of organic base pairings through hypothetical H-bonds. As clearly stated by Szent-Györgyi: *"without water and inorganic ions, organic matter just looks stupid..."* Figure IX.7 also shows how organic matter is densely packed in a cell leaving very few spaces for water molecules and ions. Thinking of intracellular water as an incoherent liquid with freely diffusing ions is thus definitively not a good idea. Starting from average data concerning molecular weight, density, volume and radius of globular proteins,[33] it is also easy to show (Figure IX.8) that a 70wt% in water translates into at most 4 water layers around each globular protein. Moreover, if the biopolymer is linear, then the same weight percentage translates into just 2 or 3 water layers depending on the anisotropy. Obviously, water molecules within these layers should be able to polarise the FZPGS as in LW and owing to the dense packing of water molecules, one may safely assume full structuration into CDs of reduced size relative to LW displaying higher stability towards thermal agitation. Four water layers having a thickness of 1.2 nm associated to a 75 nm 2D-packing of CDs at the surface of membranes or DNA lead to $V_{CD} = \pi(75)^2 \cdot 1,2/4 = 5300$ nm^3 or with $\rho_{coh} = 0.92$ g·cm^{-3} to

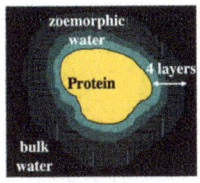

Average molecular weight = 40 kDa
Average density = 1,35 g·cm⁻³
Average volume = 50 nm³
Average radius = 2,3 nm
Water layer thickness = 0,3 nm

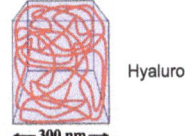

Hyaluronate

←— 300 nm —→

$$\%H_2O = 100 \times \frac{V_n - 50}{50 \times 1{,}35 + (V_n - 50)} = \frac{100 \times (V_n - 50)}{17{,}5 + V_n}$$

R(1 H$_2$O) = 2,6 nm ⇒ V$_1$ = 74 nm³ ⇒ 26%

R(2 H$_2$O) = 2,9 nm ⇒ V$_2$ = 102 nm³ ⇒ 44%

R(3 H$_2$O) = 3,2 nm ⇒ V$_3$ = 137 nm³ ⇒ 56%

R(4 H$_2$O) = 3,5 nm ⇒ V$_4$ = 180 nm³ ⇒ 66%

Macromolecule	Water monolayer (g/g)	Hydration
Globular protein	0,4	0,4/1,4 ≈ 29%
Nucleic acid	0,5	0,5/1,5 ≈ 33%
Collagen	0,6	0,6/1,6 ≈ 38%
Hyaluronate	0,7	0,7/1,7 ≈ 41%

$$A, B, C \,/\, cm^{-1} = \frac{h \cdot 10^{-2}}{8\pi^2 c \cdot I_{A,B,C}} = \frac{2{,}7993 \cdot 10^{-46}}{I_{A,B,C} \,/\, kg \cdot m^2}$$

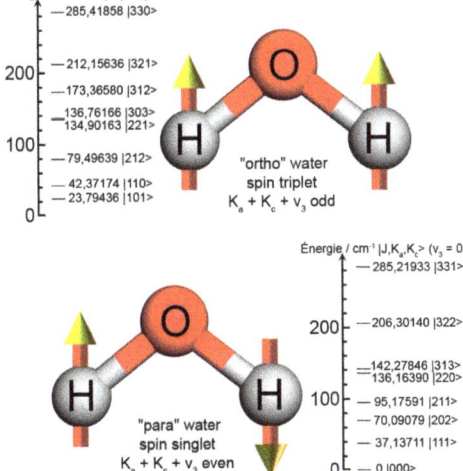

Figure IX.8. Water layering around nanometric biopolymers in a cell (top) and ortho- and para-water spin-isomers with their associated energy levels (bottom).

$N \approx 163{,}240$ water molecules. Microtubules with an inner diameter of 12 nm, should have $V_{CD} = \pi(12)^2 \cdot 75/4 = 482$ nm³, i.e. $N \approx 261{,}255$ water molecules. Proteins with a mean diameter of 4 nm may store coherent

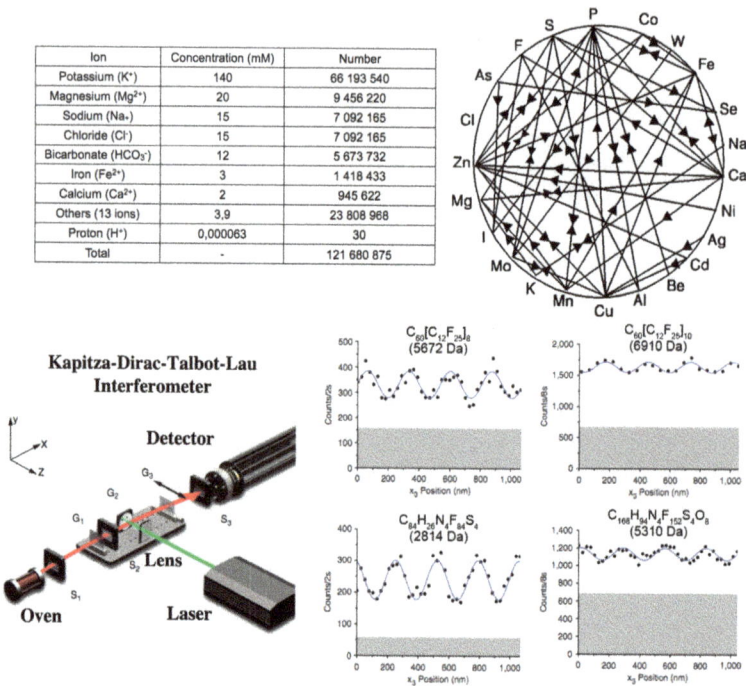

Figure IX.9. Importance and quantum entanglement of ions in the cell (top) and evidence of coherent quantum behaviour up to a molecular weight of 6 kDa.

water clusters within pockets having radii close to 1 nm, leading to $N \approx 100$ as observed in Mo_{132} nanocapsules of similar sizes.[26,27] One of the most striking consequences of water coherence in the cell is the selective enrichment in para-water around biomolecules as evidenced by four-photon Rayleigh-wing spectroscopy.[24,34] Accordingly, among the two nuclear spin isomers of the free water molecule, only the para-isomer of zero spin is able to stop rotating in its GS.

If intracellular water has 100%-coherence, this means that inorganic ions may be able or not to enter inside the CDs. Hofmeister's series in biology should then be a macroscopic consequence of LW's QFT-structure. Figure IX.9 shows how many numbers of ions may be found within a single cell such as *E. Coli*,[35] evidencing the prominent topological role played by potassium and magnesium ions, a convincing proof of their ability to penetrate water's CDs surrounding biological matter. On the other hand, the fact that sodium ions are generally found outside the cell should be clear evidence of their inability to enter these CDs. A QFT-view of the cell brings thus full support to G.N. Ling's theory,[36] where ionic pumps violating the first principle of thermodynamics are discarded, ionic gradients arising from the ability of ions to penetrate or not water layers surrounding biopolymers. Moreover, the energy liberated upon ATP hydrolysis is 0.26 eV,[37] a value very similar to the energy E_{coh} get or lost when a CD appears or disappears. This suggests that ATP's energy could be used to monitor water coherence and not to fuel ionic pumps, again bringing full support to Ling's hypotheses. Figure IX.9 also shows the antagonistic relationships between minerals upon absorption by the intestines.[38] The complexity of this diagram could suggest quantum entanglement of ions in a cell, as it was experimentally shown that interference patterns evidencing coherent behaviour could be observed up to a molecular weight of about 7 kDa.[39] This means that everything that is not a protein, RNA, DNA or polysaccharides should be treated as wave packets and not as matter-particles. The fact that such interference patterns can be surprisingly robust against a large variety of internal state transformations or against interactions with external force fields is full evidence of QFT interactions mediated through the FZPGS as in LW.

Finally, at $T\approx300K$ each CD holds about 700,000 highly polarisable electrons that could be ionised upon IR absorption. Should we be able to separate coherent and non-coherent water we would get an electric potential ΔV difference of about -100mV.[40] As such separation may occur near interfaces, the existence of a membrane potential in all cells could be a mere QFT-consequence of water's ability to capture virtual photons popping up from the FZPGS. For all theses reasons, it seems wise to give a special name to this coherent state of water involving FZPGS polarisation through a spontaneous lasering effect near the ionisation potential of water molecules that allows life sustainment. I thus propose the term "zoemorphic water", from the Greek roots "zoe" meaning life in general and "morphos" meaning shape, i.e. "water as a living shape".

Conclusion

This QFT-view of LW based on the ability of water molecules to absorb or emit spontaneously virtual photons coming from the FZPGS in order to undergo a vapour-liquid transition when the density in water molecules becomes large enough, typically $N/V > 0.3 \text{g} \cdot \text{cm}^{-3}$, is by no means new. It was first published in 1988,[41] and refined in two books.[15,16] One may thus wonder why water scientists still persist to think about LW using CP or CQM concepts and never take into account the fact that a network of continuously moving charges (HBs) should necessarily generate EMFs having at least the same fluctuation time. When this fact is explicitly taken into account at a QFT-level, LW behaves as a perfectly normal fluid without any "anomalies". It is beyond any doubts, the ignorance by condensed matter physicists of the existence of virtual photons spontaneously emerging from a FZPGS responsible for any kind of attraction or repulsion that causes troubles. Accepting the reality of a field able to create or destroy virtual or real QPs is the very first step towards a full comprehension of LW's properties. Sceptical people should realise that besides virtual photons there are also, with even lower probability, virtual electrons and virtual positrons that could be responsible for FZPGS's (or vacuum) permeability μ_0 and permittivity ε_0.[42] The most convincing evidence that the FZPGS is full of virtual radiation quanta are the static[43] or dynamical Casimir effect[44] as well as the Lamb shift in atoms.[45] Admitting the existence of VPs popping up from the FZPGS implies necessarily CDs' formation in LW following QFT-principles established during the late twenties. Those who may feel uncomfortable by seeing LW as a collection of immaterial optical cavities (CDs) able to trap EMFs and behaving as a multimode laser without prior inversion population, have to realise that this behaviour comes from non-linear effects that are usually neglected in laser physics. As explained above, spontaneous EMF amplification appears to be possible only when the oscillator strength of a given optical transition $|0> \leftrightarrow |q>$ weighted by the ratio of its plasma frequency ω_p to the self-exciting frequency ω_q, becomes larger than a critical value $g_c = [8/27 + 2\mu/3 + (4/9 + 2\mu/3)^{3/2}]^{1/2}$, where μ is a dispersive coupling constant taking into account virtual excitations involving the whole energy spectrum of the substance (6-200 eV). Such a critical behaviour is linked to the occurrence of a non-zero third derivative of the trapped EMF in the CEs. Neglecting this non-linear third order term leads to a second order equation, $p^2 + \mu \cdot p - g = 0$ having two real solutions $p = [-\mu \pm (\mu^2 + 4\mu g)^{1/2}]/2$ with absolutely no possibility of running away from the PGS (no complex solutions). It is then amazing to think that life on earth based on HBs, is linked to coherence's emergence from the FZPGS as soon as a minute third order term is different from zero. Assuming that $d^3A(\tau)/d\tau^3 = 0$, i.e. decoupling matter from the EMF created by a flickering network of continuously moving charges (for instance $q_H = +0.24$ in the TIP5P model)[46] is the main reason for failure of sophisticated MD-simulations at the CP/CQM level. On the other hand, assuming $d^3A(\tau)/d\tau^3 \neq 0$ implies associating to each optical transition of the water monomer a liquid structure having its own critical density, coherent energy gap (i.e. HB-strength), trapped EMF (A_0, ω_q), phase locking ($d\varphi/d\tau$) and mixing angle (α). Zoemorphic water involves excitation and oscillation between the ground state and a 5d-level of the O-atom located 12.07 eV above the GS that corresponds to ambient conditions on earth ($T \approx 300$ K, $P = 1$ atm). However, other excitations may be favoured by merely changing temperature and/or pressure. The occurrence of at least 15 crystalline ice polymorphs displaying HB-strengths in the range *0.17-0.25 eV*,[47] and of 3 amorphous ices would then just be the consequence of using the whole palette of excitations of the water molecule given in Figure IX.3. Biologists should also realise that life emergence on Earth may be intimately linked with the existence of zoemorphic water and the existence of a FZPGS. Most CP-reasoning found in standard biology textbooks not taking explicitly into account the intrinsically quantum aspect of all processes occurring within the cell should thus be revised. Accordingly, the most important variable in QFT is field's phases θ linked to the EM-potentials through the equations (in SI units):

$$\begin{cases} \vec{A} = \frac{\hbar}{e} \cdot \vec{\nabla}\theta \\ \varphi = -\frac{\hbar}{e} \cdot \frac{d\theta}{dt} \end{cases}$$

Consequently, any EMF variation should affect quantum phases as nicely illustrated by the celebrated Bohm-Aharanov effect.[48] But owing to the phase locking existing between matter-fields and EMFs within a water CD, biological matter (99mol% water) behaviour monitoring through pulsed EMFs becomes an alternative to hormones or drugs absorption. Recognition of the existence of water's CDs would then beyond any doubts considerably affect both physiology and medicine in a very near future.

References

1. Everts, S. *Chem. Eng. News* **2013**, *91*, 28.
2. Isaacs, E. D.; Shukla, A.; Platzman, P. M.; Hamann, D. R.; Barbiellini, B.; Tulk, C. A. *Phys. Rev. Lett.* **1999**, *82*, 600.
3. Ghanty, T. K.; Staroverov, V. N.; Koren, P. R.; Davidson, E. R. *J. Am. Chem. Soc.* **2000**, *122*, 1210.
4. Romero, A. H.; Silvestrelli, P. L.; Parrinello, M. *J. Chem. Phys.* **2001**, *115*, 115.
5. Nilsson, A.; Ogasawara, H.; Cavalleri, M.; Nordlund, D.; Nyberg, M.; Wernet, Ph.; Pettersson, L. G. M. *J. Chem. Phys.* **2005**, *122*, 154505.
6. Guo, J.-H.; Luo, Y.; Augustsson, A.; Rubensson, J.-E.; Såthe, C.; Ågren, H.; Siegbahn, H.; Nordgren, J. *Phys. Rev. Lett.* **2002**, *89*, 137402.
7. Teixeira, J.; Bellissent-Funel, M.-C.; Chen, S.-H.; Dianoux, A. J. *Phys. Rev. A*, **1985**, *31*, 1913.
8. Bertolini, D.; Cassettari, M.; Ferrario, M.; Grigolini, P.; Salvetti, G.; Tani, A. *J. Chem. Phys.* **1989**, *91*, 1179.
9. Lee, H. S.; Tuckerman, M. E. *J. Chem. Phys.* **2007**, *126*, 164501.
10. Fernandez-Serra, M. V.; Artacho, E. *Phys. Rev. Lett.* **2006**, *96*, 016404.
11. Bukowski, R.; Szalewicz, K.; Groenenboom, G. C.; van der Avoird, A. *Science* **2007**, *315*, 1249.
12. Wernet, Ph.; Nordlund, D.; Bergmann, U.; Cavalleri, M.; Odelius, M.; Ogasawara, H.; Näslund, L. Å.; Hirsch, T.K.; Ojamaë, L.; Glatzel, P.; Pettersson, L.G.M.; Nilsson, A. *Science* **2004**, *304*, 995.
13. Morrone, J. A.; Car, R. *Phys. Rev. Lett.* **2008**, *101*, 017801.
14. Guillot, B. *J. Molec. Liq.* **2002**, *101/1-3*, 219.
15. Preparata, G. *An introduction to realistic quantum physics*, World Scientific: New Jersey, **2002**.
16. Preparata, G. *QED coherence in matter*, World Scientific: Singapore, 1995.
17. Judge, D. *Il Nuovo Cimento* **1964**, *31*, 332.
18. Bono, I.; del Giudice, E.; Gamberale, L.; Henry, M. *Water* **2012**, *4*, 510.
19. Gürtler, P. ; Saile, V.; Koch, E. E. *Chem. Phys. Lett.* **1977**, *51*, 386.

20. Chan, W. F.; Cooper, G.; Brion, C.E. *Chem. Phys.* **1993**, *178*, 387.
21. Arani, R.; Bono, I.; Del Giudice, E.; Preparata, G. *Int. J. Mod. Phys. B* **1995**, *9*, 1813.
22. Craig, V. S. J.; Ninham, B. W.; Pashley R. M. *J. Phys. Chem.* **1993**, *97*, 10192.
23. Feynman, R. P.; Cohen, M. *Phys. Rev.* **1956**, *102*, 387.
24. Bunkin, A. F.; Pershin, S. M.; Nurmatov, A. A. *Laser Phys. Lett.* **2006**, *3*, 275.
25. Sliter, R.; Gish, M.; Vilesov, A. F. *J. Phys. Chem. A* **2011**, *115*, 9682.
26. Müller, A.; Henry, M. *C. R. Chimie* **2003**, *6*, 1201.
27. Henry, M.; Bögge, H.; Diemann, E.; Müller, A. *J. Molec. Liq.* **2005**, *118*, 155.
28. Higo, J.; Sasai, M.; Shirai, H.; Nakamura, H.; Kugimiya, T. *Proc. Natl. Acad Sci. USA* **2001**, *98*, 5961.
29. Shih, J.-P.; Shen, S.-Y.; Mou, C.-Y. *J. Chem. Phys.* **1994**, *100*, 2202.
30. Watson, J. D. *Molecular Biology of the Gene*, 2nd Ed. Saunders: Philadelphia, PA, 1972.
31. Neidhardt, F. C. *Escherichia coli and Salmonella: Cellular and Molecular Biology*, ASM Press, Washington DC, 1996, Vol 1. pp. 14.
32. Gerstein, M.; Levitt, M.; *Scientific American*, **1998**, November issue, 101.
33. Quillin, M. L. ; Matthews, B. W. *Acta Cryst. D* **2000**, *56*, 791.
34. Bunkin, A. F.; Lebedenko, S. I.; Nurmatov, A. A.; Pershin, S. M. *Quantum Electronics* **2006**, *36*, 612.
35. Alberts, B.; Johnson, A.; Lewis, J.; Raff, M.; Roberts, K.; Walter, P. *Molecular Biology of the Cell*, 4th Ed. Garland Science: New York, 2002.
36. Ling G. N. *Life at the cell and below-cell level*, Pacific Press: New York, 2001.
37. Podolsky, R. J.; Morales, M. F. *J. Biol. Chem.* **1995**, *218*, 945.
38. Ashmead, D. H. *The role of aminoacid chelates in animal nutrition*, Noyes Publ.: Westwood, New-Jersey, 1993, pp. 21.
39. Hornberger, K.; Gerlich, S. ; Haslinger, P. ; Nimmrichter, S. ; Arndt, M. *Rev. Mod. Phys.* **2012**, *84*, 40. Marchettini, N.; Del Giudice, E.; Voeikov, V.L.; Tiezzi, E. *J. Theoret. Biol.* **2010**, *265*, 511.

41. Del Giudice, E.; Preparata, G.; Vitiello, G. *Phys. Rev. Lett.* **1988**, *61*, 1085.
42. Urban, M.; Couchot, F.; Sarazin, X. Djannati-Atai, A. *Eur. Phys. J. D* **2013**, *67*, 58.
43. Casimir, H. B. G. *Kon. Ned. Akad. Wetensch. Proc.* **1948**, *51*, 793
44. Wilson, C. M.; Johansson, G.; Pourkabirian, A.; Simoen, M.; Johansson, J. R.; Duty, T.; Nori, F.; Delsing, P. *Nature* **2011**, *479*, 376.
45. Lamb, W. E.; Retherford, R. C. *Phys. Rev.* **1947**, *72*, 241.
46. Mahoney, M. W.; Jorgensen, W. L. *J. Chem. Phys.* **2000**, *112*, 8910.
47. Henry, M. *CHEMPHYSCHEM* **2002**, *3*, 607.
48. Aharonov, Y.; Bohm, D. *Phys. Rev.* **1959**, *115*, 485.

X
Aqua extrema and *vita incognita* deep below the waves

Shigeru Deguchi, Keigo Kinoshita, Takaaki Kubota

Japan Agency for Marine-Earth
Science and Technology (JAMSTEC),
2-15 Natsushima-cho, Yokosuka 237-0061, Japan.
Shigeru.Deguchi@jamstec.go.jp

1. Deep sea and space

Ocean covers approximately 70% of the earth's surface and stores 97% of the water on this planet. Its mean depth is 3,800 metres, and the deep sea represents the layer deeper than 200 metres. The deep sea and space are often compared as the final frontiers for human exploration. When people think about going out to space, they always think of traveling to the moon. Since the first successful moon landing by Neil Armstrong on 20 July 1969, 12 astronauts have landed the moon by the manned Apollo missions. Compared with the moon that is 384,400 kilometres away from the earth, even the deepest point of the ocean, the Challenger Deep at the southern end of the Mariana Trench, is merely 11 kilometres below the surface. Yet, going there is not at all an easy travel. On 26 March 2012, a movie director, James Cameron, successfully reached the Challenger Deep and the news made headlines around the globe. It was a very significant achievement indeed, because he was the third to reach the deepest point after the first successful attempt in 1960 by Jacques Piccard and Don Walsh.

2. Extreme environments in the deep sea

It is extreme physicochemical conditions in the deep sea that make manned exploration difficult. Hydrostatic pressure increases 0.1 MPa for every 10 metres of water depth, and reaches 110 MPa at the deepest point. It is technically challenging even today's standard to design and construct a pressure-resistant chamber that is capable of withstanding this enormous pressure. The deep sea is also a lightless world as sunlight does not reach below 200 metres. Accordingly, the temperature of the deep water is low around 2-4 °C.

As for the temperature, there is an exception. Hydrothermal vents are hot springs occurring in the bottom of the ocean where very hot water is ejected into cold deep-sea water (Figure X.1). Temperature of the vent water often exceeds 100 °C, but it remains liquid due to high hydrostatic pressure. In the Mid-Atlantic Ridge, a hydrothermal vent was discovered, where vent water exceeds the critical condition of water (T_c = 374 °C, P_c = 22.1 MPa, Figure X.2).[1] A steep thermocline is formed when hot vent water, sometimes in the supercritical state, is mixed with surrounding cold deep water. It was suggested that such

Figure X.1. Photograph (© JAMSTEC) of a hydrothermal activity at the depth of 1,492 metres near Ryukyu Islands, Japan.

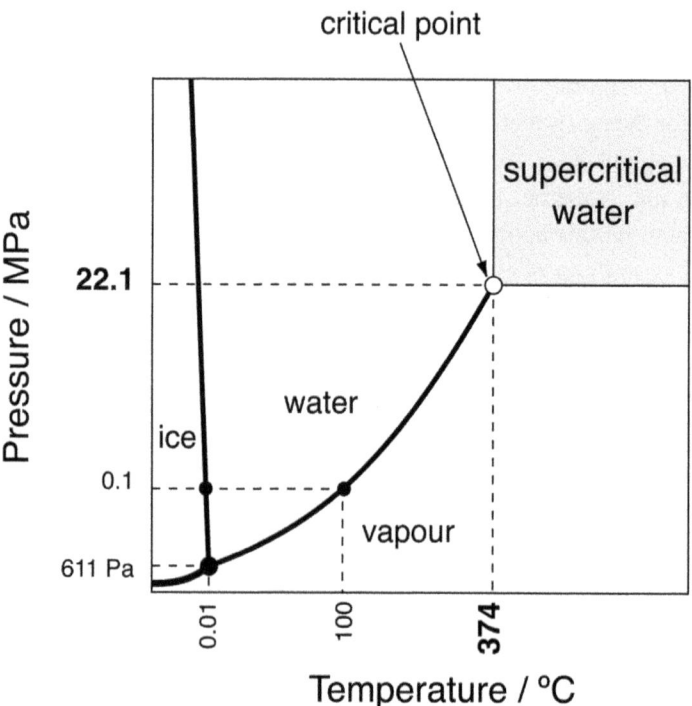

Figure X.2. Schematic phase diagram of water.

the steep thermocline from very high temperature to near-freezing temperature might have played a key role in abiotic formation of proteins by dehydration condensation of amino acids.[2,3]

3. Water under extreme conditions

Water at high temperatures and high pressures exhibits properties that are remarkably different from those of ambient liquid water, which we all are familiar with. The density of water decreases from 0.997 g/cm^3 at 25 °C and 0.1 MPa to 0.322 g/cm^3 at the critical point,[4] and various solvent properties such as dielectric constant, viscosity and refractive index change accordingly. The dielectric constant of water is 78 at 25 °C and 0.1 MPa, and is very high among liquid solvents, but decreases substantially when heated under pressure. Specifically, the dielectric constant decreases to 21 at 300 °C and 25 MPa and to 6 at the critical point. The value at the critical point is comparable to that of 1-dodecanol at ambient conditions (5.82).

This description may give an impression that supercritical water (SCW) has anomalous solvent properties, but it seems that the reverse is true. In Figure X.3, the dielectric constant is plotted

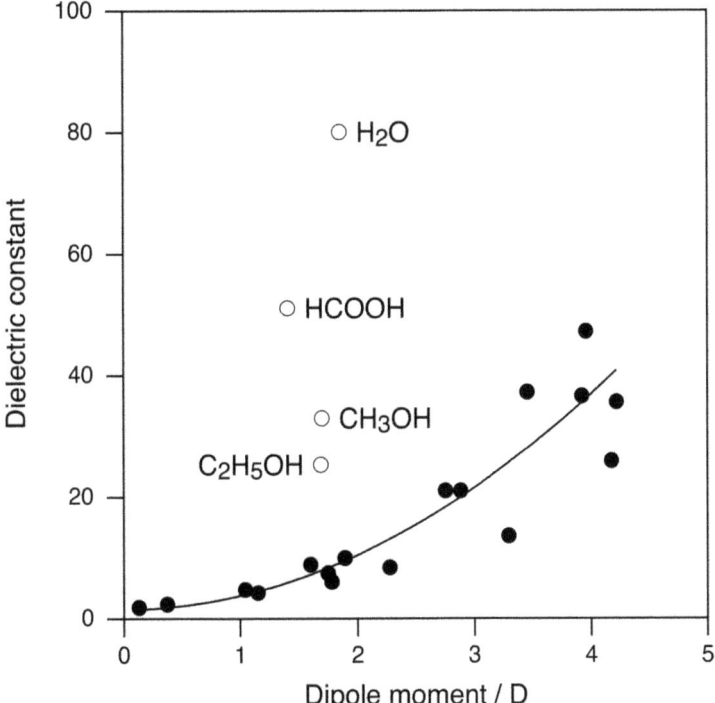

Figure X.3. Relationship between dielectric constant and dipole moment for various solvents. The curve is a guide to the eye. Reproduced with permission from Ref. 5.

against a dipole moment for various solvents.[5] The dipole moment represents a polarity of a single molecule, while the dielectric constant represents a bulk polarity. The data for most solvents fall on a single curve, indicating that a collection of polar molecules forms a polar bulk liquid. However, the data for water, formic acid, methanol, and ethanol deviate significantly from the common curve. These solvents are known as "associating solvents", and the molecules interact with each other strongly to form clusters. Very interestingly, if we take the dielectric constant of water at the critical point, the data point for water falls on the common curve. The diagram demonstrates that it is ambient water that is anomalous, and water in a supercritical state reveals its "normal" behaviour expected from its molecular structure, because high thermal energy suppresses the cluster formation through hydrogen bonding.[6]

4. Oil and water relationship

Due to remarkably different properties of SCW, physicochemical processes in water are very different at such extreme conditions. For example, oil and ambient water do not mix. However, oil and SCW freely mix because of the low dielectric constant of the latter.[4,7,8] We have utilised such unique phase behaviours to develop a novel bottom-up process for fabricating nano-sized oil droplets in water (nanoemulsions).[9]

Emulsions are widely used in such industries as food, pharmaceutical, cosmetic, chemical, agricultural, print/ink, and petroleum.[10] Nanoemulsions are those that contain droplets in the size range between 20 nm and 200 nm.[11] Owing to the small droplet size, nanoemulsions are transparent or translucent, and creaming or sedimentation is significantly slower. The small droplet size also has potential benefits for developing a new realm of applications such as pharmaceutical and cosmetic formulations as well as reactors for synthesizing nanomaterials.

Nanoemulsions are usually prepared by top-down processes, in which external forces are applied to water/oil/surfactant mixtures to deform and disrupt large droplets into smaller ones, but there are difficulties associated with down-sizing liquid droplets to nanometre range.[11,12] Nano-sized solid particles are usually prepared by bottom-up processes.[13] Unlike the top-down process, the bottom-up process starts with homogeneous solutions and solutes are allowed to assemble to form nanoparticles. It is plain to see that the bottom-up approach is also preferable for fabricating nano-sized liquid droplets, but homogeneous solutions of oil and water need to be prepared. We successfully addressed the issue by using unique properties of SCW.[9]

When coarse droplets of dodecane in water were heated under a constant pressure of 25 MPa, they started to shrink above approximately 330 °C (Figure X.4A–C). The system eventually entered into a one-phase regime at around 337 °C and the droplets disappeared completely to give a homogeneous solution (Figure X.4D). The system phase-separated again when the solution was cooled, and dodecane droplets having a fairly uniform size reappeared (Figure X.4E), which increased in size as it was cooled further (Figure X.4F).

In the process called MAGIQ (Monodisperse nAnodroplet Generation In Quenched hydrothermal solution), a fast and deep temperature-quench, ~200 °C/sec, is applied to homogeneous solutions so that they phase-separate rapidly to generate nano-sized droplets. Emulsions containing nano-sized and monodisperse oil droplets with average diameter of 61 nm were obtained in just 10 seconds.[9]

5. Cooking cellulose in hot and compressed water

Starch is polysaccharide made of glucose units that are connected via α-1-4 glycosidic linkages. It is semi-crystalline at room temperature, but undergoes crystalline-to-amorphous transformation when heated in water to 60–70 °C (known as gelatinisation).[14] The starch chains become swollen upon gelatinisation and can be attacked readily by

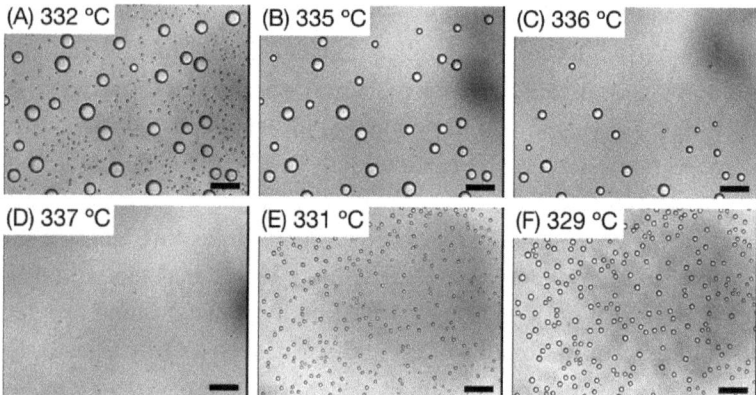

Figure X.4. A-D) *In situ* optical microscopy images of droplets of dodecane in water taken at a constant pressure of 25 MPa. Images were taken while heating the mixture at 11.6 °C/min. E and F) Formation of dodecane droplets upon cooling from D) at 4.5 °C/min. Scale bars represent 50 μm. Reproduced with permission from Ref. 9.

hydrolytic enzymes, resulting in better digestibility of cooked starchy foods like pasta. Gelatinisation is an indispensable step for processing starch in the food industry, and is also the preliminary process necessary to render starch suitable for enzyme-catalysed biomass conversion.

Cellulose is linear polysaccharide consisted of glucose units connected by β-1-4 glycosidic linkages, and the most abundant biomass on Earth with an estimated annual production of ~100 billion dry tonnes.[15] It is a highly crystalline material, and its utilisation is hampered by the recalcitrance of the crystalline domains to chemical or enzymatic hydrolysis.[16,17] A large change in the reactivity would be expected if cellulose was gelatinised, but we know from our experience that gelatinization of cellulose does not occur in water at ambient conditions.

A series of *in situ* polarized optical microscopic images in Figure X.5 show crystalline cellulose (CF1, Whatman) in hot and compressed water, which were taken by heating the specimen at 11–14 °C/min.[18]

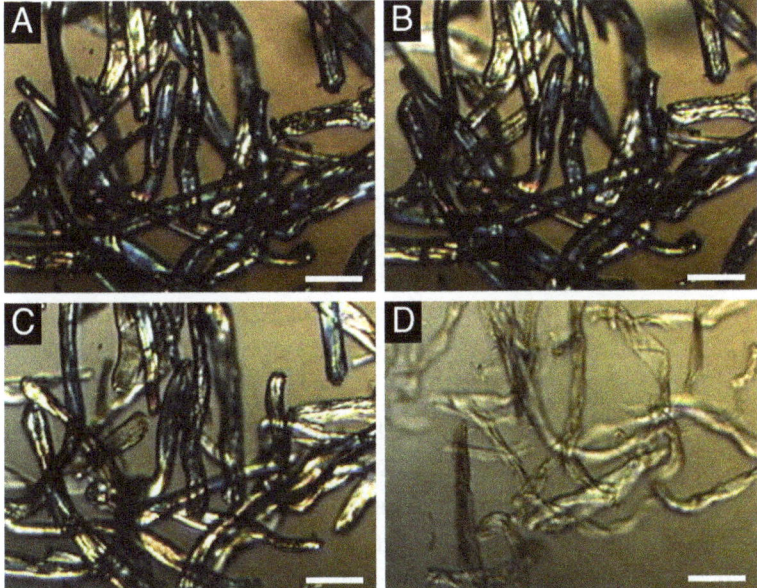

Figure X.5. In situ polarized microscopic images of crystalline cellulose in water taken between 300 and 330 °C and at constant pressure of 25 MPa. Temperature, A) 300 °C, B) 310 °C, C) 320 °C, D) 330 °C. Scale bars represent 50 μm. Reproduced with permission from Ref. 18.

Birefringence of the fibrous cellulose, which is evident from the pseudo colour under crossed polarizers, was retained up to 310 °C without any noticeable change, showing that cellulose remained crystalline up to this high temperature. However, cellulose became less birefringent at around 320 °C, and the birefringence was completely lost at 330 °C. The micrographs clearly show that cellulose undergoes crystalline-to-amorphous transformation, just like gelatinisation of starch, in water at around 320 °C and 25 MPa. Dissolution of cellulose followed the transformation, and no cellulose remains at 340 °C. Recrystallisation was not observed when the system was cooled, confirming the previous observations that cellulose was hydrolysed very rapidly under similar experimental conditions. [19-21]

Figure X.6 shows a sequence of images showing a single cellulose fibre (indicated by a white triangle in Figure X.6A) near the transformation. The fibre gradually lost birefringence as it was heated (Figures X.6A, 6B, and 6C), but no significant change was seen in the shape of the fibre. The fibre started to deform when it almost completely lost birefringence (Figure X.6D), and deformed further at higher temperatures (Figures X.6E and 6F). The large deformation suggests that cellulose becomes plastic upon transformation, and the mechanical properties change dramatically upon the transformation. Twisting and bending were observed for most of the fibrous cellulose upon the transformation.

The crystalline-to-amorphous transformation of cellulose in hot and compressed water depended very much on the crystallinity, but not on the crystalline form and surface area.[22] The transformation is not simple thermal melting, but rather interaction between water and cellulose plays an essential role, because no such transformation was observed in ethanol (T_c = 243 °C, P_c = 6.4 MPa) at 7 MPa and at temperatures up to 350 °C. Birefringence was retained throughout the observation, and char formation resulted above 330 °C.

Crystalline cellulose consists of highly ordered crystallites called

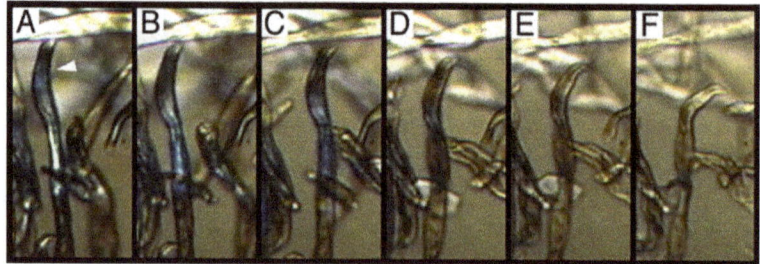

Figure X.6. *In situ* polarized micrographic images showing a cellulose fibre shown by a white triangle in A) during the crystalline-to-amorphous transformation. Temperature, A) 324 °C, B) 325 °C, C) 326 °C, D) 327 °C, E) 328 °C, F) 329 °C. Each image is 100 × 200 μm. Reproduced with permission from Ref. 18.

fringed micelles and less-ordered domains in between.[23] Under normal conditions, water only interacts with the less ordered domains.[24] The present observations clearly show that water also interacts with highly ordered domains at high temperatures under pressure, leading to a crystalline-to-amorphous transformation. It is suggested that the transformation is rather elicited by changes in the solid properties of cellulose,[22] such as glass transition,[25] transformation of cellulose crystals,[26] and a drastic change in the hydrogen-bond structure,[27,28] all of which occur above 200 °C. Such changes affect cellulose–water interactions, leading eventually to the transformation to an amorphous state. However, elucidating water–cellulose interactions under the present experimental conditions is not straightforward. On the one hand, the dielectric constant of water, which is 21 at 300 °C and 25 MPa, the value of which is comparable to that of 1-propanol. It seems unlikely that such a nonpolar solvent interacts favourably with cellulose. On the other hand, water between the crystallites might be in a supercritical state, even though the transformation takes place well below the critical temperature of water.[29] Unique solvation properties of supercritical fluids such as formation of a dense solvation shell may play an important role in the transformation.[30]

6. Black smoker

The colloidal dispersions play important roles in geological processes. For example, transportation of water-insoluble matters as colloidal dispersions is of enormous importance,[31-33] and it is estimated that 9.3–58 Gt of the suspended sediment is delivered by all rivers to the oceans annually.[33] An interesting aspect of natural colloidal systems is that they occur in a wide range of physicochemical conditions in terms of temperature and pressure. One such example is seen in a special type of deep-sea hydrothermal vent called a black smoker. The vent water from the black smoker contains large amounts of inorganic colloids of metal sulphides, typically iron sulphide, and forms a colloidal dispersion at extremely high temperatures and high pressures. It has been known for quite some time that dispersion stability of colloids in water is very different when dispersions are heated under pressure,[34] but our knowledge of colloidal phenomena under such extreme conditions was very limited.

The dispersion stability is governed by the balance of the forces acting between the surfaces of the particles. The net interaction potential usually includes electrostatic repulsion and van der Waals attraction, i.e. so-called DLVO forces.[35] The terms accounting for non-DLVO interactions such as hydration repulsion, oscillatory structural forces, hydrophobic attraction, ionic correlation, are also important in many cases.[36] All of these interactions depend not only on the surface properties of the particles but also on the properties of the dispersion medium.[36] It is plain to see that the large changes of various properties of water at high temperatures and high pressures affect the dispersion stability of the colloids.

Figure X.7 shows the change of the measured diffusion coefficients (D) of a representative model colloid, monodisperse polystyrene latex ($d_h = 206.5$ nm), as a function of temperature. The solid curve represents the theoretical prediction expected from the Stokes–Einstein relation,

$$D = \frac{k_B T}{3\pi\eta d_h}$$

where k_B is the Boltzmann constant, T is the absolute temperature, η is the viscosity, and d_h is the hydrodynamic diameter. In the calculation, it was assumed that d_h remained unchanged throughout the measurements. η of water at different temperatures was calculated according to the International Association for the Properties of Water and Steam (IAPWS) 1995 formulation by using NIST/ASME Steam. A good agreement was found between the experimental points and the theoretical prediction below 275 °C, indicating that the polystyrene latex remained well-dispersed up to this temperature. However, downward deviation was seen at 300 °C. The measured diffusion

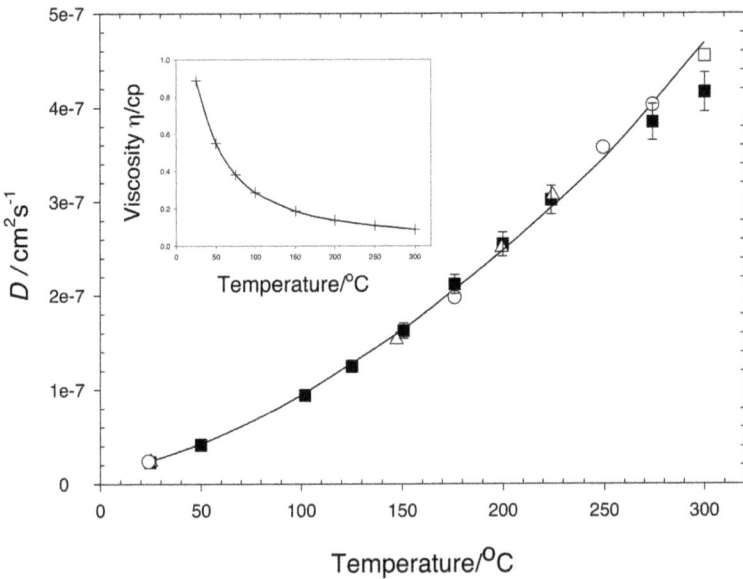

Figure X.7. Diffusion coefficient of polystyrene spheres, D, measured as a function of temperature at three different pressures: (15 MPa solid squares); 20 MPa open triangles); and (25 MPa open circles). Before reaching 300°C the pressurised sample was heated gradually and DLS measurements at various temperatures were performed (solid squares). The open square at 300°C represents the result for the diffusion coefficient, measured when the temperature is increased rapidly up to 300°C. The solid line shows the theoretical predictions for the temperature dependence of the diffusion coefficient of the particles with a mean diameter of 206.5 nm at 15 MPa. The inset shows the literature data for the viscosity of water versus temperature at 15 MPa pressure. Reproduced with permission from Ref. 37.

coefficients decreased with time at this temperature.[37] The results indicate that the polystyrene latex, which remains dispersed for a long time in water at ambient conditions, coagulates rapidly simply by heating under pressure. Similar coagulation was observed also for colloidal gold, diamond nanoparticles, clay minerals, and C_{60} nanoparticles.[38]

The results suggest that water is responsible for the coagulation at high temperatures. The properties of water where the coagulation was observed are already different from those at ambient temperatures. The DLVO forces are obviously affected by the decrease in dielectric constant and refractive index. The surface charge density would also be reduced, as counter-ion condensation should be facilitated in water of low dielectric constants. Reduced viscosity leads to increased collision frequency at high temperatures. All of these effects result in instability of the colloids. Numerical calculations indeed showed that the coagulation at high temperatures and high pressures is induced primarily by large decrease of the dielectric constant of water.[38]

7. Life deep below the waves

The biosphere on Earth is dominated by microorganisms, and so is the deep sea. One of the major driving forces for such domination is the fact that microorganisms have adapted to a wide range of extreme habitats such as high temperature, high pressure, and high gravity.[39-41] Obligately piezophilic bacterium that was isolated from the Challenger Deep grows only when the incubation pressures is above 50 MPa.[42] A hyperthemophilic archeon, *Methanopyrus kandleri* strain 116 that was isolated from a hydrothermal field in the Central Indian Ridge grows even at 122 °C, the temperature of which is the known upper limit for life.[40] Such organisms, the so-called extremophiles, play critical roles in carbon turnover even at the deepest point of the ocean.[43].

The deep sea is also a nutrient-deprived environment because sunlight does not reach and photosynthesis does not occur. It appears that the deep-sea microorganisms have acquired unique molecular mechanisms to utilise organic molecules that are otherwise not utilised. It was shown that microorganisms in the deep sea incorporate L-aspartic acid and D-aspartic acid equally well.[44,45] The results may suggest that a whole new biosphere that relies on D-amino acids may exist deep below the waves.

References

1. Koschinsky, A.; Garbe-Schönberg, D.; Sander, S.; Schmidt, K.; Gennerich, H.-H.; Strauss, H. *Geology* **2008**, *36*, 615–618.

2. Imai, E.-I.; Honda, H.; Hatori, K.; Brack, A.; Matsuno, K. *Science* **1999**, *283*, 831–833.

3. Alargov, D. K.; Deguchi, S.; Tsujii, K.; Horikoshi, K. *Orig. Life Evol. Biosph.* **2002**, *32*, 1–12.

4. Weingärtner, H.; Franck, E. U. *Angew. Chem. Int. Ed. Engl.* **2005**, *44*, 2672–2692.

5. Deguchi, S.; Tsujii, K. *Soft Matter* **2007**, *3*, 797–803.

6. Matubayasi, N.; Wakai, C.; Nakahara, M. *Phys. Rev. Lett.* **1997**, *78*, 4309–4309.

7. Brunner, E. *J. Chem. Thermodyn.* **1990**, *22*, 335–353.

8. Bröll, D.; Kaul, C.; Krämer, A.; Krammer, P.; Richter, T.; Jung, M.; Vogel, H.; Zehner, P. *Angew. Chem. Int. Ed.* **1999**, *38*, 2998–3014.

9. Deguchi, S.; Ifuku, N. *Angew. Chem. Int. Ed.* **2013**, *52*, 6409–6412.

10. Leal-Calderon, F.; Schmitt, V.; Bibette, J. *Emulsion Science – Basic Principles*; 2nd ed.; Springer: New York, 2007.

11. Solans, C.; Izquierdo, P.; Nolla, J.; Azemar, N.; Garcia-Celma, M. *Curr. Op. Coll. Interface Sci.* **2005**, *10*, 102–110.

12. Solans, C.; Solé, I. *Curr. Op. Coll. Interface Sci.* **2012**, *17*, 246–254.

13. Schmid, G. *Nanoparticles: From Theory to Application*; Wiley-VCH: Weinheim, 2004.

14. Atwell, W. A.; Hood, L. F.; Lineback, D. R.; Varriano-Marston, E.; Zobel, H. F. *Cereal Foods World* **1988**, *33*, 306–311.

15. Zhang, Y.-H. P.; Himmel, M. E.; Mielenz, J. R. *Biotechnol. Adv.* **2006**, *24*, 452–481.

16. Nishiyama, Y.; Langan, P.; Chanzy, H. *J. Am. Chem. Soc.* **2002**, *124*, 9074–9082.

17. Nishiyama, Y.; Sugiyama, J.; Chanzy, H.; Langan, P. *J. Am. Chem. Soc.* **2003**, *125*, 14300–14306.

18. Deguchi, S.; Tsujii, K.; Horikoshi, K. *Chem. Commun.* **2006**, *2006*, 3293–3295.

19. Adschiri, T.; Hirose, S.; Malaluan, R.; Arai, K. *J. Chem. Eng. Jpn.* **1993**, *26*, 676–680.

20. Sasaki, M.; Kabyemela, B.; Malaluan, R.; Hirose, S.; Takeda, N.; Adschiri, T.; Arai, K. *J. Supercrit. Fluids* **1998**, *13*, 261–268.

21. Sasaki, M.; Fang, Z.; Yoshiko Fukushima; Adschiri, T.; Arai, K. *Ind. Eng. Chem. Res.* **2000**, *39*, 2883–2890.

22. Deguchi, S.; Tsujii, K.; Horikoshi, K. *Green Chem.* **2008**, *10*, 191–196.

23. Klemm, D.; Heublein, B.; Fink, H. P.; Bohn, A. *Angew. Chem. Int. Ed.* **2005**, *44*, 3358–3393.

24. Zeronian, S. H. In *Cellulose chemistry and its applications*; Nevell, T. P.; Zeronian, S. H., Eds.; Ellis Horwood: Chichester, 1985; pp. 139–158.

25. Nishio, Y.; Roy, S. K.; Manley, R. S. J. *Polymer* **1987**, *28*, 1385–1390.

26. Horii, F.; Yamamoto, H.; Kitamaru, R.; Tanahashi, M.; Higuchi, T. *Macromolecules* **1987**, *20*, 2946–2949.

27. Watanabe, A.; Morita, S.; Ozaki, Y. *Biomacromolecules* **2006**, *7*, 3164–3170.

28. Watanabe, A.; Morita, S.; Ozaki, Y. *Biomacromolecules* **2007**, *8*, 2969–2975.

29. Brovchenko, I.; Geiger, A.; Oleinikova, A. *J. Chem. Phys.* **2004**, *120*, 1958–1972.

30. Shaw, R. W.; Brill, T. B.; Clifford, A. A.; Eckert, C. A.; Franck, E. U. *Chem. Eng. News* **1999**, *69*, 26–39.

31. Krauskopf, K. B.; Bird, D. K. *Introduction to geochemistry*; McGraw-Hill: New York, **1995**.

32. McCarthy, J. F.; McKay, L. D. *Vadose Zone J.* **2004**, *3*, 326–337.

33. Vörösmarty, C. J.; Meybeck, M.; Fekete, B.; Sharma, K.; Green, P.; Syvitski, J. P. M. *Global Planet. Change* **2003**, *39*, 169–190.

34. Frondel, C. *Econ. Geol.* **1938**, *33*, 1–20.

35. Derjaguin, B. V. *Theory of Stability of Colloids and Thin Films*; Plenum: New York, **1989**.

36. Israelachvili, J. *Intermolecular and surface forces*; Academic Press: London, **1992**.

37. Alargova, R. G.; Deguchi, S.; Tsujii, K. *Colloids Surf. A* **2001**, *183-185*, 303–312.

38. Ghosh, S. K.; Alargova, R. G.; Deguchi, S.; Tsujii, K. *J. Phys. Chem. B* **2006**, *110*, 25901–25907.
39. Rothschild, L. J.; Mancinelli, R. L. *Nature* **2001**, *409*, 1092–1101.
40. Takai, K.; Nakamura, K.; Toki, T.; Tsunogai, U.; Miyazaki, M.; Miyazaki, J.; Hirayama, H.; Nakagawa, S.; Nunoura, T.; Horikoshi, K. *Proc. Natl. Acad. Sci. U.S.A.* **2008**, *105*, 10949–10954.
41. Deguchi, S.; Shimoshige, H.; Tsudome, M.; Mukai, S.-A.; Corkery, R. W.; Ito, S.; Horikoshi, K. *Proc. Natl. Acad. Sci. U.S.A.* **2011**, *108*, 7997–8002.
42. Kato, C.; Li, L.; Nogi, Y.; Nakamura, Y.; Tamaoka, J.; Horikoshi, K. *Appl. Environ. Microbiol.* **1998**, *64*, 1510–1513.
43. Glud, R. N.; Wenzhöfer, F.; Middelboe, M.; Oguri, K.; Turnewitsch, R.; Canfield, D. E.; Kitazato, H. *Nature Geosci.* **2013**, *6*, 284–288.
44. Pérez, M. T.; Pausz, C.; Herndl, G. *J. Limnol. Oceanogr.* **2003**, *48*, 755–763.
45. Teira, E.; van Aken, H.; Veth, C.; Herndl, G. *J. Limnol. Oceanogr*: **2006**, *51*, 60-69.

XI
Water talks to water. Might we listen in?

D. James Morré and Dorothy M. Morré

Mor-NuCo, West Lafayette, IN, USA.
dj_morre@yahoo.com

Our life contains a thousand springs,
And dies if one be gone.
Strange that a harp of thousand strings
Should keep in tune so long.
Robert Boyle (1665)

Background/Hypothesis

Hypothesis. This summary report is predicated on the hypothesis that a periodic molecular disequilibrium within the structure of water is the mechanistic basis for functional cellular oscillations with a regular 24 min periodicity of certain growth regulated ECTO-NOX (ENOX) proteins which, in turn, would by necessity, link to well established transcriptional controls[1] to drive the cells' circadian 24 h clock.[1]

Time Keeping ENOX Proteins. Central to the above hypothesis was the discovery in the late 1980s of a growth-related and cyanide-insensitive family of proteins with alternating quinone oxidase and protein disulfide thiol interchange activities and a period length of 24 min also capable of oxidizing NADH classified by the HUGO Gene Nomenclature Committee as ENOX (ECTO-NOX = Ecto-Nicotinamide Dinucleotide Oxidase Disulfide Thiol Exchanger) proteins.[2]

Among the distinguishing characteristics of the ENOX proteins is that they exhibit two different activities that alternate. The first activity

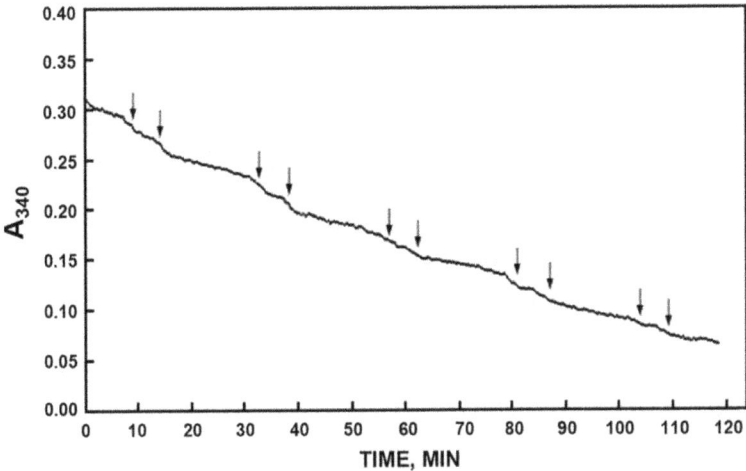

Figure XI.1. Continuous trace of the activity of recombinant ENOX1 purified by isoelectric focusing. NADH oxidation (decreases in A_{340}) was assayed over 120 min with data collected automatically and stored using a SPECTRA max 340 PC microplate reader. Reproduced from Ref. 4 with permission from ACS Publications and Ref. 2 with permission from Springer Science + Business Media.

is that of hydroquinone oxidation with NAD(P)H serving as an alternate non-physiological substrate. The second activity is that of protein disulfide-thiol interchange measured either from the restoration of activity to inactive scrambled RNase or from the cleavage of synthetic dithiopyridine substrates. Each activity generates a distinct oscillatory activity with a precise 24 min period suggestive of a role in biological time keeping.

With recombinant human ENOX1, rates of NADH oxidation, measured as a decrease in A_{340}, are seen in continuous traces (Figure XI.1) as patterns that are neither sinusoidal nor monotonic and with two major rate maxima (arrows in Figure XI.1) separated by an interval of 6 min with an overall period length of 24 min (Figure XI.1).

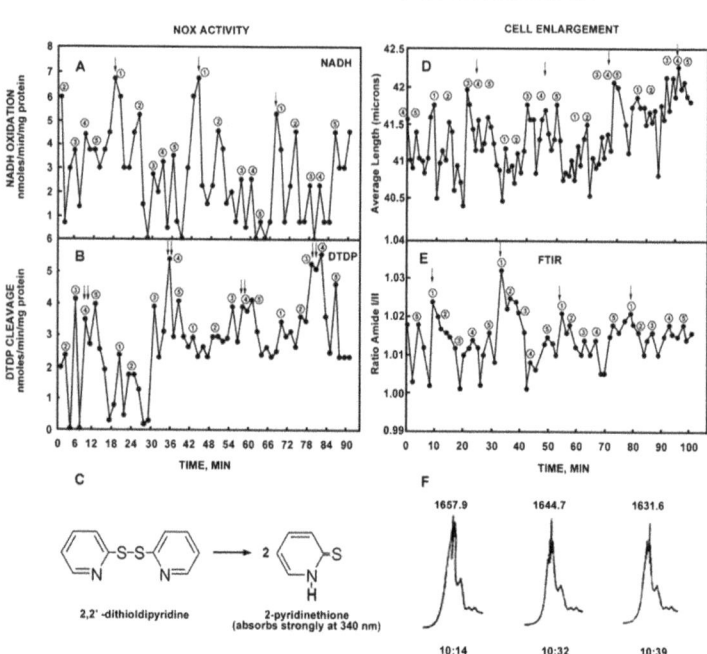

Figure XI.2. The 3 + 2 pattern of NOX activity oscillations. (a,b). NADH oxidation determined by the increase in A_{340} (upper curve). Maxima (arrows) were at 18, 42, and 66 min with secondary maxima at 24, 48 and 72 min. Three minor peaks completed each 24 min period. (b) Disulfide-thiol interchange activity measured simultaneously in parallel with NADH oxidation as an increase in A_{340} from the cleavage of dithiodiopyridine (DTDP). Major peaks are at 6, 9, and 12 min and at 24 min intervals thereafter (double arrows) with minor peaks at 18 and 24 min and at 24 min intervals thereafter. The two activities, NADH oxidation (a) and DTDP cleavage (b) alternate. (c) DTDP substrate generating 2 mol of 340 nm-absorbing 2-pyridinethionine and cleaved as a measure of the disulfide-thiol interchange activity of the NOX protein. (d) Increase in length (enlargement of a single cell as determined by

(Continued on opposite page.)

In contrast to the 24 min hydroquinone/ NADH oxidation portion of the 24 min ENOX cycle, the protein disulfide-thiol interchange activity consists of 3 maxima separated from each other and from the ① and ② oscillations by intervals of 4.5 min [2 x 3 + 4. (4.5) min] = 24 min (Figure XI.2). Normally, with ENOX proteins, the two activities alternate.

The 24 min oscillations require copper^{II}. Both the oxidative and the interchange activities illustrated diagrammatically in Figure XI.3, require copper. If the copper is removed from the protein, both oscillatory activities cease.[3]

Not only are Cu^{II} ions required for the oscillatory ENOX2 activity but solutions of copper^{I} chloride in the absence of ENOX protein exhibit an oscillatory pattern of NADH oxidation.[4]

As the copper^{II} chloride solutions do not catalyze protein disulfide thiol interchange, a function requiring the intact ENOX protein, now all five maxima of the oscillating pattern are expressed as oscillations

Figure XI.2. (continued)

image enhanced light microscopy). Cell enlargement proceeds in bursts every 12 min separated by rest periods where the cells actually shrink. As with NADH oxidase activity, each 24 min period (single arrows) is comprised, on average, of five resolvable maxima separated by minima. Three maxima are contained within the elongation phase and correspond to the protein disulfide-thiol interchange determined in parallel (b). The two maxima contained within the resting period correlate with the two maxima of NADH oxidation (a). (e) Fourier transform infrared analyses of recombinant ENOX2. Sixty-one 1 min scans taken 1.5 min apart over 100 min are illustrated. The ratio of the amide I (1,645)-amide II (1,545) absorbances varied with maxima at 22 min intervals as indicated by the arrows. (f) Within the amide I region (below), peak absorbance varied between 1,658 and 1,638 indicative of alternating α-helix-β-sheet transitions. Concanavalin A, cytochrome c, or albumin when analyzed in parallel showed no such pattern of activity fluctuations. Reproduced from Ref. 42 with permission from Taylor and Francis, and Ref. 2 with permission from Springer Science + Business Media.

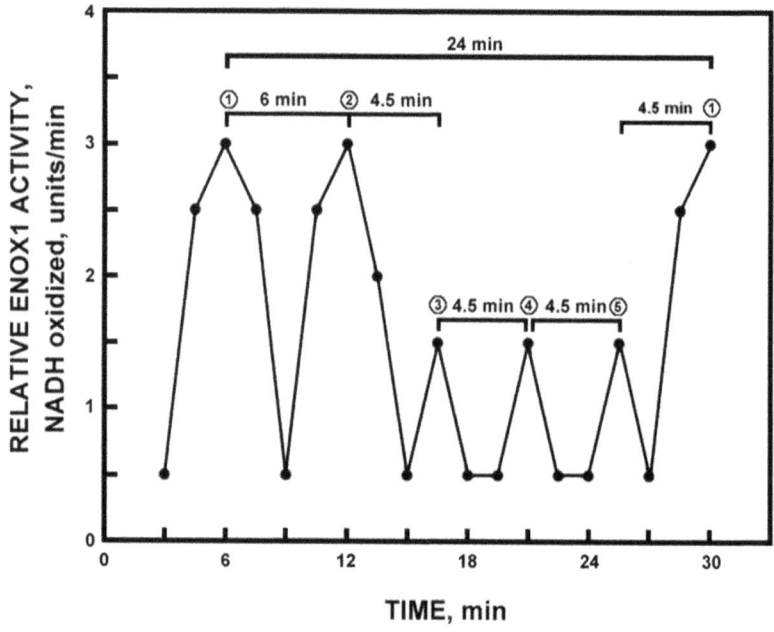

6 min + (4 X 4.5 min) = 24 min

Figure XI.3. Diagrammatic representation of the signature ENOX cycle with measurements averaged over 1 min at intervals of 1.5 min. Typically there are five maxima, two of which ① and ②, are separated by 6 min and three of which ③, ④ and ⑤ are separated from each other and from ① and ②, by 4.5 min. The asymmetry defines a 24 min period [6 min + (4 x 4.5 min)]. From Ref. 2 with permission from Springer Science + Business Media.

in the rates of NADH oxidation with two maxima separated by an interval of 6 min plus 3 maxima separated by 4.5 min as in Figure XI.3.

Oscillations in NADH oxidation are driven by oscillations in redox potential. The changing rates of NADH oxidation correlated with fluctuations in continuous traces of changes in redox potential are sufficient to drive the oxidation of NADH (Figure XI.4). Figure

Water talks to water. Might we listen in? 261

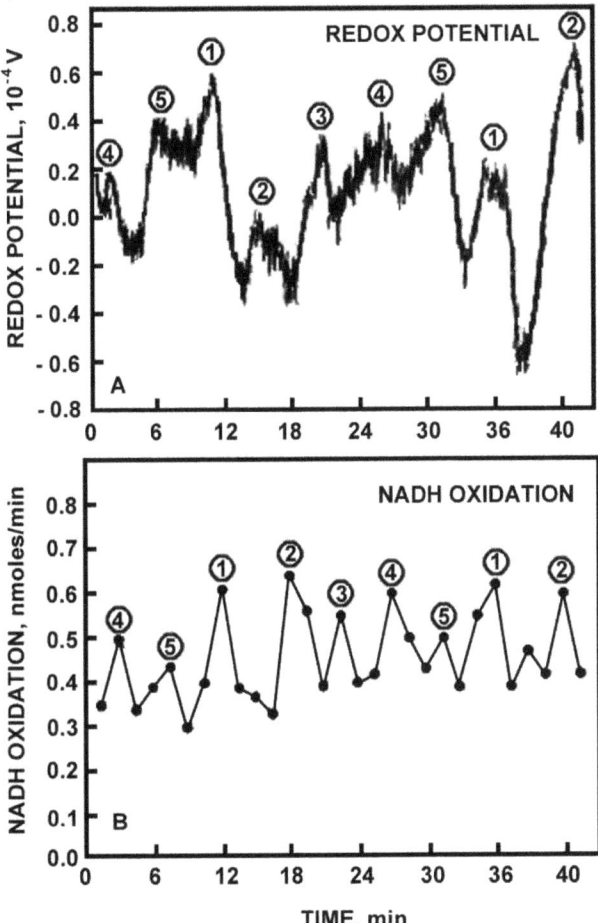

Figure XI.4. The redox potential (a) of an aqueous solution of copper chloride measured continuously showing a 2 + 3 pattern of oscillation with NADH oxidation measured over 1 min at intervals of 1.5 min (b) in parallel. The period length of both is 24 min. Results with pure water are similar except that the period length is now 18 min.[2]

XI.4 illustrates simultaneous measurements of redox potential and of NADH oxidation for a solution of copper chloride with a pattern consisting of five maxima. The redox measurements are from a continuous trace (Figure XI.4A) whereas the rates of NADH oxidation were measured over 1 min every 1.5 min in parallel (Figure XI.4B) to improve resolution. Two of the maxima labeled ① and ② in the figures are separated by 6 min. The remaining three maxima are separated from each other and from maxima ① and ② by 4.5 min. These intervals confer a characteristic asymmetry to the pattern of oscillations, and for solutions of $Cu^{II}Cl_2$ repeat every 24 min to impart a time-keeping aspect to the activity oscillations.

Oscillations not unique to copper^{II}. Chlorides, in solution, of other cations also exhibited periodic oscillations but the period lengths were longer or shorter than those of copper. Some property of these different cations was sought that might correlate with the period length of the oscillations. Ionic radii were found to do so (Figure XI.5). Period lengths of the oscillations of an aqueous solution were directly proportional to the ionic radius of the cation present and independent of the cation concentration. In the data of Figure XI.5 the cation concentrations were 10 μM.

With pure water redox potential and rate of NADH oxidation still oscillate. Surprisingly, even in the absence of cations pure water exhibited periodic oscillations in redox potential and NADH oxidation.[2] However, the average time between maxima was approximately 3.5 min to generate an overall period length of about 18 min.

If water is the fundamental source of the periodic oscillations, what property of water might be responsible? Water would be an ideal medium to serve as a universal time-keeper. It permeates every region of the cell, and is capable of interaction with both small and large molecules.

Heavy water, deuterium oxide, is the only substance known that consistently changes the length of the circadian day.[5] Whatever aspect of water might be responsible for the oscillations, it must be a physical rather than a chemical phenomenon due to the temperature independence of period length which is also a universal characteristic of the circadian day.[6] However, the oscillations were far too slow to correlate with any physical phenomena at the molecular or atomic level known to us at the time.

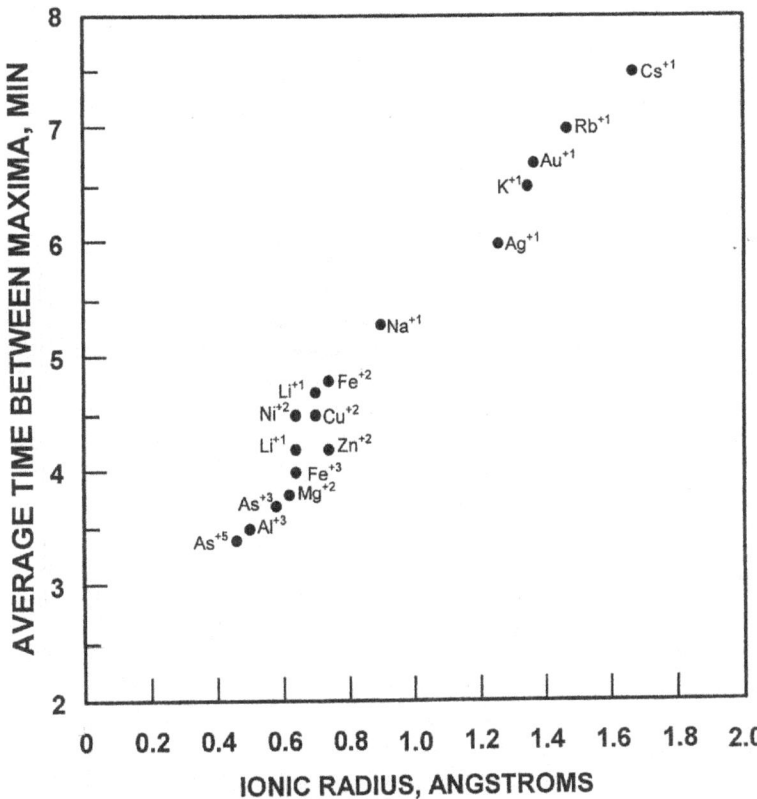

Figure XI.5. Period length of the oscillations of an aqueous solution is directly proportional to the ionic radius of the cation present and independent of cation concentration. All solutions were tested as the chloride at a final concentration of 10 μM. Only with Cu^{2+} (replaceable by Ni^{2+}) was the asymmetric period length of 5.8 min x 5 = 24 min observed. From Ref. 2.

The physical basis for water oscillations. Periodic disequilibration of ortho to para spin pair isomers of water protons

It was our Purdue University friend and physics colleague, the late Dr. Albert Overhauser, that drew our attention to the work of Tikhonov and Volkov published in *Science* in 2002 (Figure XI.6a).[8] For equilibration of pure ortho water vapor to reach a ca 3:1 mixture of ortho to para

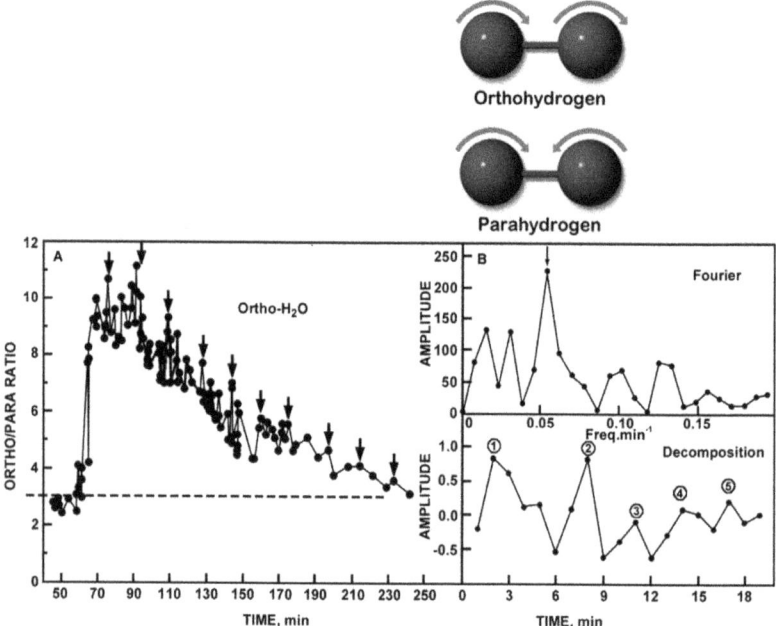

Figure XI.6. (a) Time dependence of equilibrium of the ratio of the *ortho-para* spin isomers determined from far infrared spectra starting with *ortho* enriched (1:10) water vapor derived from the original data in Ref. 8. The time to reach equilibrium was estimated to be 55 ± 5 min. During equilibration, the *ortho-para* ratio oscillated with a ca. 19-min period length (arrows). (b) Period length determined by Fast Fourier Analysis (frequency arrow = 18.6 min). The amplitude of the effect exceeded the instrumental error (± 5%) by an order of magnitude. (c) Decomposition fit of the data of (a) for a 19-min period to show the five peak (2 + 3) pattern characteristic of the ENOX-related time keeping. From Ref. 2.

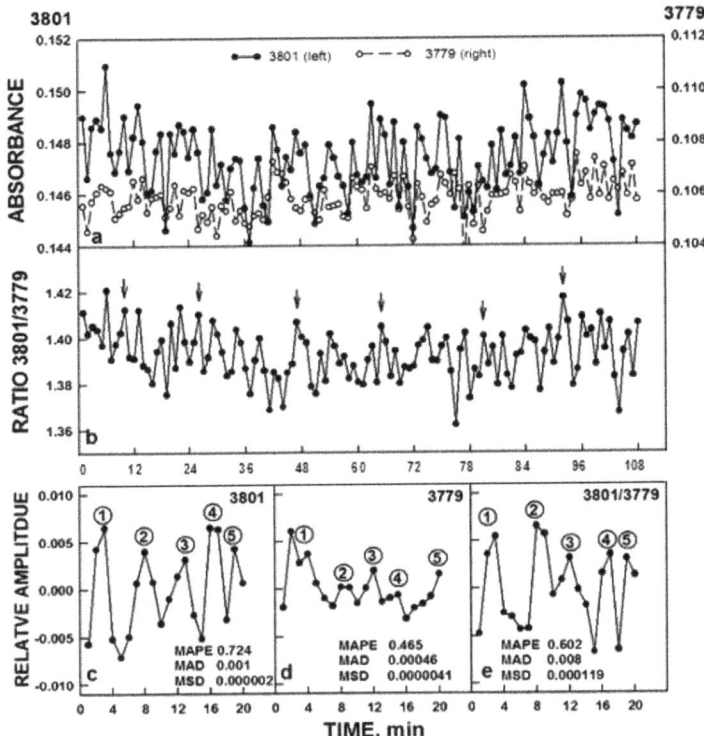

Figure XI.7. FTIR spectroscopic measurement of the ratio of *para*-H_2O above a water sample surface determined at 3,801 and 3,779/cm^{-1}, respectively (a). The ratio of the two wave-lengths exhibited a repeating pattern of oscillations of five maxima at intervals of about 18 min (arrows) (b). Decomposition fits using an imposed period length of 18 min of data collected at 2,801/cm (c), and at 3,779/cm (d) as well as the ratio of the two (e revealed the oscillatory pattern typical of water with five recurrent maxima, two of which, labeled ① and ② were separated in time by about 6 min and two to three additional maxima separated in time by about 4 min labeled ③, ④ and ⑤. The accuracy measures, MAPE, MAD and MSD are indicative of a close fit between the original and the fitted data. From Ref. 21.

water not only required 4 h but appeared to oscillate with a period length of about 18 min (Figs. 6b, c). The latter corresponded to the period length of the oscillations of pure water based on redox changes using NADH oxidation as the indicator.

Other works have reported spin selective interactions with porous surfaces[7,8] or in biological solutions.[9,10] Spin enrichment was achieved most recently by means of matrix deposition[11] whereas Veber et al.[12] failed to reproduce previous results possibly due to differences in their methods for determination of ortho/para ratios. None-the-less, these studies have contributed conclusive evidence for ortho/para spin conversions which, prior to the studies of Tikhonov and Volkov,[8] were missing from our information.

Spectral evidence. In our own work, FTIR spectroscopic measurements of ortho-H_2O/para-H_2O interconversion were carried out in the middle infra-red spectra region above a water sample surface determined at 3.801 and 3.779/cm^{-1} respectively in a manner similar to those for NAD(P)H oxidation and redox potential (Figure XI.6).

The ratios of the two wavelengths exhibited a repeating pattern of oscillations (Figure XI.7a, b) which when analyzed by decomposition fits using an imposed period length of 18 min revealed the typical oscillatory pattern of water with five recurrent maxima, two of which labeled ① and ②, were separated in time by about 4.5 min and three additional maxima separated in time by 3.4 min labeled ③, ④ and ⑤ (Figure XI.7).

Two independent states of liquid water. Evidence for ortho-para spin isomer disequilibrium in liquid water may be deduced as well from the studies of Pershin (Figure XI.8).[13,14] The spectrometric measurements were of backscattered Raman signal of water excited by the second-harmonic radiation of an Nd: YAG laser (a single 10-ns pulse or a train of such pulses with a repetition period of 1 s). The laser radiation was focused in a cell with room-temperature water. The measurements were at laser intensities of 2, 15, 35 and 350 MW/cm². The Raman signal was focused on the entrance slit of a polychromator

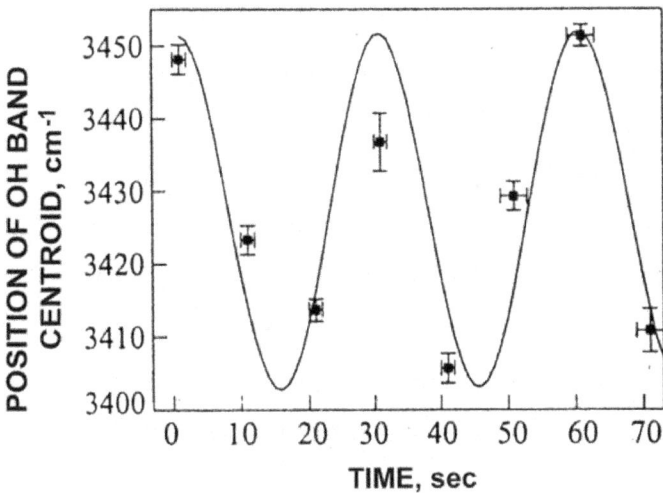

Figure XI.8. Harmonic oscillations of the OH centroid band determined from Raman scattering as a measure of exchange between two water states related to *ortho/para* ratios. The period length was determined to be 35 ± 13s. Reproduced from Ref. 13 with permission.

interfaced with an image intensifier and a diode array. Low-intensity measurements were carried out with a cooled diode array.

Spectral manifestation of a relatively high mobility of "hot" molecules was observed to result in the formation of a high frequency wing of the OH band and a shift of the center of the band with a coefficient of about 1/cm/grad upon heating. Such a shift was interpreted as a decrease in the concentration of strong H-bonds as a result of disordering through destruction of polymer water and an increase in the number of complexes whose molecules can rotate.[13,14]

The result is the experimental validation of the hypothesis of the independent existence of two liquids in water. Both liquids consist of hydrogen-bonded complexes of H_2O molecules, which exhibit characteristic frequency distributions of OH oscillators and form an integrated envelope of the OH-stretching vibration band.

These findings, plus evidence that water vapor is a mixture of independent fractions of ortho and para modifications, led Pershin to assume the existence of two independent states of liquid water with long lifetimes and two hydrogen bond types differing in energy.[13] Specifically "the energy of the hydrogen bond between ortho isomers of molecules which always rotate seems to be lower than between para molecules, a part of which cannot rotate at room temperature."

The conclusion was reached that this overheating-overcooling process is a fundamental property of water, which manifests itself at any temperature and is not the result of perturbation of overcooled water by an optical pulse. His findings show "that such an evolution of the band center is steadily observed in water and at room temperature. Moreover, this overheating-overcooling process was approximated by a harmonic function with a period of 35 ± 13 s without noticeable damping for ~ 16 min" (Figure XI.8).

Attributing the overheating-overcooling process observed by Pershin to different energy levels associated with ortho and para spin states of water, respectively, now requires a model whereby the ratio of ortho to para water might be maintained in a steady oscillation as opposed to simply reaching the equilibrium ratio and remaining there. Such phenomena are not without precedent in biology. An appropriate model being that of a limit oscillation with the classical example being the limit oscillator that controls heartbeat.

The limit oscillator model. That certain materials may preferentially adsorb para water[11] due to its non-rotation ground state implies that populations of water molecules might spontaneously form some sort of collective order whose structure subtly favor formation of para water.

In the manner of a limit oscillator, as para water forms, it might locally add to the field generated by the collective order which would further favor the accumulation of para water for some period of time (Figure XI.9). But also in keeping with the limit oscillator model, the initial conditions favoring para water formation would produce

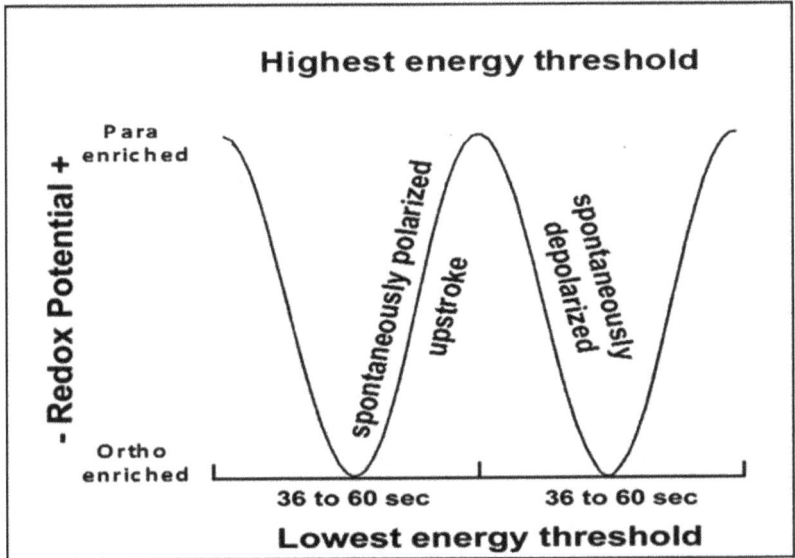

Figure XI.9. Model based on the concept of limit oscillations to explain *ortho-para* oscillations and their failure to reach a steady-state equilibrium. *Para* water reaches some high energy threshold and then spontaneously converts to *ortho* water to reach some low energy threshold at which time the conversion reverses back to *para* water. From Ref. 2.

an increased potential that might not be released or discharged until some minimum threshold of activation energy is achieved. Once the minimum threshold was achieved, a discharge of potential and conditions favorable to ortho water formation would result and the process would now run for a time in the opposite direction until some threshold level of ortho water was restored and the cycle would reverse.

However, information from biological limit oscillators also teach us that complex composite waves with relatively long well synchronized periods may derive from much shorter less well synchronized oscillations. For example, heart beat rate, the classic example of a limit oscillator, when averaged over 1 min every 1.5 min as for NADH oxidation, oscillates with a ca. 24-min period similar to that of the ortho/para oscillations of water with five maxima two of which are separated by 6 min and three of which are separated by 4.5 min recapitulating the asymmetry of the copper II hexahydrate measurements (Figure XI.3). However, the primary oscillations for

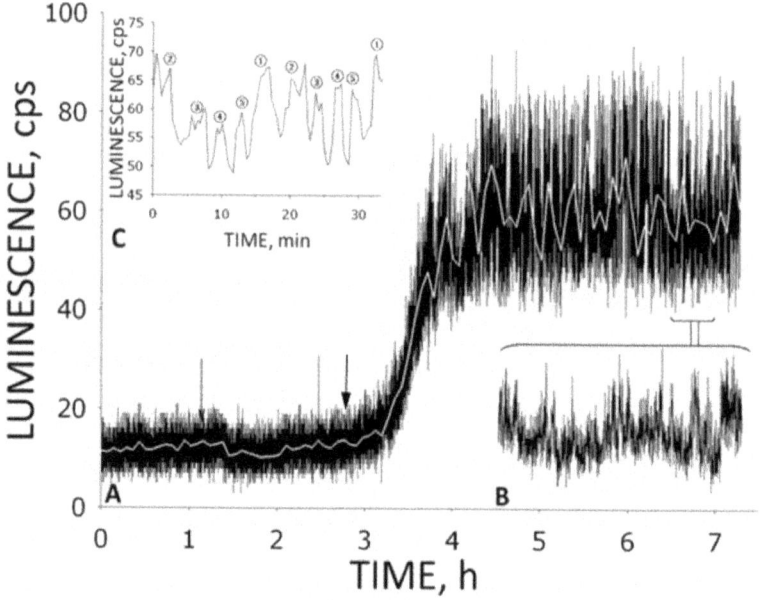

Figure XI.10. Effect of laser radiation on water luminescence. (a) Data of a representative experiment. Water was exposed to an infrared laser (A = 1,264 nm, power 5 mW) for 5 min. The time of exposure is shown by a *vertical arrow*. To the *left* of the *arrow* is a record of the background water luminescence before the onset of laser irradiation. The *white line* on the basic plot represents the macrostructure of a SigmaPlot revealing the characteristic 2 + 3 pattern of oscillations with a period length of 18 min in a form of "carrier wave" generated from *ortho/para* water oscillations with period lengths of 30-50 s as shown in (c). (b) The microstructure of changes in water luminescence in the bracketed data portion. (c) The integral intensity of water luminescence corresponding to that bracketed in (b). Maxima labeled ①-⑤denote an 18 min period length. Modified from Ref. 18 with permission.

heart beat rate, on average, occur at intervals of about 6 sec. Thus, the 2 + 3 pattern in this example must represent a form of composite wave generated by interaction of one to several parallel sets of primary oscillations of much greater frequency such as the ortho/para water

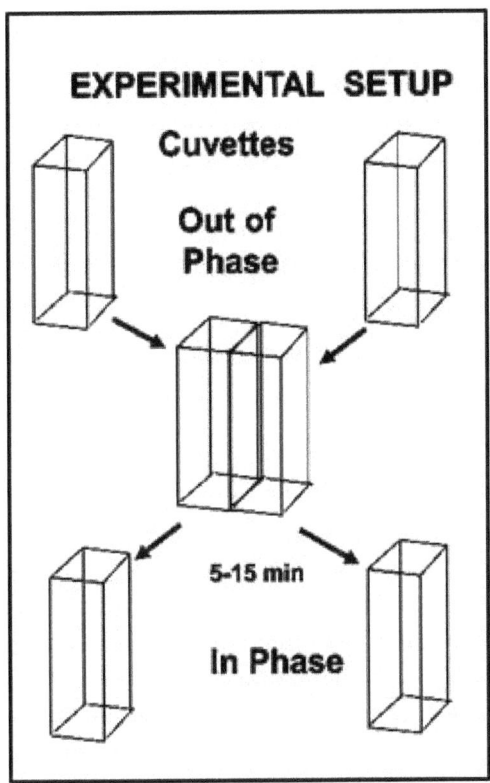

Figure XI.11. Experimental set up for autophasing of two out-of-phase water samples.

oscillations with period lengths of 30 to 50 sec reported by Pershin (Ref. 13, Figure XI.8) or even multiples of 12 sec or less indicated from measurements of water luminescence determined at 1 sec intervals (Figure XI.10).[15,16]

Luminescence data. An approach to determination of the fundamental period length. The luminescence data of Figure XI.10 present both unenhanced (Figure XI.10, left of arrow) and enhanced by laser illumination (Figure XI.10, right of arrow). The data were characterized by sinusoidal oscillations with a period length in multiples of 1.2 sec when evaluated by Fourier analysis.[17] When summed algebraically the values generated a composite (carrier) wave that was periodic but no longer sinusoidal. When values were summed over 1 min every 1.5 min in the manner of analysis of water-related changes in redox potential as measured by changes in NADH oxidation, the result was a repeating pattern of 4 or 5 maxima with two of the maxima separated by an interval of 6 min and an overall period length of 18 min. This agrees with the previously found oscillatory period of 18 min in the ratio of ortho to para spin pair isomers and redox potential in pure water.

In the original study of Gudkov *et al.*,[18] air-saturated double distilled water was exposed for 5 min to low-intensity laser infrared radiation at the wavelength of the electronic transition of dissolved oxygen to the singlet state (1264 nm). After a latent period of more than 2 h auto-oscillations of water luminescence in the blue-green region were observed over several h without indications of damping. The assumption was that the oscillations were the result of an alternation of two energetic water states with luminescence augmented auto catalytically by the laser-induced steady state formation of singlet oxygen.[19] Two such different energetic states are provided by alternation of the ratios of ortho and para nuclear spin isomers of water between two distinct energy states with the water luminescence representing the release of energy in the form of light as the highly synchronized ortho-para disequilibration returns from the high energy to the low energy state. A similar phenomenon occurs in pure water in the absence of excitation by laser radiation although at an intensity ca. 25% of that following laser excitation and presumably does not require involvement of singlet oxygen in the excitation process. The latter supposition is supported by observations that oscillatory changes in redox potential and NADH oxidation in pure water occur in solutions purged of dissolved oxygen and occur in a manner where both amplitude and period length are independent of the concentration of dissolved oxygen and are unresponsive to infrared irradiation.[17]

The 2 + 3 oscillatory pattern is likely a composite wave. The findings do not directly answer the question of what is the fundamental period length of the ortho/para oscillations of water that appear to generate the ultradian rhythm suggested to underlie the biological clock. While consistent with the observations of Pershin for a period length in the order of 36 sec,[13] the actual period of the individual oscillations may be shorter perhaps in the order of 12 sec or less which, by a process of algebraic summation, would generate higher order oscillations the sum of which were no longer sinusoidal and give rise to "super" periods in multiples of 12 sec (including a period length of 36 sec) up to at least 96 sec or longer. The resultant series of sine waves were combined algebraically to generate a "composite wave". Averaging over 1 min then generated the 18 min asymmetric pattern of water oscillations derived by other methods.[17]

Synchronous water oscillations are attained and maintained

One issue concerning the oscillatory properties ascribed to water is a need to explain how their high level of synchrony is attained and maintained. In order to even measure the protein/copper hexahydrate/water oscillations, the systems must have been in near perfect synchrony. To quote the famous chemist, Robert Boyle, in his musing on the marvels of the human body from 1665 "Strange that a harp of a thousand strings should keep in tune so long".[20]

Life requires synchrony. Day-night cycles, albeit important, lack the precision required and do little to serve those without sight. Totally blind persons still maintain normal biological rhythms. Our search for the ultradian oscillator of the biological clock led to the discovery of the ECTO-NOX (ENOX) proteins whose activities oscillate with a precise temperature-independent period length of 24 min.[2] The oscillations are likely counted like those of the balance wheel of a spring-driven watch with the hands represented by diurnal activity cycles. The emergent question then was how was it possible for all members of a single population of protein molecules to oscillate indefinitely with absolute precession as monitored by changes in redox potential, enzymatic activity and conformational (α-helix $\leftrightarrows$ β sheet) change (Figure XI.2E)? That the underlying basis was physical rather than chemical was implicit in the lack of temperature dependence of period length.

Role of low frequency electromagnetic fields. A logical extension of our hypothesis was that the basis for water synchrony lay in the energetics of the ortho-para disequilibrium functioning as a classic limit oscillator and the fact that water molecules might communicate with each other via low frequency electromagnetic fields (LFEMF) or very low frequency electromagnetic fields (VLFEMF).

That water oscillations are phased by LFEMF supports the suggestion that water molecules communicate via LFEMF or VLMEF.[21,22] Even pure water is diamagnetic. It generates a field that opposes an applied magnetic field although substantially weaker by several orders of magnitude (e.g. < 10 microgauss) than the usual forces associated with ferromagnetic materials. These fields might be modulated by the energetics of synchronous ortho-para interconversion of the nuclear spin pairs of the paired water hydrogens. To test this hypothesis, the earlier observation that oscillatory changes in redox potential correlate with rates of NADH oxidation measured spectrophotometrically, was employed to monitor synchrony achieved

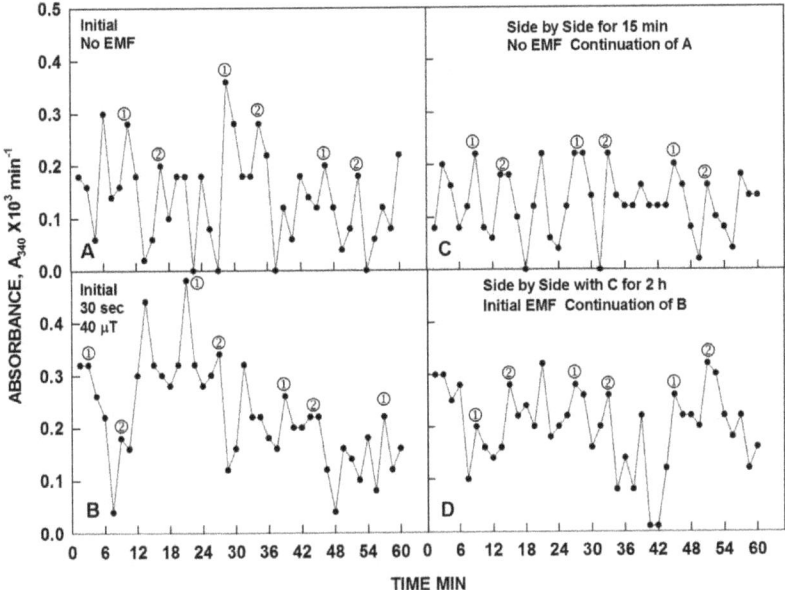

Figure XI.12. Low frequency EMF phasing of water samples through a thin plastic barrier. A. HPLC grade water. No EMF. B. A sample of HPLC grade water identical to that in A was phased using low frequency EMF (30 sec 40 µT) and shown to be out of phase with the original water sample (A). C, D. When the two plastic cuvettes containing the samples were placed adjacent to each other for 15 min both samples then began to oscillate in synchrony.

in contiguous water samples without mixing. That the energy difference between the two states (ortho and para) may generate a redox potential sufficient to catalyze the oxidation of NADH has been demonstrated experimentally. Such measurements also provide a biochemical basis for monitoring the degree of synchrony of the water oscillations. The experimental design is illustrated in Figure XI.11. Synchronization of asynchronous water samples was achieved through communication through the thin plastic barrier. A sample of HPLC grade water was phased using LFEMF (30 sec 40 µT) in a manner to be out of phase with the original water sample (Figure XI.12A,B). When the two plastic cuvettes containing the samples were placed adjacent to each for only a few min, both samples began to oscillate in parallel (Figure XI.12C,D). Thus, synchronization of asynchronous water samples was achieved by communication through the thin plastic barrier. When the two samples placed side by side were separated by a thin barrier of sheet copper sufficient to block LFEMF, the phasing of the two water samples was prevented. To test the prediction that contiguous water samples oscillate in phase through LFEMF communication, water was sampled in a shallow pond at two locations at a distance of 100 ft. apart. Also, water was sampled at opposite ends of a 20 acre lake. In both instances, the samples were synchronous in their oscillatory pattern. Two samples of West Lafayette, Indiana tap water collected simultaneously at the same time at two different locations separated

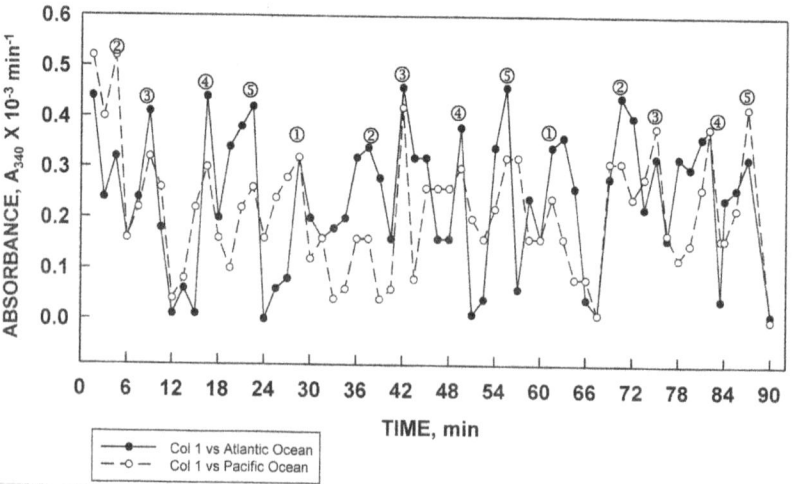

Figure XI.13. Water sampled simultaneous from the Atlantic ocean at Ocean City, Maryland (solid line, closed symbols) and from the Pacific ocean at San Diego, California (dashed line, open symbols) exhibited similar oscillatory patterns when analyzed in parallel indicative of synchrony.

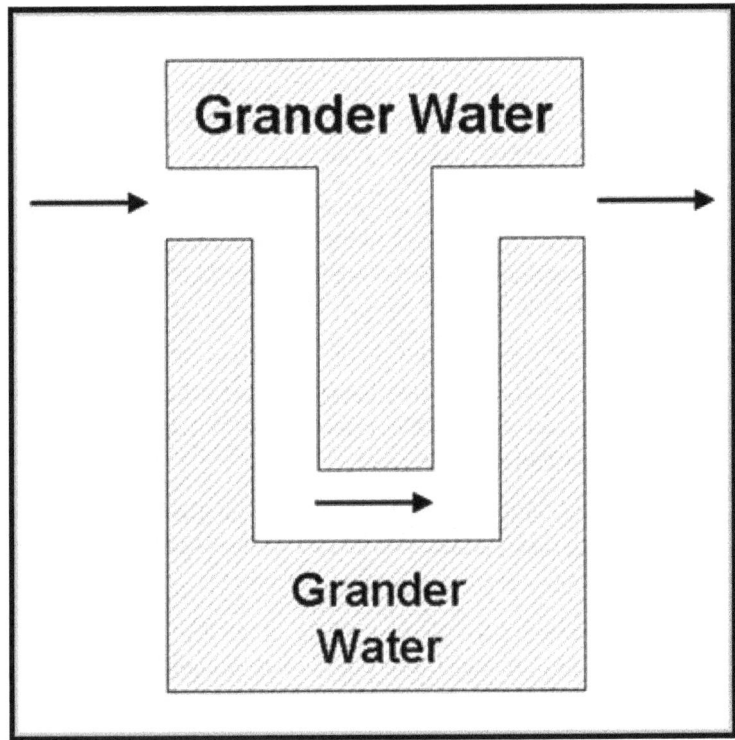

Figure XI.14. Diagrammatic representation of the Grander water purification device constructed of stainless steel. A transit time through the device of less than 1 min is sufficient to effect the transformation.

by a linear distance of 5 miles (8.3 km) also were synchronous. Water samples from a flowing stream at $t = 0$ and then 15 min later were synchronous. It was estimated that at least 50,000 gallons of water had passed the test site between sample collections.

The largest bodies of fresh water sampled have been from Lake Ontario and from the Niagara River in New York with sampling points separated by approximately 20 miles for each. Samples from both were found to be synchronous. Simultaneous sampling of water from the Pacific Ocean at San Diego, California and of the Atlantic

Ocean at Ocean City, Maryland, also revealed the water at the two sampling points to be in phase with each other (Figure XI.13). These findings were confirmed in two separate experiments and support the concept of water coherence advanced by others that translates into highly correlated water populations but to an extent much greater than that which may have been previously anticipated.[23,24]

Frequency of the EMF associated with coherent water is estimated to be 1.2 μm in the infrared.[25] One coherent domain "speaks" to another coherent domain to align the contained water molecules. The frequency of transmission might be a few kHz. However, the waves are trapped within the water (generated and absorbed by contiguous water molecules) and for the most part, do not escape (Marc Henry, personal communication).

A seeming exception is across a plastic or glass surface where transmission apparently does take place to effect synchrony of water in an adjacent plastic or glass container.

A more puzzling exception was subsequently encountered in our attempts to understand how a certain water purification system developed by Grander Wasserbelebung of Jochburg, Austria could function.

Trans-metal direct contact water communications

The Grander water purification device is based on the principle that water inside a metal chamber causes a fundamental alteration in the properties of the water passing through the device (Figure XI.14). Industrial water passing through the device containing the pure (Grander) water on the inside of the stainless steel encased chamber comes out with properties of pure water after only a few seconds of transit though the device.

Some cited benefits of Grander-treated water include: More palatable to drink, bacteria will not grow, reduction in scale buildup,

Water talks to water. Might we listen in?

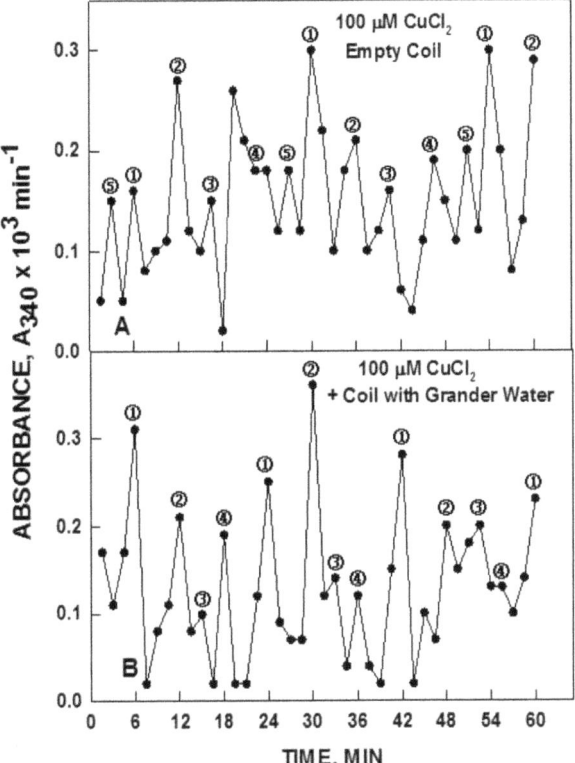

Figure XI.15. NADH, 120 μM, was prepared in water or 100 μM $CuCl_2$ with a standard copper coil filled with Grander water. The pattern shown in A was observed. The empty coil had no effect. With the coil containing Grander water, the result was the 4-peak pattern with the 18 min period of water shown in B and not the 5 peak pattern with a 24 min period characteristic of 100 μM $CuCl_2$ shown in A. Results with HPLC water in the coil were equivalent.

increased service life of equipment, reduced allergies and skin problems, use of less soap or detergents when washing clothes and elimination of a need for a water softener.

In the experiments below, a copper coil was used with no liquid inside (empty) or with water or copper chloride (100 μM) inside. The test sample over the coil was in a plastic container and contained 120 μM NADH as the redox indicator to measure the length and phase of the period.

In the first experiment illustrated with 120 μM NADH prepared in water or 100 μM $CuCl_2$, the copper coil filled with Grander water generated the pattern shown by Figure XI.15. With a coil without liquid inside, there was no effect. With the coil containing Grander water (Figure XI.15), the result was the 4-peak pattern with the 18 min period of water not the 5 peak pattern with a 24 min period characteristic of 100 μM $CuCl_2$ (Figure XI.16). Whatever water, either Grander or HPLC water, was placed in the container on top of the coil assumed the characteristics of the water inside the coil with no change in ionic or solute composition. Impure water exposed to pure water inside the coil behaved as pure water. The exposure time in these experiments was 15 min but less than 1 min exposure was required to affect the transformation.

The metal needed not be copper nor was it necessary for the container be in the form of a coil. Identical results were obtained with a water-filled chamber of stainless steel (Figure XI.17). With the $CuCl_2$ solution placed over the stainless steel chamber containing Grander water (Figure XI.17), the $CuCl_2$ solution now oscillated with the 4 peak pattern and 18 min period typical of water instead of the 5 peak pattern and 24 min period characteristic of the $CuCl_2$ solution.

As measured with a shielded DC magnetometer, water inside the coil or chamber generated a ca 7 milligaus LFEMF at the surface of the coil or chamber. A thin sheet of metal interposed between the coil or chamber and the external sample blocked the communication consistent with a LFEMF signal. It would appear that the LFEMF at the coil or chamber surface is of the same frequency and phase as the aqueous liquid inside the coil or chamber and is capable of transmitting that frequency to aqueous liquids external to the coil or chamber.

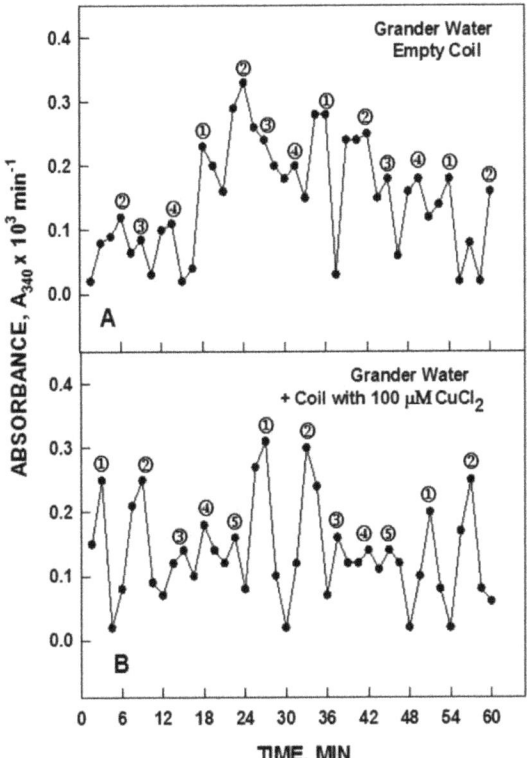

Figure XI.16. The reverse experiment gave the reverse result. With HPLC water (4 maxima, 18 min period (A) placed over the coil now filled with 100 µM CuCl$_2$ caused the HPLC water to oscillate with the 5 peak 24 min period characteristic of the 100 µM CuCl$_2$ (B).

Question 1: How does water at the surface of a metal barrier communicate with the metal of the barrier to generate the measured LFEMF at the outer barrier surface?

A possible answer is that the electromagnetic carrier wave from the water in direct contact with a conducting metal surface conducts that signal via the conductor to a suitable detector in exactly the same way

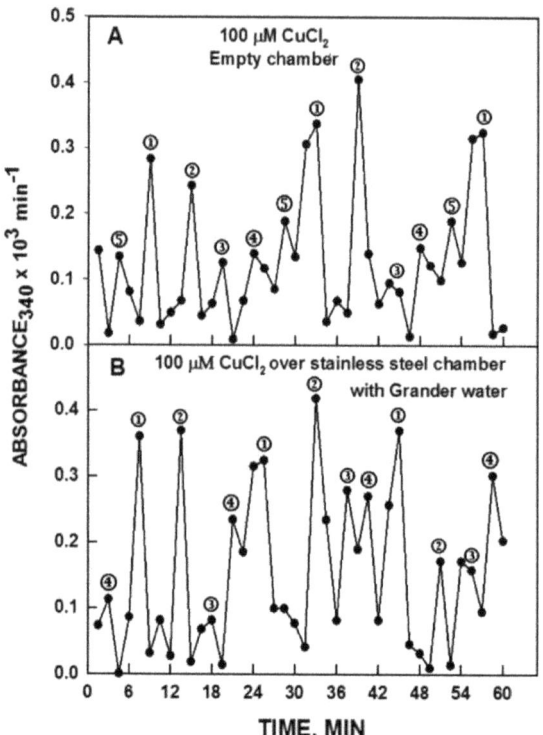

Figure XI.17. The metal need not be copper nor the container in the form of a coil. Identical results were obtained with a water filled chamber of stainless steel. A. 120 μM NADH prepared in 100 μM CuCl$_2$. The typical 5 peak 24 min period characteristic of CuCl$_2$. B. With the CuCl$_2$ solution placed over the stainless steel chamber containing Grander water, the CuCl$_2$ solution now oscillated with the 4 peak pattern and 18 min period typical of water.

as do electrodermal sensors (see below).

Question 2: Why is the frequency and phase of the oscillations inside the device not influenced by the frequency and phase of the aqueous solution at the outer surface of the device?

A possible answer to Question 2 is that water in direct contact with the outer metal surface, partly as a result of surface tension, will exhibit properties of the inside water continuum such that any signal returning from the outside water layer to the inside water will be only that of the inside water. If there is no outside water to metal contact, the LFEMF signal coming from the outside non-metal water container is blocked by the metal of the coil and the properties of the water inside the coil are unaffected.

Complex (carrier) wave formation for water solutions

In most forms of communication that involve the transmission of information from one location to another, modulation is used to encode the information into some form of carrier wave. The characteristics of the carrier wave will determine how the signal will propagate over any significant distance and the wave will carry information as a function of the changing property of the wave. Three key parameters, amplitude (volume), phase (timing) and frequency (pitch) characterize all periodic wave forms. Waves can be superimposed upon each other without limit. For electromagnetic waves, their electric and magnetic fields simply add at each point. So why not for water?

Water associated with Cu^{II} hexahydrate generates a specific carrier wave perceived by other copper hexahydrates so all are synchronized. However, Cu^{II} hexahydrate is just one example. What if every molecule in aqueous solution, every micro environment, encountered by water would contribute in a similar manner to the unique composite or carrier wave generated by water for any particular environment? For example, a sample of body fluid might then contain information on levels of drugs, metabolic disorders, toxins, sources of chronic pain or cancer all contained within the pattern of water oscillations. With the appropriate receiver system, it then might be possible for water messages to be heard as a means to obtain useful diagnostic information. Something to this effect known as electrodermal screening may already be in use in naturopathy. Electrodermal screening both of diseases and of the efficacy of drugs is often unexpectedly accurate. Based on our findings above with trans metal direct contact water communication, electrodermal measurements may act as receivers, at least in part, for water-derived signals implying that meridians are water communication lines.

Electrodermal screening. A means to listen in?

Electrodermal screening is used to measure the electrical resistance on the skin's surface (Figure XI.18). The purpose is to detect "energy" imbalance along invisible lines of the body described by acupuncturists as meridians. The testing device sends a tiny electrical current, too small to be detected by the patients, through a probe. The patient may hold a probe in one hand, while a second probe is touched to another part of the body. This completes a low-voltage electrical circuit and a computer screen or a needle on a gauge reads out a number between 0 and 100. This may be repeated at many different places on the skin. These numbers are used to decide if the patient's "energy" is in or out of balance.

If the patient is being tested for a type of treatment, samples of various remedies may be tried as the probe is touched to the problem area. Remedies may include homeopathic liquids and dietary or vitamin supplements. Different substances are tested until one is found that "balances" the energy disturbance.

What do electrodermal diagnostic acupuncture instruments really measure? How might such measurements provide an early warning system of body pathology as claimed? Up until now, the connection between these skin measurements and the functioning of internal organs has not been obvious. Water-generated electromagnetic fields may hold the answer.

From a theoretical modeling basis, William Tiller wrote "one expects that cooperative cellular oscillations in an organ will lead

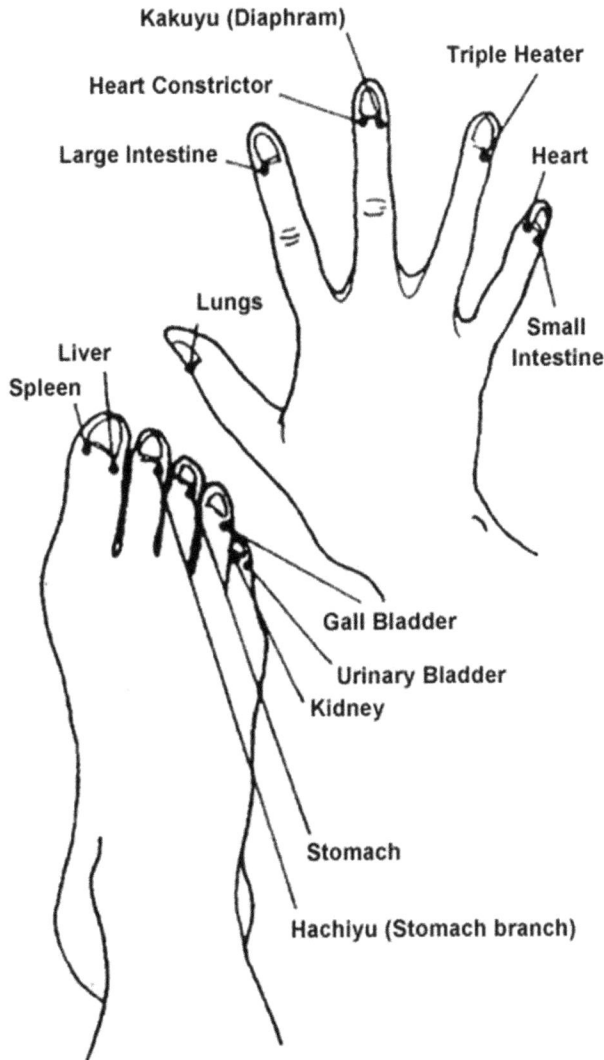

Figure XI.18. Acupuncture points at the ends of the meridians fingers and toes). From Tiller with permission.[43] See also Ref. 44 for role of water in defining meridians.

to the generation of electromagnetic radiation in a broad frequency band from above the infrared to the KHz range.[26] If meridians are indeed present as a type of conductance channel in the body, then electromagnetic radiation waves of the appropriate wavelength from this generated organ spectrum will be guided away from the organ environs and out to the skin through specific acupuncture points. Because the dimension of the meridian channel is expected to be a fraction of a millimeter, the guided electromagnetic radiation waves will be in the microwave portion of the spectrum. Such a guided microwave beam will have an added value in the modeling because it is expected to cause the acupuncture point to be of higher conductance than the surrounding tissue ...".

Early detection of cancer by electrodermal screening. One meaningful example recently brought to our attention is the early detection of cancer by electrodermal testing. To date, sera from 15 subjects where electrodermal screening has indicated early cancer in advance of clinical symptoms have been received by Mor-NuCo's ONCOblot® Tissue of Origin Test for early detection of cancer presence.[2,27] Of the 15 serum samples received, all but 2 reacted positively by the ONCOblot® Tissue of Origin Cancer Test for the presence of cancer-specific ENOX2 protein transcript variants in patient sera (Table 1). The cancer-specific and potentially clock-related ENOX2 proteins, studied mostly in man, oscillate with a 21-22 min period. ENOX2 proteins are inhibited by a variety of quinone site inhibitors all with anticancer activity.[2] In contrast, the growth and clock-related ENOX1 proteins are constitutive, largely drug resistant, are widely distributed among animals, plants and yeast and oscillate with a 24 min period. Moreover, unlike ENOX1 proteins which are a single species, ENOX2 proteins produce transcript variants . Each cancer tissue of origin generates transcript variants of unique molecular weight and isoelectric points and are shed into sera to serve as molecular markers both of cancer presence and tissue of cancer origin.[27]

Tiller in his 1987 paper on electrodermial testing, attached considerable importance to a single earlier screening as one example of early detection of cancer using an electrodermal screening device.[28] Greater than 85% accuracy of electrodermal screening in disclosing cancer in advance of clinical symptoms, as summarized in Table 1, is not only worthy of note but also deserving of elucidation of the mechanistic basis for the level of accuracy attainable.

Table 1. Cancer presence indicated from electro-dermal screening (No clinical symptoms) confirmed by ONCOblot® Tissue of Origin Cancer Test.[2]

Age (y)/Gender	Cancer Present and Tissue of Origin
62 M	Malignant melanoma
40 M	Prostate
62 F	Generic ENOX2
27 F	Endometriosis
60 F	Negative
27 F	Blood Cell Cancer
38 M	Colorectal
62 F	Ovarian
67 F	Colon
35 M	Prostate
32 F	Generic ENOX2*
73 M	Generic ENOX2*
63 F	Breast
19 M	Negative
44 M	Colon

*Generic ENOX2 is a fully processed ENOX2 transcript variant indicative of cancer presence but unspecific as to tissue of orgin.

Why should water of a cancer patient exhibit unique oscillatory properties? Evidence that water from a subject with cancer would exhibit a pattern of oscillations different form non-cancer is limited but compelling. The cancer-specific ENOX2 proteins shed from cancer cells oscillate with a ca. 21-22 min period compared to the 24 min period of the constitutive ENOX1.[29] Since both ENOX1 and ENOX2 oscillations do not cross entrain and remain synchronous in the presence of each other, it seems reasonable to assume that their effect on the water microenvironment might, just of themselves, be sufficient to signal cancer presence via electrodermal screening.

As the dominant ENOX of cancer cells, ENOX2 or tNOX, having a period length of 21 to 22 min might be expected to generate a 21 to 22 h circadian day as observed in cultured cells overexpressing ENOX2.[30] Activity patterns in cancer patients have also been observed to differ from those in non-cancer patients as the basis for chronotherapy, which in oncology, strives to take advantage of asynchronies in cell proliferation and drug metabolism rhythms that distinguish normal and malignant tissues, to time chemotherapy to where the cancer cells and tissues are maximally sensitive to a drug, for example.[31,32] Such therapy has been accomplished on an empirical basis with patients and offers promise to minimize chemotherapeutic damage to host tissue and maximize drug toxicity to malignancies.[33,34]

Several carefully designed studies have been published to document that tumor development is affected by disruption of circadian rhythms.[35] Breast cancer is accelerated by constant light exposure (as in people that work predominantly at night).[36] Robustness of diurnal rhythmicity in patients may favor survival in patients with breast cancer.[37] In a two-year study of patients with metastatic colorectal cancer, those with clearly defined rest and activity rhythms had a fivefold higher survival rate and experienced less fatigue than patients with diminished rest/activity rhythms.[32] Increased cancer risk to pilots, flight attendants, and airline cabin crews subjected to repeated circadian disruptions involving jet lag is also an area of concern.[38]

Water as target for biological responses to external electromagnetic fields. Within the water molecule, nuclear spins of protons of water have been proposed as the primary target in biological systems to external electromagnetic fields.[39] There is extensive literature documenting the ability of weak electromagnetic fields to affect human cells and human health including generation of reactive oxygen species, immunological responses, inflammation, cell proliferation, wound healing, developmental processes, nerve regeneration, and cancer.[2] As a result of the wide range effects of EMF on biological processes, concern has been expressed that exposure to low frequency magnetic fields, particularly those generated by electrical transmission lines, distribution lines, and electrical appliances, have complications to human health including increased incidence of cancer. Beneficial effects might accrue as well with water as the target such as the possibility of lowering the therapeutic dose of drugs by combination with EMF.[39, 40]

Water as a long range communication channel and molecular time keeper

If water coursing through the body in various body fluids is in fact, as it seems to be, a long range communication channel, the opportunities that arise are profoundly important. Not only might it be possible to obtain wellness information electrodermally but also transmit information back to the body to repair the reported damage and restore balance to targets of imbalance. Opportunities afforded may be not only to listen in but to talk back as well.

At the cellular level, the mammalian circadian clock has been largely ascribed to a series of transcriptional feedback loops. However, recent studies,[41] no longer fully support the prevailing models where generation of biological rhythms is controlled primarily by transcriptional mechanisms. A water based time keeping system would serve the need for a new paradigm of cellular time keeping based on physical principles rather than chemical reactions with period lengths truly independent of temperature rather than "temperature compensated" and both synchronous and entrainable within an entire cell or organism. A water based molecular clock would readily link to the myriad of outputs currently known to be under circadian control within the cell and help explain the characteristic of robustness of circadian oscillations in general.

References

1. Morré, D. J.; Chueh, P.-J.; Pletcher, J.; Tang, X.; Wu, L.-Y.; Morré, D. M. *Biochemistry* **2002**, *40*, 11941-11954.
2. Morré, D. J.; Morré, D. M. *ECTO-NOX Proteins*, Springer, New York, 2013.
3. Tang, X.; Chueh, P.-J.; Jiang, Z.; Layman, S.; Martin, B.; Kim, C.; Morré, D. M.; Morré, D. J. *J. Bioenerg. Biomembr.* **2010**, *42*, 355-360.
4. Jiang, Z.; Morré, D. M.; Morré, D. J. *J. Inorg. Biochem.* **2006**, *100*, 14028-14038.
5. Pittendrigh, C. S.; Caldarola, P. C.; Cosbey, E. S. *Proc. Natl. Acad. Sci. U.S.A.* **1973**, *70*, 2037-2041.
6. Edmunds, L.N., Jr. *Cellular and molecular basis of biological clocks*. Springer, New York, Berlin, 497 pp. 1998.
7. Potekhin, S. A.; Khusainova, R. S. *Biophys. Chem.* **2005**, *118*, 84-87.
8. Tikhonov, V. I.; Volkov, A. A. *Science* **2002**, *296*, 2363.
9. Bunkin, A. F.; Pershin, S. M.; Nurmatov, A. A. *Laser Phys. Lett.* **2006**, *3*, 275-277.
10. Bunkin, A. F.; Lebedenko, S. L.; Nurmatov, A. A.; Pershin, S. M. *Quantum Electron.* **2006**, *36*, 612-615.
11. Sliter, R.; Gish, M.; Vilesov, A.F. *J. Phys. Chem. A* **2011**, *115*, 9682-9688.
12. Veber, S. L.; Bagryanskaya, E. G.; Chapovsky, P. L. *J. Exp. Theor. Phys.* **2006**, *102*, 76-83.
13. Pershin, S. M. *Physics Wave Phenom.* **2005**, *13*, 192-208.
14. Pershin, S. M. *Laser Phys.* **2006**, *16*, 1140-1190.
15. Morré, D. J.; Morré, D. M. In: *Water and society*, Pepper, D. W., Brebbia,

C. A. (eds.), WIT Press, Southampton, Boston, pp. 13-23, 2012.

16. Morré, D. J.; Morré, D. M. In: *Cancer Prevention*, Georgakilas, A. G. (ed.), InTech, Rijeka, pp. 389-402, 2012.

17. Morré, D. J.; Morré, D. M.; Pendleton, R. L.; Gudkov, S. V.; Zakharov, S. D. *J. Phys. Chem. A*, submitted.

18. Gudkov, S. V.; Brushov, V. I.; Astrshey, M. E.; Chemikov, A. V.; Yaguzhinsky, L. S.; Zakharov, S. D. *J. Phys. Chem. B* **2011**, *115*, 7693-7697.

19. Zakharov, S. D.; Ivanov, A. V. *Quantum Electron.* **1999**, *29*, 1031-1053.

20. Davis, E. B. *J. Congregational Song* **2002**, *53*, 46-47.

21. Morré, D. J.; Orczyk, J.; Hignite, H.; Kim, C. *J. Inorg. Biochem.* **2008**, *102*, 260-267.

22. Morré, D. J.; Jiang, Z.; Marjanovic, M.; Orczyk, J.; Morré, D. M. *J. Inorg. Biochem.* **2008**, *102*, 1812-1818.

23. Pollack, G. H.; Clegg, J. *Unexpected linkage between unstirred layers, exclusion zones, and v. Phase Transitions in Cell Biology*, eds. G.H. Pollack and W.C. Chin, Springer Science Business Media, Berlin, pp. 143-152, 2008.

24. Del Giudice, E.; Spinetti, P. R.; Tedeschi, A. *Water* **2010**, *2*, 566-586.

25. Bono, I.; Del Giudice, E.; Gamberale, L.; Henry, M. *Water* **2012**, *4*, 510-532.

26. Tiller, W. A. *Amer. J. Acupuncture* **1987**, *15*, 15-23.

27. Hostetler, B.; Weston, N.; Kim, C.; Morré, D. M.; Morré, D. *J. Clin. Proteomics* **2009**, *5*, 46-51.

28. Kobayashi, T. *Amer. J. Acupuncture* **1985**, *13*, 63-68.

29. Wang, S.; Morré, D. M.; Morré, D. J. *Cancer Lett.* **2003**, *190*, 135-141.

30. Chueh, P.-J.; Wu, L.-Y.; Morré, D. M.; Morré, D. J. *BioFactors* **2004**, *20*, 235-249.

31. Mormont, N. C.; Levi, F. *Int. J. Cancer* **1997**, *70*, 241-247.

32. Mormont, M. C.; Waterhouse, J.; Bleuzen, P.; Giacchetti, S.; Jami, A.; Bogdan, A.; Lellouch, J.; Misset, J.; Touitou, Y.; Levi, F. *Clin. Cancer Res.* **2002**, *6*, 3038-3045.

33. Levi, F. *Chronobiol. Int.* **2002**, *25*, 459-461.

34. Levi, F.; Schibler, U. *Annu. Rev. Pharmcol. Toxicol.* **2007**, *47*, 593-628.

35. Fu, L.; Lee, C. C. *Natl. Rev. Cancer* **2003**, *3*, 350-361.
36. Schemhammer, E. S.; Laden, F.; Speizer, F. E.; Willetti, W. C.; Hunter, D. J.; Kawachi, I.; Colditz, G. A. *J. Natl. Cancer Inst.* **2006**, *93*, 1563-1568.
37. Sephton, S. E.; Sapolsky, R. M.; Kraemer, H. C.; Spegal, D. *J. Natl. Cancer Inst.* **2000**, *92*, 994-1000.
38. Wartenberg, D. A.; Stapleton, C. P. *Br. Med. J.* **1998**, *316*, 1902-1916.
39. Binhi, V. N. *Magnetobiology: underlying physical problems*. Academic, San Diego, 2002.
40. Chakkalakal, D. A; Mollner, T. J.; Bogard, M. R.; Fritz, E. D.; Novak, J. R.; McGuire, M. H. *Cancer Biochem. Biophys.* **1999**, *17*, 89-98.
41. Doherty, C. J.; Kay, S. A. *Science* **2012**, *338*, 338-340.
42. Morré, D. J.; Morré, D. M. *Free Rad. Res.* **2003**, *37*, 795-808.
43. Tiller, W. A. *J. Holistic Medicine* **1982**, *4*, 105-127.
44. Lo, Y. *Acupuncture Today* **2009**, *10*, 1-4.

XII
Aqua incognita: liquor aquae superficies

Richard J. Saykally

Department of Chemistry, University of California and Chemical Sciences Division, Lawrence Berkeley National Laboratory, Berkeley, CA 94720, USA.
http://www.cchem.berkeley.edu/rjsgrp/
saykally@berkeley.edu

In 2007, the author was among a group of scientists who were asked by noted science writer Michael Schirber to identify for an article appearing on LiveScience (*http://www.livescience.com*) what they viewed as "the most important unsolved mystery in science," to which he responded "The greatest mystery in science is understanding why, after literally centuries of tireless research and endless debate, we remain unable to accurately describe and predict the properties of water-the ostensibly simple, third most abundant molecule in The Universe, and the basis of all life as we know it" In this article, we briefly review the status of a few of these current "mysteries" involving water that our research group has recently addressed. Most of these explicitly or implicitly involve the liquid water surface.

Note that his article is not intended to be a comprehensive review, but rather to provide a general outline. The reader is referred to references provided in the cited literature.

Selective Adsorption of Ions to the Liquid Water Surface

The last decade has witnessed remarkable progress in our understanding of the interfaces of aqueous solutions-a subject of critical importance in broad areas of contemporary science. Initially motivated by observations in the study of atmospheric aerosol chemistry and subsequent computer simulations, results from numerous research groups are producing a new view of aqueous interfaces that is expanding rapidly in both detail and scope. Several recent reviews survey this progress.[1-3] While initial reports of positive surface adsorption of simple ions were met with considerable skepticism, the accumulated experimental evidence in support of such predictions from simulations has now convinced most doubters, and current efforts have turned to elucidating the mechanism responsible for this counterintuitive behavior.

Temperature dependent UV-SHG experiments have shown that ion adsorption is driven by enthalpy and impeded by entropy.[4] Accompanying model calculations (from Pat Schaffer and Phill Geissler) demonstrated that the displacement of weakly interacting water molecules into the bulk solution by anions moving to the air-water interface produces a negative (favorable) enthalpy change, while the suppression of capillary waves by the adsorbed ions yields a negative (unfavorable) entropy change of similar magnitude. The fact that a point dipole solvent model (Stockmayer fluid) yielded results essentially identical to those employing a widely used explicit water model (SPC-E) testifies to the generality of this ion adsorption mechanism for the air/water interface. We have also begun to explore the nature of ion adsorption to more complicated interfaces. Notably, the adsorption of bromide ions to the interface formed by a monolayer of dodecanol on liquid water was shown to exhibit an essentially identical free energy to that found for the air/water interface.[5]

But both the experimental and theoretical studies were performed on high (molar) concentration solutions. Quite different results have been reported for dilute (millimolar) solutions. For example, molar solutions of potassium iodide and potassium ferrocyanide exhibit "normal" ion adsorpion behavior, viz adsorption free energy changes of ca. -1 kJ/mol, whereas much higher values are found for millimolar concentrations.[2,6] This is reminiscent of the controversial "Jones-Ray experiments" of the 1930s, wherein 13 salts were found to exhibit negative surface tension increments (and thus positive ion adsorption) at sub-millimolar concentrations, while exhibiting maxima at ~ 1 millimolar before becoming positive and behaving normally (~proportional to salt concentration) in the molar region.[6]

We have analyzed these Jones-Ray data with the simple adsorption models used in our UV-SHG experiments, showing them to exhibit free energy changes that are ~10x larger than our "usual" molar concentration values.[2,6] Another study in the molar region surprisingly revealed very different behavior for the adsorption of $NaNO_3$ and $NaNO_2$ salts. While *nitrate* exhibited "normal" behavior, following the independent adsorption model used for all systems we have studied in the molar region,[7] sodium *nitrite* data could not be fit with that model, but instead fit well to an ion pair adsorption model, with a much higher free energy change.[8]

As described above, our temperature dependent studies of thiocyanate and associated calculations establish that it is the displacement of weakly interacting water molecules (from the interface and the ion solvation shells) into the strongly hydrogen bonding bulk solution that is the driving force for the electrostatically unfavorable ion adsorption to the air-water interface. There are competing models for this driving force, including a large surface field produced by the asymmetric interfacial distribution of water dipoles.[9] Future studies will clarify the nature of this vitally important process.

Cation-Cation pairing in aqueous solutions

Another surprising phenomenon that is seemingly forbidden by electrostatics has recently been shown to be enabled by a similar displacement of weakly interacting waters into the more strongly interacting bulk-formation of contact pairs between two *positively* charged ions.[10] In this case, guanidinium ions, which form strong N-H donor hydrogen bonds with water in the plane of the molecule, but form weak acceptor bonds with water into the pi electron network perpendicular to the plane, expel the weakly interacting waters as they are drawn together by fluctuations and strong dispersion forces. The net interaction is favorable for the cations to form a stacked contact pair, in defiance of the like-charge repulsion rule. This finding, which was predicted by several theoretical calculations and verified by our recent x-ray spectroscopy experiments and first principles' theory (with David Prendergast), may have important general implications for biological systems, since guanidinium is the side chain of the amino acid arginine, which is known to form unusual arrangements in polypeptides, and other biological molecules have structures similar to guanidine.

Water evaporation

The evaporation of water is likewise a subject of both consequence and controversy. The competition between evaporation and condensation determines the particle size distribution in clouds, which, in turn, determines their optical properties-viz. absorption, transmission, reflection, and scattering of sunlight. Accordingly, there has been much effort directed at characterizing the kinetics of evaporation (and condensation). The standard description of evaporation proceeds in terms of the simple Hertz-Knudtson equation for the maximum rate of collision of a gas molecule with a surface. Equilibrium (reversibility) is assumed between condensation and evaporation, and the inevitable molecular-level complexity is embodied in the "fudge factors" called the sticking coefficient (for condensation) or evaporation coefficient (for evaporation). Many experiments have addressed the determination of the evaporation coefficient, with published results spanning three orders of magnitude. The most recent determinations cover a more restricted range of 0.05 to 1.0, however, while most cloud models employ a value of unity.

We have combined liquid microjet technology with Raman thermometry for measuring evaporation coefficients.[11-14] By simply employing jet diameters that are smaller than the mean free path of evaporating molecules from the jet in a vacuum, collisions between gaseous molecules and between evaporating molecules and the liquid jet are minimized, and evaporation rates free from contamination by attending condensation, ostensibly the principal uncertainty in previous determinations, are easily obtained. With this approach, we (with Ron Cohen) obtained an evaporation coefficent of 0.62 +/-0.09 for H_2O and 0.57+/-0.06 for D_2O (independent of temperature over the range [245-298K], implying the presence of a small barrier in the evaporation reaction coordinate. The change from the value of unity used in the cloud models to our value of 0.6 is not expected to cause any dramatic changes in the models, whereas values of 0.1 or less would indeed engender significant effects.

In addition to determining an accurate value for use in climate models, there is also much interest in establishing the detailed molecular mechanism by which water molecules evaporate from the liquid surface. While this has been attempted via a number of molecular dynamics simulations, the results have been unconvincing and uncertain. The underlying reason for this is the fact that the evaporation event is a very rare occurrence (ca. one event from a 1nm^2 area every 10 nanoseconds) and thus difficult to model. Very recently, Varrily *et al.* addressed this problem with the use of Chandler's powerful Transition Path Sampling (TPS) approach for simulating rare events.[15] Two salient results emerged from this study: 1) no barriers were evident in the evaporation coordinate, implying an evaporation coefficient of unity 2) the critical feature of the reaction coordinate was found to be the amplitude of the thermally excited capillary waves on the liquid surface. Water molecules evaporated when an anomalously large amplitude capillary appeared (as a rare event) due to constructive interference of the many thermally excited wavelengths. The appearance of such a large amplitude protrusion weakened the water hydrogen bonds, allowing the molecule to dissociate from other interfacial molecules. The relatively small difference between the current experimental value of the evaporation coefficient (0.6) and the TPS result (unity) remains to be resolved, but could be due to factors such as quantum effects not considered in the simulations, the use of an effective water model (SPC-E), or to aspects of the experiment and interpretation that were not considered in detail, e.g. extrapolation of the cooling curve in the supercooled region. Meanwhile, essentially the same experimental value of the evaporation coefficient has been obtained for 3M solutions of ammonium sulphate (the dominant inorganic component of atmospheric aerosols),[16] while a 25% reduction in this value was found for molar solutions of sodium perchlorate-comprising a highly surface active anion.[17]

Water Clusters in the Atmosphere: Contributions to the "Water Continuum"

Water is the third most abundant molecule in the atmosphere and the principal absorber of both incoming sunlight and reradiated blackbody radiation. However, models of atmospheric absorption that only take into account the water monomer rotational and vibration-rotation transitions fail to match with measurements of the atmosphere's absorption spectrum.[18] Water continuum absorption extending from the microwave to the infrared was first discovered in the microwave region by Hettner in 1918 and elaborated by Elsasser shortly afterward, and has since been studied by many groups, but still no consensus exists as to its origin. It has been determined that for ambient atmospheric conditions, this continuum absorption scales quadratically with the H_2O number density and has a strong, negative temperature dependence. Three possible mechanisms have been proposed to account for these observations: Far wing absorption of allowed monomer transitions, absorption by water dimers, and collision-induced absorption. No consensus exists as to which mechanism is primarily responsible [http://www.met.reading.ac.uk/caviar/water_continuum.html], and the recent detailed studies by Leforestier and coworkers have elucidated the difficulties in ascertaining the origin of the continuum.[19-22] However, they were able to deduce some important conclusions based on the new information available, as we discuss below.

The possibility that water dimers and perhaps larger water clusters act as sources of excess absorption has been intensely debated for over four decades, with recent work suggesting that water dimers are indeed involved in both the radiation balance and chemistry of the atmosphere.[23] However, definitive detection of water dimers in the atmosphere has remained elusive. Towards the goal of elucidating the abundance and the spectroscopy of dimers in the atmosphere, we (with Claude Leforestier) have employed successively more accurate and sophisticated water dimer potential surfaces and computational methods that utilize the available terahertz and IR data to compute the dimer equilibrium constant, that in turn permits abundance and spectra to be calculated as a function of atmospheric conditions (temperature, pressure, humidity).[24-26] Qualitatively, these results do indeed indicate that the dimer abundance can be high enough to engender important effects, as proposed, especially at high relative humidity.

Using the new potential surfaces and spectroscopic information, Leforestier and coworkers were able to deduce some important information regarding the origin and properties of the water continuum.[19-22] In particular, they could assess the relative importance of each of the three proposed continuum mechanisms.[22] In the 800–1150 cm^{-1} region, which is the window between the pure rotational band and the bending vibrational band, contributions from far wings of allowed H_2O lines are the dominant source of the continuum, whereas contributions from dimer absorption are small or even negligible, and those from the collision induced absorption are negligible. In the "pure rotational band" between 30 and 500 cm^{-1}, the calculated dimer absorption could account for approximately 20%–35% of the predicted self-continuum at room temperature and even more at lower temperatures. Above 500 cm^{-1} dimer contributions fall off very rapidly as the frequency increases. In the microwave region, dimer absorptions could make even larger contributions to the self-continuum. The three continuum mechanisms were found to have completely different temperature dependences. The dimer spectra exhibit a very strong negative T dependence, the far-wing theory exhibits a moderately strong negative dependence, while collision induced absorption has a weak and predominantly negative dependence. The nature of these temperature dependences was also generally complicated.

Electrokinetic Energy Conversion and Hydrogen Generation from Liquid Water Microjets

The exploitation of water as an essentially limitless source of clean energy has been a perennial dream of scientists and engineers- and science fiction writers- particularly noting the "Cold Fusion" debate of two decades past. On a more modest scale, we have found that by simply squirting water through a small hole under the proper conditions, it is possible to generate both electrical power and gaseous hydrogen. Building on the work of Faubel and coworkers,[27] we investigated the use of liquid microjets for "electrokinetic energy conversion", finding that 10-20% of the kinetic energy of flowing water could be easily converted to electrical power, with the accompanying generation of hydrogen.[28,29] No serious attempts were made to optimize the efficiency of either hydrogen generation or power conversion in these initial experiments that used the same liquid microjet metal orfice apparatus employed in our initial X-ray experiments.[30] However, accompanying modeling supported the intuition that electrokinetic current generation resulted from the overlap of the interfacial charge distribution with the hydrodynamic velocity profile, assuming that hydroxide ions initially adsorbed to the metal nozzle, with hydrated protons forming a mobile counterion layer, such that the hydrodynamic flow swept more protons than hydroxide out of the nozzle region. A patent was filed for this process, with suggestions for improving the efficiency [patent: *Method and Apparatus for Electrokinetic Co-Generation of Hydrogen and Electric Power from Liquid Water Microjets,* Inventor(s): Richard J. Saykally; Andrew M. Duffin; Kevin R. Wilson; Bruce S. Rude].

Towards a Predictive Universal Model for Water

The essential missing ingredient necessary (but perhaps still not sufficient) for a definitive resolution of the above issues and the many other unresolved "mysteries" of water is a model capable of describing water in all its forms and under all conditions with predictive accuracy. While literally hundreds of water models have been developed, and some of them can describe a considerable number of properties under a limited ranges of conditions, none come close to the "universal first principles' "model described above, despite impressive recent progress. For over two decades, our group has pursued this goal (with Claude Leforestier) via the highly detailed study of small water clusters by Terahertz Vibration-Rotation-Tunneling (VRT) spectroscopy.[31-41] The plan is to understand every detail of the water dimer, trimer, tetramer, pentamer, etc through the combination of terahertz VRT spectroscopy and state of the art quantum chemistry. Initially, we refined Stone's highly detailed (72 parameter) ASP model for the dimer via least squares regression analyses of VRT spectral parameters(intermolecular vibrational frequencies, rotational constants, tunneling splittings, ...), with the philosophy being that the dimer contains the most essential components of a complete water potential energy surface, since the latter is dominated by pairwise intermolecular interactions, while the trimer, tetramer, etc will provide quantification of the three-body, four-body, etc terms that account for the famous cooperative nature of hydrogen bonding in condensed water that underlies its unusual behavior. The subsequently refined models (called VRT-ASP) were shown to reproduce the structures of the small water clusters as well as the properties of liquid water reasonably well,[42,43] but ultimately require a full quantum mechanical treatment of the dynamics with this very complex intermolecular potential energy surface, which is expressed in molecular coordinates. This is currently beyond the state of the art for the liquid.

Recently, important advances have been made, both in the ab initio description of water potential surfaces and in the treatment of the VRT dynamics in the small water clusters. Several groups have reported new ab initio-based water models that yield improved descriptions of both clusters and bulk properties.[44,45] Leforestier and coworkers[46] have developed a new formalism for computing the VRT dynamics of clusters with a relatively simple model, but requiring monomer flexibility. This enables calculation of redshifts in the OH bond vibrations from a global potential surface, which is a difficult problem.[47] Using the flexible monomer CCpol-8sf potential of Salewicz *et al.*, excellent agreement with Terahertz VRT spectra was achieved for the dimer.[48] Van der Avoird, Leforestier, and Salewicz describe the progress towards a universal first principles' water model in a recent review, and Bowman and co-workers have developed a new model that also yields impressive agreement with experiments [*J. Chem. Phys.* 134, 094509 (2011)].[49] Finally, a dramatic advance in microwave spectroscopy of water clusters was recently reported by the Brooks Pate group,[50] who were able to measure microwave rotation-tunneling spectra for several low-energy forms of the water hexamer, heptamer, and nonamer. Rapid progress is indeed being made toward the stated goal of this section, which will ultimately enable the resolution of many "aqua incognita" issues by a straightforward calculation.

Acknowledgements

The Terahertz spectroscopy projects on water clusters are supported by the Experimental Physical Chemistry Program of the National Science Foundation. The synchrotron X-ray spectroscopy, evaporation, electrokinetic energy conversion studies, and the nonlinear laser spectroscopy efforts are supported by the Basic Energy Sciences Division of the Department of Energy through the Lawrence Berkeley National Laboratory.

References

1. Petersen, P. B.; Saykally, R. J. *Ann. Ref. Phys. Chem.* **2006**, *57*, 333-364.
2. Petersen, P. B.; Saykally, R. J. *J. Phys. Chem. B* **2006**, *110*, 14060-14073.
3. Baer, M. D.; Mundy, C. J. *J. Phys. Chem. Lett.* **2011**, *2*, 1088-1093.
4. Otten, D. E.; Shaffer, P.; Geissler, P.; Saykally, R. J. *Proc. Natl. Acad. Sci. USA* **2012**, *109*, 701-705.
5. Onorato, R. M.; Otten, D. E.; Saykally, R. J. *J. Phys. Chem. C* **2010**, *114*, 13746-13751.
6. Petersen P. B.; Saykally, R. J. *J. Am. Chem. Soc.* **2005**, *127*, 15446-15452.
7. Otten, D. E.; Petersen, P. B.; Saykally, R. J. *Chem. Phys. Lett.* **2007**, *449*, 261-265.
8. Otten, D. E.; Onorato, R.; Michaels, R.; Goodnight, J.; Saykally, R. J. *Chem. Phys. Letters* **2012**, *519-520*, 45-48.
9. Vácha, R.; Marsalek, O.; Willard, A. P.; Bonthuis, D.J.; Netz, R. R.; Jungwirth, P. *J. Phys. Chem. Lett.* **2012**, *3*, 107-111.
10. Shih, O.; England, A.; Dallinger, G.; Smith, J.; Duffey, K.; Cohen, R.; Prendergast, D.; Saykally, R. J. *J. Chem. Phys.* **2013**, *139*, 035104.
11. Cappa, C. D.; Drisdell, W.; Smith, J. D.; Saykally, R. J.; Cohen, R. C. *J. Phys. Chem. B* **2005**, *109*, 24391-24400.
12. Smith, J. D.; Cappa, C. D.; Drisdell, W. S.; Cohen, R. C.; Saykally, R. J. *J. Am. Chem. Soc.* **2006**, *128*, 12892-12898.
13. Cappa, C. D.; Smith, J. D.; Drisdell, W. S.; Saykally, R. J.; Cohen, R. C. *J. Phys. Chem. C* **2007**, *111*, 7011-7020.

14. Drisdell, W. S.; Cappa, C. D.; Smith, J. D.; Saykally, R. J.; Cohen, R. C. *Atmos. Chem. Phys. Discuss.* **2008**, *8*, 6699-6706.
15. Varilly, P.; Chandler, D. *J. Phys. Chem. B* **2013**, *117*, 1419-1428.
16. Drisdell, W. S.; Saykally, R. J.; Cohen, R. C. *Proc. Natl. Acad. Sci. USA* **2009**, *106*, 18897-1890.
17. Drisdell, W. S.; Saykally, R. J.; Cohen, R. C. *J. Phys. Chem. C* **2010**, *114*, 11880-11885.
18. Duffey, K.C.; Shih, O.; Wong, N.L.; Drisdell, W.S.; Saykally, R.J.; Cohen, R.C. *Phys. Chem. Chem. Phys.* **2013**, *15*, 11634-11639.
19. Saykally R. J. *Physics* **2013**, *6*, 22.
20. Scribano, Y.; Leforestier, C. *J. Chem. Phys.* **2007**, *126*, 234301.
21. Ma, Q.; Tipping, R. H.; Leforestier, C. *J. Chem. Phys.* **2008**, *128*, 124313.
22. Ma, Q.; Tipping, R. H.; Leforestier, C. *J. Chem. Phys.* **2010**, *132*, 164302.
23. Anglada, J. M.; Hoffman, G. J.; Slipchenko, L. V.; Costa, M. M.; Ruiz-Lopez, M. F.; Francisco, J. S. *J. Phys. Chem. A* **2013**, *117*, 10381-10393.
24. Goldman, N.; Fellers, R. S.; Leforestier, C.; Saykally, R. J. *J. Phys. Chem. A* **2001**, *105*, 515-519.
25. Goldman, N.; Leforestier, C.; Saykally, R. J. *J. Phys. Chem. A* **2004**, *108*, 787-794.
26. Scribano, Y.; Goldman, N.; Saykally, R. J.; Leforestier, C. *J. Phys. Chem. A* **2006**, *110*, 5411-5419.
27. Faubel, M,; Steiner, B. *Ber. Bunsen-Ges., Phys. Chem. Chem. Phys.* **1992**, *96*, 1167.
28. Duffin, A. M.; Saykally, R. J. *J. Phys. Chem. C* **2007**, *111*, 12031-12037.
29. Duffin, A. M.; Saykally, R. J. *J. Phys. Chem. C* **2008**, *112*, 17018-17022.
30. Nakayama, Y.; Pauzauskie, P. J.; Radenovic, A.; Onorato, R. M.; Liang, W.; Saykally, R. J.; Liphardt, J.; Yang, P. *Nature* **2007**, *447*, 1098-1102.
31. Busarow, K. L.; Cohen, R. C.; Blake, G. A.; Laughlin, K. B.; Lee, Y. T.; Saykally, R. J. *J. Chem. Phys.* **1989**, *90*, 3937-3943.
32. Saykally, R. J.; Blake, G. A. *Science* **1993**, *259*, 1993, 1570-1575.
33. Braly, L. B.; Cruzan, J. D.; Liu, K.; Fellers, R. S.; Saykally, R. J. *J. Chem. Phys.* **2000**, *112*, 10293-10313.
34. Braly, L. B.; Liu, K.; Brown, M. G.; Keutsch, F. N.; Fellers, R. S.;

Saykally, R. J. *J. Chem. Phys.* **2000**, *112*, 10314-10326.

35. Fellers, R. S.; Leforestier, C.; Braly, L. B.; Brown, M. G.; Saykally, R. J. *Science* **1999**, *284*, 945-948.
36. Goldman, N.; Fellers, R. S.; Leforestier, C.; Saykally, R. J. *J. Phys. Chem. A* **2001**, *105*, 515-519.
37. Saykally, R. J. *Proc. Nobel Symp. 117 on "The Physics and Chemistry of Clusters"*, Visby, Sweden, E.E.B. Campbell and M. Larsson (eds.); 2001, pp. 206-218.
38. Keutsch, F. N.; Saykally, R. J. *Proc. Natl. Acad. Sci. USA* **2001**, *98*, 10533-10540.
39. Goldman, N.; Fellers, R. S.; Brown, M. G.; Braly, L. B.; Keoshian, C. J.; Leforestier, C.; Saykally, R. J. *J. Chem. Phys.* **2002**, *116*, 10148.
40. Leforestier, C.; Fellers, R. S.; Saykally, R. J. *J. Chem. Phys.* **2002**, *117*, 8710-8722.
41. Richardson, J. O.; Wales, D. J.; Althorpe, S. C.; McLaughlin; Shih, O.; Saykally, R. J. *J. Phys. Chem. A.* **2013**, *117*, 6960-6966.
42. Keutsch, F. N.; Goldman, N.; Harker, H. A.; Leforestier, C.; Saykally, R. J. *Mol. Phys.* **2003**, *101*, 3477-3492.
43. Petersen, P. B.; Saykally, R. J. *Chem. Phys. Lett.* **2004**, *397*, 51-55.
44. Hamm, P.; Fanourgakis, G. S.; Rao, R. F.; Xantheas, S. S. *J. Chem. Phys.* **2013**.
45. Babin, V.; Medders, G. R.; Paesani, F. *J. Phys. Chem. Lett.* **2012**, *3*, 3765-3769.
46. Leforestier, C.; van Harrevelt, R.; van der Avoird, A. *J. Phys. Chem. A* **2009**, *113*, 12285-12294.
47. Leforestier, C. *Philos. Trans. A Math. Phys. Eng. Sci.* **1968**, *370*, 2675-2690.
48. Leforestier, C.; Szalewicz, K.; van der Avoird, A. *J. Chem. Phys.* **2012**, *137*, 014305.
49. Szalewicz, K.; Leforestier, C.; van der Avoird, A. *Chem. Phys. Lett.* **2009**, *482*, 1-14.
50. Liu, K.; Cruzan, J. D.; Saykally, R. J. *Science* **1996**, *271*, 929-933.

XIII

Helen Keller problem: tactile texture of water isn't necessarily favourable

Yoshimune Nonomura,[†] Rina Saito,[†] Takashi Maeno[‡]*

[†]: Department of Biochemical Engineering, Graduate School of Science and Engineering, Yamagata University, 4-3-16, Jonan, Yonezawa, Japan.
nonoy@yz.yamagata-u.ac.jp

[‡]: Graduate School of System Design and Management, Keio University, 4-1-1 Hiyoshi, Kohoku-ku, Yokohama, Japan.

1. Introduction

The mechanism of water perception is an interesting topic because obtaining this liquid is one of the most important activities of life.[1-5] Almost all studies have focused on water perception by taste; however, humans can also recognize water by their tactile sense. For example, Helen Keller described water as 'the wonderful cool something' when Miss Sullivan poured water on her hands.[6] On the basis of a study that compared the perception of liquidity and solidity, the tactile texture of water changes with temperature.[7] Kajimoto and co-workers found that mechanical stress on skin hair plays a major role in the perception of a water surface.[8] Guest *et al.* showed that fluids were perceived as more watery if high normal and tangential vibrations and less friction were applied on a skin surface.[9]

The point of issue for understanding the physical origin of the tactile texture of water is the clarification of frictional phenomena on wet human skin because tactile texture is caused by factors such as roughness, hardness and friction.[10-12] Comaish *et al.* evaluated friction forces on skin surface and found that the behaviour of friction forces does not follow the simple Amonton's laws, but follows a more complex power law $F = \mu W^n$, in which F = friction force, W = vertical force and μ and n are constants.[13] This is probably because skin is an object that exhibits viscoelastic rather than purely plastic deformation.[14] Application of water drastically changes frictional properties of skin. For example, water swells human skin, increases the contact area and friction force between skin and solid surfaces and causes the stick–slip phenomenon, which is the spontaneous jerking motion that can occur while two objects are sliding over each other.[15] Nacht et al. showed that the application of oils such as petrolatum, glycerin and baseline increased skin friction with skin hydration.[16] On the basis of the frictional measurement on a human finger, André et al. reported that fingertip moisture is optimally modulated during object manipulation.[17]

In previous studies, we have evaluated the tactile texture and frictional properties of water and oils on solid surfaces to show the mechanism of water perception by touch. The results of frictional analysis predicted that the stick–slip feel is caused by a drastic change in frictional resistance.[18,19] High-speed observation and finite-element analysis predicted that periodic shears with acceleration seven times greater than gravitational acceleration and characteristic activation of tactile receptors occurred during the application process.[20] When such stimuli were applied to fingertips by an ultrasonic vibrator, a water-like tactile texture was perceived by some subjects, even though no liquid was present between the fingertip and vibrator surface. In these previous studies, we have succeeded in building a macroscopic model of water perception process by touch.

In this study, we evaluated the tactile texture (i.e. favourability of touch), frictional properties and fingertip movement when 20 subjects applied small amounts of water and five thickener aqueous solutions to the artificial skins that mimicked the structure and mechanical properties of human skin.[21] The technical improvements in the present study are as follows. First, a novel tactile evaluation system, in which a friction evaluation unit and a high-speed camera were connected through a data logger, was developed to achieve simultaneous evaluation of tactile feel, friction on human skin and moving behaviour of a fingertip (Figures XIII.1(a) and (b)). Second, the subjects executed our original task, i.e. a two-step contact method (Figure XIII.1(c)). In

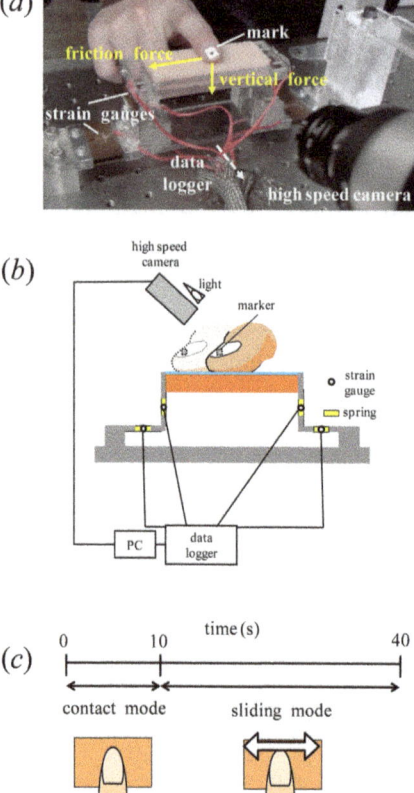

Figure XIII.1. Tactile and friction evaluation method. (a, b) Image and setup of a friction evaluation system that simultaneously evaluated tactile sensation, friction properties and moving behaviour of a fingertip. (c) Task of the two-step contact method consisting of the contact and sliding modes.

the first step, subjects placed down their forefinger on the artificial skin from a finger holder at a height of 2 cm and held it there for 10 s. In the second step, they moved their forefinger side to side for 30 s and applied liquid materials on the artificial skin. This original task effectively achieves active touch under controlled conditions and removes the effect of differences among individuals.

2. Tactile texture of water.

A favourability score for tactile texture was evaluated when 20 subjects applied 0.1 mL of water and 0.15 ~ 5.00 wt% thickener solutions to the artificial skin. Figure XIII.2 shows the favourability scores for the six liquids. The favourability score of water was 3.4 ±

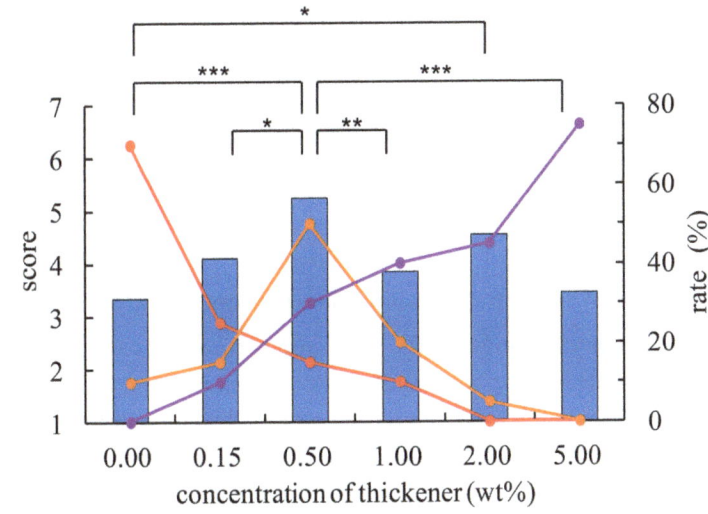

Figure XIII.2. Tactile feel of water and thickener aqueous solutions on artificial skin: bar chart; favourability score of water and thickener aqueous solutions. All data are averages of all subjects (± standard deviation). Line chart: incidence rates of stick–slip (red), slimy (purple) and smooth feel (orange).

1.5, which was the lowest among the six liquids. The score increased in proportion with thickener concentration and attained the maximum value of 5.3 ± 1.2 when the concentration was 0.50 wt%. The score for the liquid containing excessive thickener was also low; the score for 5 wt% thickener solution was 3.5 ± 1.7. The statistical significance was measured between the scores and we predicted that the p-values between the score for 0.5 wt% and water (or 5.00 wt% solution) were less than 0.01. We analysed the comments that the subjects gave as reasons for their evaluations. On the touch of water, 0.50 and 5.00 wt% thickener solutions, many subjects commented 'stick–slip feel', 'smooth feel' and 'slimy feel', respectively. Figure XIII.2 shows the incidence rate of these tactile feel for each liquid. The favourability score was linked to the incident rate of smooth feelings; the correlation coefficient between these two scores had the highest value (0.75), indicative of a positive correlation. The score of stick–slip (slimy) feel decreased (increased) with increasing thickener concentration. Surprisingly, these results predict that the tactile texture of water is not necessarily favourable because of its strong stick–slip feel. Many subjects favoured the liquid with smooth tactile texture such as 0.50 wt% thickener solution.

3. Frictional stimuli in the application process of water.

What type of friction induces the stick–slip feel, which causes water to be unfavourable? In previous studies, we have shown that water creates a stick–slip feel when a small amount is rubbed using a fingertip on an artificial skin or glass.[18-20] The results of frictional analysis predicted that this stick–slip feel is caused by a drastic change in frictional resistance. In the present study, we analysed friction profiles in the sliding mode to find the mechanical origin of the tactile texture of water. Figure XIII.3(a) shows the average of the friction coefficient in the sliding process and the incidence ratio of the drastic change in 20 subjects. Both the coefficient and ratio were the highest for water

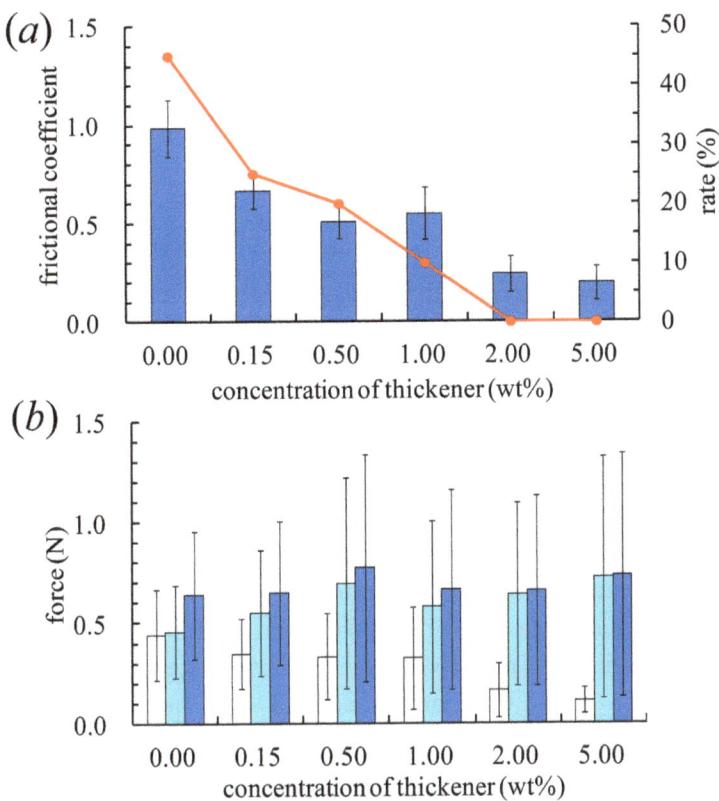

Figure XIII.3. Frictional stimuli in the sliding mode of water and thickener solutions. (a) Average of the friction coefficient (bar charts) and the incidence ratio of the drastic change in frictional resistance (red line). (b) Average friction force F (white), vertical force W (light blue) and their resultant force. $\sqrt{F^2+W^2}$ (blue).

in the six liquids, 0.98 ± 0.15 and 45%, respectively. They decreased with increasing thickener concentration and were linked to the score of stick–slip feel, shown in Figure XIII.2 and 3(a). The correlation coefficient between the incidence rate of the stick–slip feel and friction coefficient (the incidence ratio of the drastic change) were 0.94 (0.96), indicative of strong positive correlations. These results were similar with those noted in our previous studies.[18-20]

In the present analysis, we found some significant trends in the data concerning mechanical forces applied on the fingertips of the subjects. Figure XIII.3(b) shows the average friction force F, vertical force W and their resultant force $\sqrt{F^2+W^2}$ in the sliding mode. The average friction force F decreased with increasing thickener concentration, and the average vertical force W increased 1.5 times, 0.45 N for water and 0.73 N for 5.00 wt% thickener solution. The statistical significance predicted that the p-values between F of water and that of 2.00 or 5.00 wt% aqueous solutions were less than 0.01. In consequence of the decrease in F and the increase in W, their resultant forces were almost constant values, approximately 0.7 N, regardless of the thickener concentration. There is a clear reason why F decreased with the increase in thickener concentration: some films of thixotropic liquid induce a lubrication effect.[22-24] On the other hand, the mechanism of the increase in W is non-trivial: the feedback system, which senses the situation on the basis of the external force and control W, is necessary to induce such systematic change. The results shown in Figure XIII.3(b) predict that our neurological system modulates W to converge $\sqrt{F^2+W^2}$ to a steady value.

4. Tribological analysis of the water application process.

To quantify the contact condition between the finger surface and the artificial skin covered with water or thickener solutions, we analysed the change of F with W during the sliding mode. Figure XIII.4(a) shows the relationship between F and W for water and five thickener aqueous solutions. The log–log plot for water shows positive tendency, which could follow the power law $F = \mu W^n$. Here, the proportional constant μ and multiplier n are 0.89 and 0.91, respectively. The results of the power-law fittings agree with that published by Adams *et al.* when wet human skin was rubbed by glass or polypropylene balls: $\mu = 1.15$–2.84 and $n = 0.83$–0.85.[25] These parameters decreased with increasing thickener concentration: $\mu = 0.39$ and $n = 0.27$ for 0.5 wt% solution and $\mu = 0.17$ and $n = -0.17$ for 5 wt% solution. These results are indicative of the fact that the contact condition between the fingertip and wet artificial skin changed with the addition of thickener to water.

Friction behaviour on solid substrate changes with the thickness of the lubricant layer, which is reflected in the Sommerfeld number: viscosity η multiplied by the sliding velocity v divided by the vertical force W. Figure XIII.4 shows the Stribeck curves, i.e. the plot of the friction coefficient every 2 ms against the Sommerfeld number, obtained in the sliding mode for water and five thickener solutions. The friction evaluation system and high-speed camera were used to evaluate the mechanical forces (F and W) and the sliding velocity v. In these curves, when the effects of increased speed, increased viscosity or reduced vertical force are shown, there is a rightward shift on the horizontal axis.[26] The combination of low speed, low viscosity and high load will produce boundary lubrication, characterized by a small quantity of fluid at the interface, large surface contact and high friction. As the speed and viscosity increase or the vertical force decreases, the surfaces begin to separate and a fluid film begins to form. Mixed lubrication (caused by the fluid film) is the result, and is easily seen on the Stribeck curve as a sharp drop in friction coefficient. The drop in friction is a result of decreasing surface contact and more fluid lubrication. The friction coefficient will reach its minimum and there is a transition to hydrodynamic lubrication. At this point, the vertical force on the interface is entirely supported by the fluid film. There is low friction and no wear in hydrodynamic lubrication since there is a full fluid film and no solid–solid contact. In Figure XIII.4, the average friction coefficient decreased from 0.9 (water) to 0.2 (5 wt% thickener solution) with the increase of thickener concentration, and the average of the Sommerfeld number increased from 2×10^{-5}

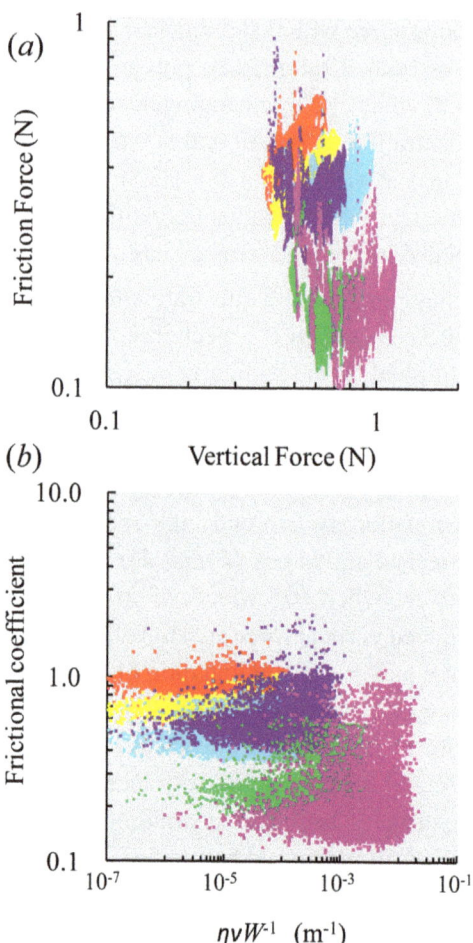

Figure XIII.4. Friction analysis of the water application process: relationship between friction and vertical forces (a) and Stribeck curves (b) for water (red), 0.15 (yellow), 0.50 (purple), 1.00 (blue), 2.00 (green) and 5.00 (pink) wt% thickener solutions.

m^{-1} to 4×10^{-3} m^{-1}. The Sommerfeld number and friction coefficient were 10^{-7}–10^{-3} m^{-1} and approximately 1 for water, and they were 10^{-5}–10^{-1} m^{-1} and approximately 0.2 for 5 wt% thickener aqueous solution.

The Sommerfeld number (friction coefficient) increased (decreased) with increasing viscosity. These results predict that boundary regions between the finger and artificial skin surfaces are in a mixed lubrication state and the fluid film forms with increasing thickener concentration.

5. Why is the tactile texture of water unfavourabile?

The present psychophysical experiments show that the tactile texture of water is not necessarily favourable. The favourability score of water was the lowest among the six liquids. The unfavourable score of water was caused by characteristics in smooth, stick–slip and slimy feel. Smooth feel is one of the most important structural factors of tactile feel. Hollins *et al.* and Yoshioka *et al.* evaluated tactile feel of a wide variety of solid materials.[10,11] Okamoto *et al.* analysed 17 previous psychophysical experiments and determined a common structure for these dimensions.[12] In all of the reports, researchers concluded that roughness/smoothness factor is one of the most prominent psychophysical dimensions for tactile textures. Previous studies have also shown that the smooth feel is favourable for many people; some experimental investigations on fabric materials suggested soft and smooth materials as pleasant, and those that were stiff, rough or coarse as unpleasant.[27-29] Both stick–slip and slimy feel were also evoked when we touched liquid materials. The authors showed that the stick–slip feel was remarkably induced when we touched water.[18,19] To our knowledge, the present paper is the first report on the effect of this feel on favourability, although there are a few reports on the effect of slimy feel on favourability. For example, Guest *et al.* showed that pleasant and harsh were related to textured and silken and were weakly associated with viscous feel.[9]

The origins of some tactile feel—smooth, stick–slip and slimy—can be described by friction phenomena on wet skin. Water swells human skin and increases the contact area and friction force between skin and solid surfaces.[24,30] Skin hydration causes tenderization of the stratum corneum because water is an efficient plasticizer of the proteins consisting of the stratum corneum.[31] Evans and Hyde predicted that swelling of the stratum corneum beyond the free volume accessible to the straightened rod packing of keratin fibers will lead to dramatic weakening, since further swelling of the pattern can only occur by losing contacts between fibres.[32] This tenderization of the stratum corneum can significantly increase the friction on human skin. The smooth and rigid sphere in contact with a smooth elastic planar surface was considered as a model for the friction of wet skin. When wet skin surface is rubbed by a spherical probe, the friction coefficient is described by the following equation:[25]

$$\mu = \pi \tau_0 \left(\frac{3R}{4E^*}\right)^{\frac{2}{3}} \frac{1}{W^{\frac{1}{3}}}$$

in which t_0 is the intrinsic interfacial shear strength, R is the radius of the sphere, W is the vertical force and E^* is $\left(\frac{1-p_1^2}{E_1} + \frac{1-p_2^2}{E_2}\right)^{-1}$. Here, E and p are Young's modulus and Poisson's ratio, respectively, and the subscripts 1 and 2 refer to the sphere and skin, respectively. When the stratum corneum is immersed in water, the increase in moisture content and the considerable reduction in Young's modulus occur rapidly.[33] Tensile experiments show that Young's modulus decreases from 1 GPa at 50 % RH to 3 MPa in water.[34] The reduction in the elastic modulus would be expected to lead to extensive smoothing of wet skin frictional contacts. Next, we consider the mechanism of the drastic changes of friction coefficient during the application of water. In the case of water, a sufficiently thick fluid film does not form between the fingertip and artificial skin because of the lowest viscosity (approximately 1 mPa s) and hv/W (10^{-7}–10^{-3} m^{-1}). The rupture in the application process can induce the stick–slip phenomenon. Recently, Nakano et al. reported similar tribological phenomena: the friction coefficients were scattered in a wide range without any regularity when the amount of the liquid material was insufficient for forming a fluid film.[35] On the other hand, in the case of thickener solution, these oscillating phenomena are inhibited because of the formation of a complete fluid film.

The friction phenomena observed in the present study contain interesting findings on our subconscious tactile behaviour. As shown in Figure XIII.3(b), the friction force F and vertical force W decreased and increased with increasing thickener concentration, and their resultant forces were constant values regardless of the thickener concentration. These results predict a feedback control system, which reduces the vertical force in the water application process with the stick–slip phenomenon. One of the most important sensory feedback mechanisms for manipulations is the cutaneous feedback from our fingertips.[36,37] Cutaneous afferent feedback provides critical information regarding the current state of the system and enables the central nervous system to modify the motor plan accordingly. Different types of cutaneous receptors provide feedback about the shape, texture, pressure, temperature and other physical properties of objects. Almost all previous studies on feedback behaviour were contact process with solid materials; there are few reports on the contact with liquids or wet surfaces. The present study is the first report on our unconscious feedback behaviour when we contact a wet surface.

6. Conclusions

In the present study, we show that tactile texture of water is not necessarily favourable. This unfavorability was caused by the non-smooth and stick–slip feel, which reflect the increase of friction coefficient and their drastic changes on wet skin. The friction data show that the rupture of a fluid film in the mixed lubrication state can induce the stick–slip phenomenon. The frictional data predict a feedback control system, which reduces the vertical force in the water application process with the stick–slip phenomenon and enhances the force in the application process of the thickener aqueous solutions with lubricant properties. The analysis of friction dynamics predicts that the feedback control occurs in stage 3 of the sliding mode; the vertical force is controlled based on the friction force in the process of reciprocal motion of a fingertip on the second timescale. These findings add to our understanding of the emotional arousal induced on contact with water, which is the main component of the human body and an essential material for life. In day-to-day life, we frequently perceive water through our tactile senses. These findings will be useful when designing virtual reality systems to mimic the sensation of the texture of liquids.

References

1. Zotterman, Y.; Diamant, H. *Nature* **1959**, *183*, 191–192.
2. Bartoshuk, L. M.; Harned, M. A.; Parks, L. H. *Science* **1971**, *171*, 699–701.
3. Liu, L.; Li, Y.; Wang, R.; Yin, C.; Dong, Q.; Hing, H.; Kim, C.; Welsh, M. J. *Nature* **2007**, *450*, 294–299.
4. Cameron, P.; Hiroi, M.; Ngai, J.; Scott, K. *Nature* **2010**, *465*, 91–96.
5. Greif, S.; Siemers, B. M. *Nat. Comm.* **2010**, *1*, 107.
6. Keller, H. *The Story of My Life*, Signet Classic, 2002, Chapter 4.
7. Sullivan, A. H. *Am. J. Psychol.* **1923**, *34*, 531–541.
8. Sato, M.; Miyake, J.; Hashimoto, Y.; Kajimoto, H. Tactile perception of a water surface: contributions of surface tension and skin hair In *Haptics: generating and perceiving tangible sensations*, Springer, 6192, 2010, 58–64.
9. Guest, S.; Mehrabyan, A.; Essick, G.; Phillips, N.; Hopkinson, A.; Mcglone, F. *J. Texture Stud.* **2012**, *43*, 77–93.
10. Hollins, M.; Bensmaïa, S.; Karlof, K.; Young, F. *Atten. Percept. Psychophys.* **2000**, *62*, 1534–1544.
11. Yoshioka, T.; Bensamaïa, S. J.; Craig, J. C.; Hsiao, S. S. *Somatosens. Mot. Res.* **2007**, *24*, 53–70.
12. Okamoto, S.; Nagano, H.; Yamada, Y. *IEEE Transactions on Haptics* **2013**, *6*, 81–93.
13. Comaish, S.; Bottoms, E. *Br. J. Dermatol.* **1971**, *84*, 37–43.
14. Sivamani, R. K.; Goodman, J.; Gitis, N. V.; Maibach, H. I. *Skin Res. Technol.* **2003**, *9*, 227–234.

15. Highley, D. R.; Coomey, M.; DenBeste, M.; Wolfram, L. J. *J. Invest. Dermatol.* **1977**, *69*, 303–305.

16. Nacht, S.; Close, J. A.; Yeung, D.; Gans, E. H. *J. Soc. Cosmet. Chem.* **1981**, *32*, 55-65.

17. André, T.; Lefévre, P.; Thonnard, L. *J. Neurophysiol.* **2010**, 103, 402–408.

18. Nonomura, Y.; Fujii, T.; Arashi, Y.; Miura, T.; Maeno, T.; Tashiro, K.; Kamikawa, Y.; Monchi, R. *Colloids. Surf. B* **2009**, *69*, 264–267.

19. Nonomura, Y.; Arashi, Y.; Maeno, T. *Colloids. Surf. B* **2009**, *73*, 80–83.

20. Nonomura, Y.; Miura, T.; Miyashita, T.; Asao, Y.; Shirado, H.; Makino, Y.; Maeno, T. *J. Roy. Soc. Interface* **2012**, *9*, 1216-1223.

21. Saito, R.; Suzuki, M.; Maeno, T.; Mayama, H.; Nonomura, Y. *Trans. Soc. Inst. Control Eng.* 2014, 50, 2-8.

22. Boxer, C. J. *J. Phys. Chem.* **1948**, *52*, 1383–1390.

23. Hahn, S. J.; Ree, T.; Eyring, H. *Ind. Eng. Chem.* **1959**, *51*, 856-857.

24. Dintenfass, L. *Nature* **1963**, *197*, 496–497.

25. Adams, M. J.; Briscoe, B. J.; Johnson, S. A. *Tribol. Lett.* **2007**, *26*, 239–253.

26. Persson, B. N. J. *Sliding Friction: Physical Principles and Applications*, Springer, 2000, Chapter 7.

27. Major, D. R. *Am. J. Psychol.* **1895**, *7*, 57–77.

28. Ripin, R.; Lazarsfeld, P. F. *J. Appl. Psychol.* **1937**, *21*, 198–224.

29. Essick, G. K.; McGlone, F.; Dancer, C.; Fabricant, D.; Ragin, Y.; Phillips, N.; Jone, T.; Guest, S. *Neurosci. Biobehav. Rev.* **2010**, *34*, 192–203.

30. Gerhardt, L. C.; Strässlel, V.; Lenz, A.; Spencer, N. D.; Derler, S. *J. Roy. Soc. Interface* **2008**, *5*, 1317–1328.

31. Park, A. C.; Baddiel, C. B. *J. Soc. Cosmet. Chem.* **1972**, *23*, 3-12.

32. Evans, M. E.; Hyde, S. T. *J. Roy. Soc. Interface* **2011**, *8*, 1274-1280.

33. Kasting, G. B.; Barai, N. D. *J. Pharm. Sci.* **2003**, *92*, 1624-1631.

34. Park, A. C.; Baddiel, C. B. *J. Soc. Cosmet. Chem.* **1972**, *23*, 471-479.

35. Nakano, K.; Kobayashi, K.; Nakao, K.; Tsuchiya, R.; Nagai, Y. *Tribol. Int.* **2013**, *63*, 8–13.

36. Shim, J. K.; Karol, S.; Kim, Y. S.; Seo, N. J.; Kim, Y.; Kim, Y. S.; Yoon, B. C. *J. Biomechanics* **2012**, *45*, 415–420.

37. Johansson, R. S.; Hager, C.; Riso, R. *Exp. Brain Res.* **1992**, *89*, 192-203.

XIV
Recognition and patterning of molecules on water surface: reconstruction of some fundamental features of biomembrane

Toyoki Kunitake

Kitakyushu Foundation for the Advancement of Industry,
Science and Technology, and Kyushu University, Japan.
kunitake@ruby.ocn.ne.jp

Abstract

Molecular recognition is a process in which a given molecule is bound to another molecule in a highly specific way. It is an essential function to maintain living systems as living. An artificial counterpart of this process is molecular recognition at the air-water interface, since the biological molecular recognition is found most often at the cell surface. The unique nature of molecular interfaces are common in both systems. Our research in this field started by producing an artificial self-organising alternative of biomembrane from simple organic amphiphiles. It was then extended to ultrathin films and giant nanomembranes. As for interfacial molecular recognition, guanidinium monolayer spread on water shows very effective binding with aqueous phosphate molecules, due to the multi-dielectric effect of the interface. Interesting other examples of interfacial molecular recognition such as specific binding of peptides have been developed by using this technique. Patterning of individual molecules at interface is achieved by taking advantage of multi-site molecular recognition.

Biomembrane Mimics

Cell membranes are formed through spontaneous assembly of lipids and proteins. The lipid bilayer thus formed is in the fluid liquid crystalline state, which is required to maintain the living function. Proteins are buried in the bilayer and/or attached to the bilayer surface, and polysaccharide chains on the membrane surface play essential role in cell recognition. This picture of biological membrane was proposed as "fluid mosaic model" by Singer and Nicholson in 1972.[1] The major driving force for lipid bilayer formation from the physicochemical viewpoint is a. ordered assembly of hydrophobic alkyl chains of lipid molecules and b. stabilisation of the membrane-water interface by hydrophilic head groups.

Such stable bilayer assembly in water is not unique to natural lipid molecules. Kunitake and Okahata showed in 1977 that aqueous bilayer organization is formed readily from simple double-chain ammonium salts.[2] This finding triggered widespread research into amphiphiles that could form vesicle-like bilayer assemblies. These new materials are composed from hydrophobic tails and hydrophilic head groups. Cationic, anionic, nonionic, and zwitterionic units can be used as polar heads, and dialkyl, trialkyl, tetraalkyl chains are useful for molecular alignment in the assembly. In addition, single-chain combination of rigid segments and flexible alkyl groups can be effective to produce spontaneous hydrophobic alignment.[3]

Giant Nanomembrane

Much efforts are subsequently devoted to apply these synthetic bilayer assemblies to practical uses. Medical and clinical applications of water-dispersed synthetic bilayer membranes have been tested, while industrial extension of such synthetic analogs was difficult to realise, as their nanometer-thickness was an obstacle against facile, macroscopic handling. Recently, this problem was solved partially by invention of giant nanomembrane.[4] In a typical case, thermosetting resins such as epoxy resins and melamine resins are highly cross-linked after curing, and therefore, are rigid and infusible. Such macroscopic mechanical properties do not necessarily apply in the regime of nanometer thickness. When precursor mixtures of typical thermosetting resins are spread on proper underlayers at nanometer-thickness and cured, the resulting nanomembrane with thicknesses of 20-200 nm are self-supporting and defect-free. Their basic mechanical properties are analogous to their macroscopic counterparts.

Thus-prepared "giant nanomembrane" must satisfy the following requirements : a. thickness of less than 100 nm, b. aspect ratio (size/thickness) of over one million (10 cm/100 nm), c. free-standing, defect-free and robust. Giant nanomembranes are highly permeating for many gases, but not so for liquids, as expected. The gap between biomembrane and artificial membrane is narrowing, but intrinsic properties that are still lacking in artificial membranes is dynamic nature and molecular thickness. Unique functional characteristics of biomembranes are supported by molecular events occurring on fluid mosaic structures. Giant nanomembranes, unlike biomembranes, are still too thick to provide functions driven by single-molecular events, and their dynamic nature must be created by additional approaches.

Molecular Recognition at Interface

Hydrogen bonding in aqueous systems is central to maintenance and operation of the biological systems, and it is decisive in controlling the biological molecular recognition, as seen in replication of nucleic acids, maintenance of tertiary structures of proteins, and substrate recognition of enzymes. In contrast, it is not effective in the presence of water with artificial systems, and has been utilised only in non-aqueous solutions. The situation is totally different for molecular interaction at the air-water interface. As shown in Figure XIV.1, dilute thymine and adenine dissolved in the aqueous phase are effectively bound to interfacial monolayers containing diaminotriazine and orotic

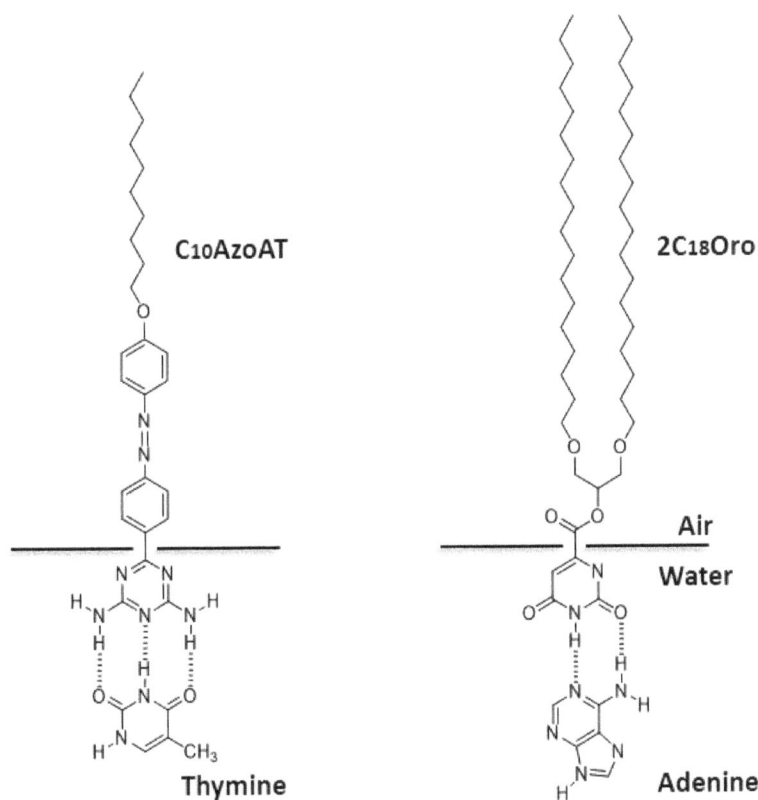

Figure XIV.1. Binding of thymine and adenine molecules in water to complementary interfacial monolayer

acid units, respectively. Spectral evidence of transferred monolayers indicates that complementary hydrogen bonding play an essential role for the binding, as illustrated in the figure. The binding constant for thymine is ca. 3×10^2 and is comparable to that of a similar functional combination in aprotic organic solvents.[5] Such binding is not observed for non-complementary adenine substrate.

Enormous difference in binding efficiency is found between

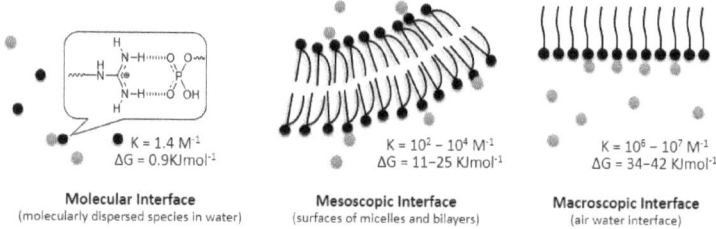

Figure XIV.2. Remarkable differences of binding of the quanidinium and phosphate functions at molecular, mesoscopic and macroscopic (air-water) interfaces.

the functional pair of guanidinium and phosphate in bulk water, in aqueous aggregates, and at molecular interfaces. These units are known to provide a strong binding pair in the biological system due to the combined electrostatic interaction and hydrogen bonding. In the case of artificial bilayer dispersed in water, the Langmuir analysis with varied guest concentrations gives a binding constant of 10^2 to 10^4 M^{-1} for AMP substrate. This value is significantly larger than that between molecularly dispersed guanidinium and phosphate in water (1.4 M^{-1}). Similar analysis conducted for a guanidinium monolayer at the air-water interface gives an even larger value of 10^6 to 10^7 M^{-1}.[6] See Figure XIV.2.

The enormous enhancement in binding energy must be related to the unique feature of the interface. To answer this question, a quantum chemical calculation was performed on the basis of a multi-dielectric model for the guanidinium-phosphate system placed at an interface,[7] as illustrated in Figure XIV.3. Dielectric constants for aliphatic and water phases are set to be 2 and 80, respectively. The binding profile was obtained by calculating the free energy of the whole system as a function of separation distance, d, of the two functions, and the binding energy was estimated as the difference in free energy between that at the potential minimum and that at infinite distance. When d value close to zero is adopted, the binding energy is estimated as ca. 30 kJ/

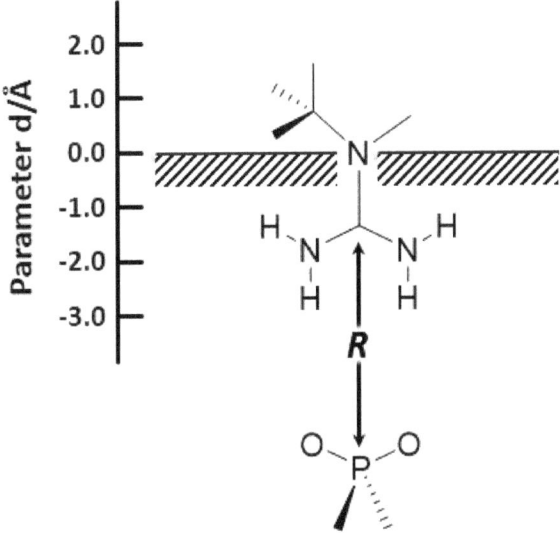

Figure XIV.3. Calculation model of the guanidinium-phosphate interaction at multi-dielectric interface.

mol, corresponding fairly closely to the binding constant observed for AMP at the air-water interface. The multi-dielectric effect of the air-water interface is outstanding.

Similarly enhanced binding was observed also between aqueous peptide guests and Langmuir monolayers with peptide head groups. Specific hydrogen bonding among peptide chains leads to selective binding of given peptides.[8]

Multi-Site Recognition and Two-Dimensional Molecular Patterns

Cooperative functional interaction is often observed in the interfacial molecular recognition. For example, aqueous UMP is bound to 2 molecules of a guanidinium monolayer through simultaneous recognition by two types of interaction. This phenomenon is also observed in interfacial recognition of FAD with four monolayer molecules of three kinds. See Figure XIV.4. At the fully bound state, the most advantageous alignment of the four monolayer molecules must be created for the recognition at the interface. This effect can provide unique molecular ups and downs.

An equimolar mixture of guanidinium amphiphile and orotate amphiphile (both with C18 alkyl groups) gives a uniform monolayer mixture on water, as evidenced by atomic force microscopy, probably because of alignment effect of the identical alkyl chains. When

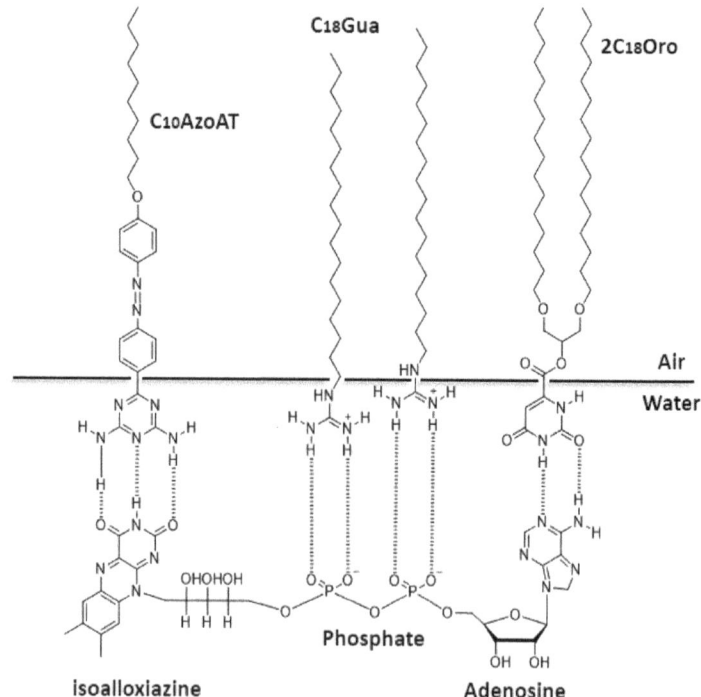

Figure XIV.4. Cooperative recognition between FAD molecule in water and four functional molecules in monolayer at the air-water interface.

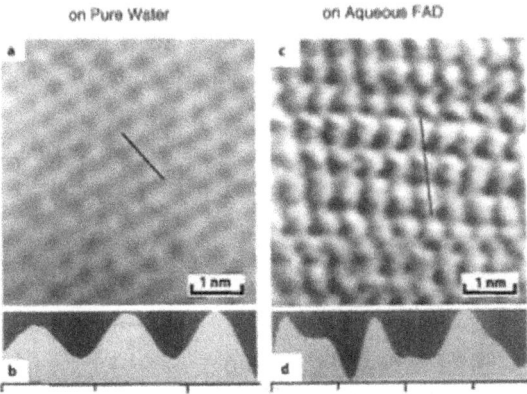

Figure 5-1

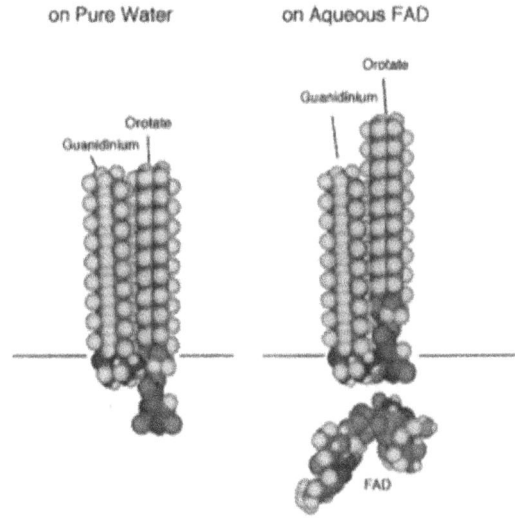

Figure 5-2

Figure XIV.5. The top surface of guanidinium-orotate monolayers on pure water and on aqueous FAD. Molecular images by atomic force microscopy (5.1) and the corresponding interaction scheme (5.2).

this mixture recognises FAD molecule in the subphase, interfacial molecular alignment is governed by convenience of the double molecular recognition, so that alkyl chain alignment is slipped by one carbon. Atomic force microscopy of the transferred monolayer shows alternate ups and downs of the alkyl chain, in accordance with this picture of molecular recognition,[9] as shown in Figure XIV.5.

Conclusion

Reconstruction of intricate molecular organisation of biomembranes is still far beyond reach of man-made chemistry. Aqueous vesicles, nanomembranes, and molecular recognition can be separately realised artificially. *Aufheben* of these partial functions to more elaborate molecular systems of higher order should be a great target of further research. The role of water in defining the biological structure is most important. The mystery of water at the molecular interface and at the macroscopic interface must be elucidated to construct better artificial alternatives of the biological molecular recognition.

References

1. Singer, S.J.; Nicholson, G.L. *Science* **1972**, *175*, 720-731.
2. Kunitake, T.; Okahata, Y. *J. Am. Chem. Soc.* **1977**, *99*, 3860.
3. Kunitake, T. *Angew. Chem. Int. Ed.* **1992**, *31*, 709-726.
4. Vendamme, R.; Onoue, S.; Nakao, S.; Kunitake, T. *Nat. Mater.* **2006**, *5*, 494-501. Watanabe, H.; Kunitake, T. *Adv. Mater.* **2007**, *19*, 909-912.
5. Kurihara, K.; Ohto, K.; Honda, Y.; Kunitake, T. *J. Am. Chem. Soc.* **1991**, *113*, 5078-5079.
6. Ariga, K.; Kunitake, T. *Acc. Chem. Res.* **1998**, *31*, 371-378.
7. Sakurai, M.; Tamagawa, H.; Furuki, T.; Inoue, Y.; Ariga, K.; Kunitake, T. *Chem. Lett.* **1995**, 1001-1002.
8. Cha, X.; Ariga, K.; Kunitake, T. *J. Am. Chem. Soc.* **1996**, *118*, 9545-9551.
9. Oishi, Y.; Torii, Y.; Kato, T.; Kuramori. M.;Suehiro, K.; Ariga, K.; Taguchi, K.; Kamino, A.; Koyano, H.; Kunitake, T. *Langmuir* **1997**, *13*, 519-524.

XV
Interfacial osmotic pressure

Martin Chaplin

London South Bank University.
martin.chaplin@btinternet.com

Abstract

The generally accepted view of osmotic pressure is that it is a colligative property, along with freezing point depression, boiling point elevation and vapour pressure lowering. These properties ideally depend on the concentration of dissolved solute molecules. Osmotic pressure, however, is also generated, without any solute, at hydrophilic surfaces and water-gas interfaces. This pressure may oppose gas exit from nanobubbles and so stabilise them. Here is presented a rationale and explanation for these phenomena.

Background

Although there is some contention amongst physical chemists over how to describe osmotic pressure, it is generally agreed that it is a reversible thermodynamic colligative property and correlates well with other colligative properties such as freezing point depression. As such, the osmotic pressure is thought ideally to depend on the number of dissolved 'particles' (e.g., molecules) present in a volume of liquid. It is, perhaps, unsurprising that ion exchange surfaces can generate very high osmotic pressures of over 100 MPa in water,[1] as they create high surface concentrations of counter-ions. Poly-ionic nanoparticles, with high surface area, produce such a great osmotic pressure that they can be used in practical desalination processes.[2,3] Likewise; polyelectrolytes may also produce high osmotic pressures between molecules.[4] However, hydrophilic uncharged molecules, without any counterions, also produce similarly high osmotic pressures to polyelectrolytes between molecules.[4] Also, it has been experimentally verified that uncharged hydrophilic particle surfaces generate high osmotic pressure, without the presence of counter-ions or solutes.[5,6] It is clear, therefore, that relatively small numbers of particles, far less than required by the 'conventional' colligative law, can generate high osmotic pressures so long as they possess extensive hydrophilic surfaces.

Much work on the effect of hydrophilic surfaces on the mesoscopic properties of the adjoining aqueous solutions has been done in the laboratory of Gerald Pollack,[7,8] and the phenomena they discuss have been confirmed by many other independent workers at other laboratories.[9-11] In essence, it has been found that the interfacial water next to ionic charged (e.g. polyacrylate) or neutral uncharged (e.g. polyvinyl alcohol) hydrophilic surfaces expels solutes to the bulk of the solution that may be several hundred microns away. These exclusion zones (named as EZ-water) can be visualized when low-molecular weight dyes, proteins, micron-sized microspheres or other solutes are used. Also the EZ-water seems to possess other physical properties such as absorption at 270 nm,[12] greater density, greater viscosity and negative charge compared with the bulk water. There is no generally-accepted explanation for this osmotic pressure phenomenon or these properties of EZ-water. A new explanation covering all these phenomena is given below, involving the generation of osmotic pressure, that is very simple, easily understood and potentially very important in a number of related fields.

Proposal for the generation of osmotic pressure at aqueous interfaces

Wherever water is present in solution, it may be considered as being either 'bound' or 'free', although there will be transitional water between these states. When considering the colligative properties, 'water' is considered bound to any solute when it has a lower entropy compared with pure liquid water. Such water may be considered part of the solute and not part of the dissolving 'free' water.[13] As pure liquid water consists of a mixture containing low-density water, made up of extensively hydrogen bonded structures, and higher density water,[14-18] consisting of much smaller less extensive clusters, the proportions of 'bound' or 'free' water in pure liquid water can vary with more strongly-bound larger clusters with behaviour approaching that of 'bound' water. In bulk liquid water, the relative concentrations of the two aqueous forms is of no consequence as all the water behaves the same throughout. If fluctuations occur, volumes of the solution contain different proportions of strongly and weakly hydrogen-bonded water molecules (or put even more simply that there is more extensive clustering present), then these different volumes will show a difference with respect to their water activity and chemical potential. Normally any such instantaneous differences in water activity and chemical potential between different volumes within the same mass of liquid would rapidly cause liquid movement from high to low water activity in order to equalize these states and so remove the chemical potential differences. However, where there are surfaces interacting with the liquid water, the concentration of the more extensive hydrogen-bonded clusters within the surface layer may differ from the bulk values with the surface interactions preventing the potential equalization between bulk and surface volumes. When this occurs, the surface water has a lower water activity and chemical potential to the bulk, leading to differences in osmotic pressure, and other colligative properties (Figure XV.1). The change in the chemical potential (μ_w) is $-\{RTln(x_{ws}) - RTln(x_{wb})\}$ (that is, a negative energy term is added to the chemical potential when $x_{ws} < x_{wb}$) where x_{ws} is the mole fraction of the 'free' water ($0 < x_{ws} < 1$) in the surface layer and x_{wb} is the mole fraction of the 'free' water ($0 < x_{wb} < 1$) in the bulk liquid. The generated osmotic pressure (Π) is given by, $\Pi = -(RT/V_M).ln(x_{ws}/x_{wb})$

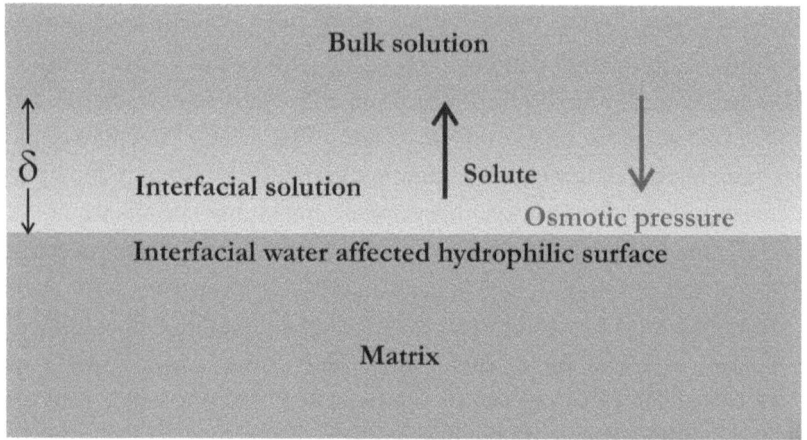

Figure XV.1. Diagram showing the effect of surface on the interfacial water. The depth of the 'unstirred' layer (δ) depends on the stillness of the bulk solution and the extent of the matrix surface.

where V_M is molar volume of water (0.0000180685 m^3 mol^{-1} at 25 °C and 0.1 MPa) (Figure XV.2).

At hydrophilic surfaces, interactions between the surface and neighbouring water molecules fix the localised hydrogen bonding and this, together with steric factors, increases the cluster extent and lifetime.[19] As the 'free' water reduces as compared with its bulk value when the formation of longer-lived and more extensive hydrogen bonded clusters increases,[19] so the water activity reduces and the osmotic pressure increases. Solutes next to the surface will move to equalise the water activity throughout the liquid. In other words, the increase in osmotic pressure next to the surface will displace solutes from the surface towards the bulk until its effect is equalled by the osmotic pressure of the surrounding solution or the system reaches a steady state. As the first effect of this solute expulsion is naturally the formation of an increased concentration band as expelled solute mixes with the prior solute concentration, the extent of the expulsion will affect the whole of the unstirred layer (~1-100 μm). Where

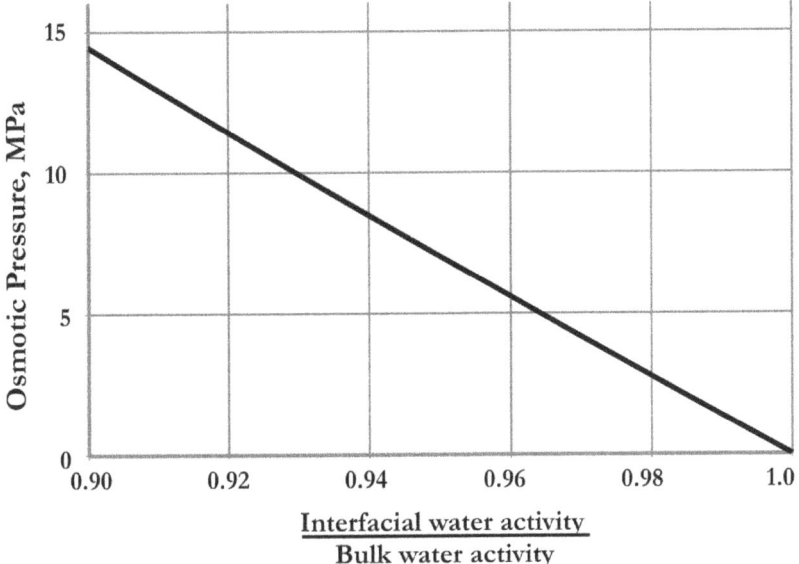

Figure XV.2. Small reductions in the interfacial water activity give rise to large osmotic pressures.

hydrophilic microparticles or nanoparticles are suspended in aqueous solutions, their surfaces will necessarily cause mutually repulsive osmotic pressure effects that may result in the ordering of the particles within small volumes of the liquid.[20-21]

It should be noted that osmotic drive does not require a membrane to separate the two solutions,[22] provided there are two phases.[23] Here the two phases consist of the unstirred and stirred layers. In this context, the affected aqueous layer behaves similarly to that described for exclusion zone (EZ) water by Pollack and here we put this forward as a simple explanation of his experimental data.[8, 12, 24] This hypothesis also helps explain the experiments on autothixotropy,[25-30] where the viscosity of unstirred solutions increase with time. The increase in density at the interface, as found in EZ-water, has been explained previously by the increase in clustering causing the water to behave as though it is at a lower temperature and this has also been used to explain the ease with which this surface layer freezes. The presence of 270 nm absorption in the interfacial water, as described for EZ-water,[12] may be attributed to the delocalization of electrons within the extended clustering as hydrogen bonding is known to involve electron delocalisation.[31] It is clear that electron delocalisation will be stabilised by the addition of electrons but not by protonation, so likely causing the charge separation seen at these interfaces.[32] Other experimental properties of the EZ layers, such as the reduction in its thickness with particle size[33] and with increased bulk ionic strength[34] are easily explained by the presented hypothesis, as the unstirred layer around particles is known to depend on the particle size and high bulk osmotic pressures will oppose the osmotic pressure at the hydrophilic surface.

Another effect of interfaces is the formation of evanescent waves due to the internal reflection of electromagnetic radiation (Figure XV.3). The standing electromagnetic wave produced will interact with water molecules to stabilise a standing wave of hydrogen

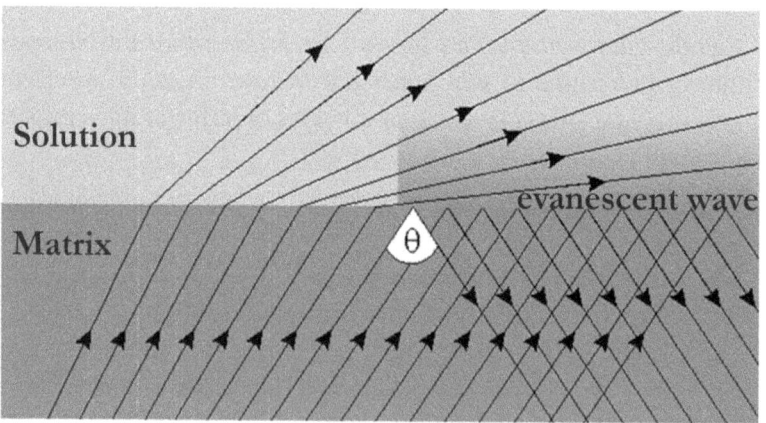

Figure XV.3. The standing electromagnetic evanescent wave within the interfacial water is caused by impinging electromagnetic radiation and the angle at which total internal reflection occurs (θ).

bonded clusters that will increase the local concentration and extent of hydrogen bonded clusters so increasing the above osmotic effect, in agreement with the experimental data.[35-38]

It appears that a similar effect on solutes to the one described for water may occur in other polar solvents that can form hydrogen bonds,[39] thus reinforcing the likelihood that a mechanism is acting that does not depend on the specific properties of water, such as the here-described osmotic pressure generation.

Osmotic pressure stabilises nanobubbles

There is now much evidence that sub-micron-sized gas-filled cavities (often called nanobubbles) can exist for significant periods of time both in bulk aqueous solution [40-45] and at submerged aqueous-hydrophobe interfaces.[46-47] Their contained gas is in constant flux with gas molecules both leaving and entering continuously. The cavities are under excess pressure given by the Laplace equation ($2\gamma/r$, where γ is the surface tension and r is the cavity radius) as the surface tension causes a tendency to minimize their surface area, and hence volume. Nanobubbles grow or shrink by diffusion according to external pressure and the degree of over- or under-saturation in the surrounding solution, with the dissolved gas relative to the raised cavity pressure. As the solubility of gas is proportional to the gas pressure and this pressure is exerted by the surface tension in inverse proportion to the size of the bubbles, there is increasing tendency for gasses to dissolve as the bubbles reduce in size so accelerating the bubble-dissolution process. Such size-reduction is increased by the bubble›s movement and contraction during this activity, which aids the removal of any gas-saturated solution around the cavities. Calculations show that nanobubbles should only persist for a few microseconds,[48] in contrast to their long lifetimes (hours to days) as detected by light scattering or resonant mass measurement [49] (bulk nanobubbles) or tapping mode atomic force microscopy [50] (surface nanobubbles). Interestingly, these bubbles are subject to Brownian motion, so behaving as though they have solid shells similarly to solid nano-particles. Nanobubbles are commonly found on solid hydrophobic surfaces in solutions open to the air, where they appear to be quite stable.[51] Surface and bulk-phase nanobubbles can both give rise to the otherwise difficult to explain long range attraction between hydrophobic surfaces. due to bridging nanoscopic gas cavities [52-53] or the osmotic effect due to local nanobubble depletion.[43]

Contributing to the stability of nanobubbles is the slow rate of gas diffusion to the bulk liquid surface from both surface and bulk-phase nanobubbles.[54-55] This is apparently due to a thick interfacial layer; a phenomenon experimentally supported by the higher forces required to penetrate greater depths of surface nanobubbles. [55-56] Here we hypothesise that the cause of this layer is the high osmotic pressure produced beneath at gas liquid interface so both preventing the gas dissolving and driving dissolved gas near the interface back into the nanobubble (Figure XV.4). This surface layer can be seen

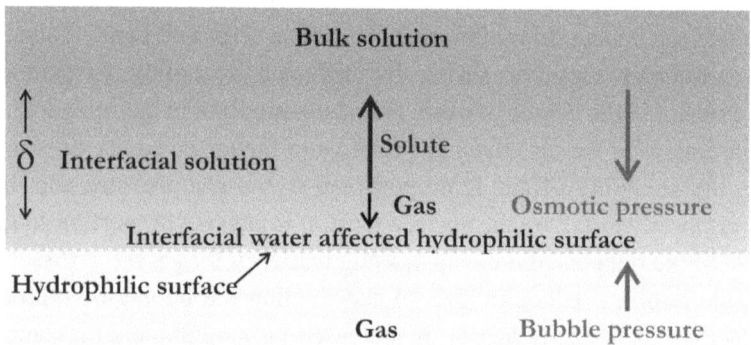

Figure XV.4. Osmotic pressure generated at nanobubble surface is responsible for nanobubble stability. The more structured interfacial water is generated from the hydrophilic gas-liquid interface. The depth of the 'unstirred' layer (δ) is approximately proportional to the size of the nanobubbles. This thickness will be substantially reduced by the mixing caused by any surface changes in the bubbles as they rise or change size.

experimentally. The generation of the osmotic pressure at the surface is due to the necessarily under-coordinated water molecules at the gas-liquid surface forming an ice-like, low-density phase [57] that has lower water activity than the bulk water. Hydrogen bonding in the surface is stronger than in the bulk,[58] due to the reduced competition from neighbouring water molecules, lower anticooperativity and compensation for the increased chemical potential on the loss of some bonding. A further reason for at least some of the stability of nanobubbles is that the nanobubble gas/liquid interface is charged,[59] expanding the surface and introducing an opposing force to the surface tension, so slowing or preventing the nanobubble dissipation. Curved aqueous surfaces may introduce a surface charge due to water's molecular structure or its ionization. It is clear that the presence of like charges at the interface will reduce the effect of the surface tension, with charge repulsion acting in the opposite direction to the surface minimization due to surface tension. Any effect may be increased by the presence of additional charged materials that favour the gas-liquid interface, such as OH⁻ ions at neutral or basic pH. [59] Recently, it has been proposed that the surface tension is reduced by the degree of supersaturation within the surface. [60] As this supersaturation is prevented from equilibrating away from the bubbles by the imposed osmotic pressure, this introduces a further stabilisation effect on the nanobubbles.

Further background information on the properties of water, interfaces and nanobubbles has been reviewed.[61]

References

1. Budinski, M. K.; A. Cook, A. *Tsinghua Sci. Technol.* **2010**, *15*, 385-390.
2. Ling, M. M.; Chung, T.-S. *Desalination* **2011**, *278*, 194-202.
3. Ling, M. M.; Wang, K. Y.; Chung, T.-S. *Ind. Eng. Chem. Res.* **2010**, *49*, 5869-5876.

4. Parsegian, V. A.; Zemb, T. *Curr. Op. Coll. Interface Sci.* **2011**, *16*, 618-624.
5. Ling, M. M.; Chung, T.-S.; Lu,X. *Chem. Commun.* **2011**, *47*, 10788-10790.
6. Ling, M. M.; Chung, T.-S. *J. Membrane Sci.* **2011**, *372*, 201-209.
7. Zheng, J.-M.; Pollack, G. H. *Phys. Rev. E.* **2003**, *68*, 031408.
8. Zheng, J.-M.; Chin, W.-C.; Khijniak, E.; Khijniak, E. Jr.; Pollack, G. H. *Adv. Coll. Interface Sci.* **2006**, *127*, 19-27.
9. Mollenhauer, H. H.; Morré, D. J. Structural compartmentation of the cytosol: zones of exclusion, zones of adhesion, cytoskeletal and intercisternal elements. *In* Roodyn D. B. (ed.) *Subcellular Biochemistry*, vol. 5 (Plenum Press, 1978) pp. 327-62.
10. Green, K.; Otori, T. *J. Physiol.* **1970**, *207*, 93-102.
11. Yoshida, H.; Ise, N.; Hashimoto, T. *J. Chem. Phys.* **1995**, *103*, 10146.
12. Chai, B.; Zheng, J.; Zhao, Q.; Pollack, G. H. *J. Phys. Chem. A* **2008**, *112*, 2242-2247.
13. Zavitsas, A. A. *Chemistry* **2010**, *16*, 5942-5960.
14. Maréchal, Y. *J. Mol. Structure* **2011**, *1004*, 146-155.
15. Mishima, O. *J. Phys. Chem. B* **2011**, *115*, 14064-14067.
16. Nilsson, A.; Pettersson, L. G. M. *Chem. Phys.* **2011**, *389*, 1-34.
17. Mallamace, F. *Proc. Nat. Acad. Sci.* **2009**, *106*, 15097-15098.
18. Banerjee, D.; Bhat, S. N.; Bhat, S. V.; Leporini, D. *Proc. Nat. Acad. Sci.* **2009**, *106*, 11448-11453.
19. Ruan, C.-Y.; Lobastov, V.A.; Vigliotti, F.; Chen, S.; Zewail, A. H. *Science* **2004**, *304*, 80-84.
20. Samal, S.; Geckeler, K. E. *Chem. Commun.* **2001**, *21*, 2224-2225.
21. Sadakane, K.; Seto, H.; Endo, H.; Shibayama, M. *J. Phys. Soc .Japan* **2007**, *76*, 113602.
22. Zhao, Q.; Ovchinnikova, K.; Chai, B.; Yoo, H.; Magula, J.; Pollack, G. H. *J. Phys. Chem. B* **2009**, *113*, 10708-10714.
23. Pollack, G. H.; Clegg, J. Unexpected linkage between unstirred layers, exclusion zones, and water, In *Phase Transitions in Cell Biology*, G.H. Pollack and W.-C. Chin (eds.), (Springer Science+Business Media B.V., 2008) pp. 143-52.

24. Zheng, J.-M.; Wexler, A.; Pollack, G. H. *J.Coll. Interface Sci.* **2009**, *332*, 511-514.
25. Vybíral, B.; Vorácek, P. "Autothixotropy" of water – an unknown physical phenomenon, *arXiv.org Physics e-Print archive* physics/0307046 (2003).
26. Vybíral, B. The comprehensive experimental research on the autothixotropy of water, In *Water and the cell*, Ed. G. H. Pollack, I. L. Cameron and D. N. Wheatley (Springer, Dordrecht, 2006) pp 299-314;
27. Vybíral, B.; Voráček, P. *Homeopathy* **2007**, *96*, 171-182.
28. Gaylard, A. P. *Homeopathy* **2008**, *97*, 46.
29. Vybíral, B.; Voráček, P. *Homeopathy* **2008**, *97*, 47.
30. Verdel, N.; Jerma, I.; Krasovec, R.; Bukovec, P.; Zupancic, M. *Int. J. Mol. Sci.* **2012**, *13*, 4048-4068.
31. Arunan, E.; Desiraju, G. R.; Klein, R. A.; Sadlej, J.; Scheiner, S.; Alkorta, I.; Clary, D. C.; Crabtree, R. H.; Dannenberg, J. J.; Hobza, P.; Kjaergaard, H. G.; Legon, A. C.; Mennucci, B.; Nesbitt, D. J. *Pure Appl. Chem.* **2011**, *83*, 1637-1641.
32. Yoo, H.; Baker, D. R.; Pirie, C. M.; Hovakeemian, B.; Pollack, G. H. Characteristics of water adjacent to hydrophilic interfaces. In *Water: the Forgotten Molecule*, D. LeBihan, and H. Fukuyama (eds), (Pan Stanford, 2011) pp 123-36.
33. Nhan, D. T.; Pollack, G. H. *Int. J. Des. Nat. Ecodyn.* **2011**, *6*, 139-144.
34. Zhang, J.; Pollack, G. H. Solute exclusion and potential distribution near hydrophilic surfaces, In *Water and the Cell*, Ed. G. H. Pollack, I. L. Cameron and D. N. Wheatley, Springer, pp. 165-174.
35. Shimokawa, S.; Yokono, T.; Yokono, M.; Yokokawa, T.; Araiso, T. *Jap. J. Appl. Phys.* **2007**, *46*, 333-335.
36. Yokono, T.; Shimokawa, S.; Yokono, M.; Hattori, H. *Water* **2009**, *1*, 29-34.
37. Ovchinnikova, K.; Pollack, G. H. *Phys. Rev. E* **2009**, *79*, 036117.
38. Chai, B.; Yoo, H.; Pollack, G. H. *J. Phys. Chem. B* **2009**, *113*, 13953-13958.
39. Chai, B.; Pollack, G. H. *J. Phys. Chem. B* **2010**, *114*, 5371-5375.
40. Kikuchi, K.; Tanaka, Y.; Saihara, Y.; Maeda, M.; Kawamura, M.; Ogumi, Z. *J. Coll. Interface Sci.* **2006**, *298*, 914-919.

41. Jin, F.; Ye, J.; Hong, L.; Lam, H.; Wu, C. *J. Phys. Chem. B* **2007**, *111*, 2255-2261.
42. Bunkin, N. F.; Suyazov, N. V.; Shkirin, A. V.; Ignatiev, P. S.; Indukaev, K. V. *J. Chem. Phys.* **2009**, *130*, 134308.
43. Jin, F.; Gong, X. J.; Yea, J.; Ngai, T. *Soft Matter* **2008**, *4*, 968-971.
44. Yount, D. E. *J. Acoust. Soc. Am.* **1979**, *65*, 1429-1439.
45. Ohgaki, K.; Khanh, N. Q.; Joden, Y.; Tsuji, A.; Nakagawa, T. *Chem. Eng. Sci.* **2010**, *65*, 1296-1300.
46. Attard, P. *Adv. Coll. Interface Sci.* **2003**, *104*, 75-91.
47. Zhang, X. H.; Quinn, A.; Ducker, W. A. *Langmuir* **2008**, *24*, 4756-4764.
48. Ljunggren, S.; Eriksson, J. C. *Coll. Surf. A* **1997**, *129-130*, 151-155.
49. Burg, T. F.; Godin, M.; Knudsen, S. M.; Shen, W.; Carlson, G.; Foster, J. S.; Babcock, K.; Manalis, S. R. *Nature* **2007**, *446*, 1066-1069.
50. Tyrrell, J. W. G.; Attard, P. *Phys. Rev. Lett.* **2001**, *87*, 176104.
51. Borkent, B. M.; Dammer, S. M.; Schönherr, H.; Vancso, G. J.; Lohse, D. *Phys. Rev. Lett.* **2007**, *98*, 204502.
52. Ishida, N.; Sakamoto, M.; Miyahara, M.; Higashitani, K. *J. Coll. Interface Sci.* **2002**, *253*, 112-116.
53. Pashley, R. M.; Francis, M. J.; Rzechowicz, M. *Curr. Op. Coll. Interface Sci.* **2008**, *13*, 236-244.
54. Weijs, J. H.; Lohse, D. *Phys. Rev. Lett.* **2013**, *110*, 054501.
55. Wang, S.; Liu, M.; Dong, Y. *J. Phys.: Condens. Matter* **2013**, *25*, 184007.
56. Walczyk, W.; Schön, P. M.; Schönherr, H. *J. Phys.: Condens. Matter* **2013**, *25*, 184005.
57. Sun, C. Q.; Zhang, X.; Zhou, J.; Huang, Y.; Zhou, Y.; Zheng, W. *J. Phys. Chem. Lett.* **2013**, *4*, 2565-2570.
58. Gan, W.; Wu, D.; Zhang, Z.; Guo, Y.; Wan, H. *Chinese J. Chem. Phys.* **2006**, *19*, 20-24.
59. Chaplin, M. *Water* **2009**, *1*, 1-28.
60. Attard, P. The stability of nanobubbles, *Eur. Phys. J. Special Topics* (2013) Article in press, doi: 10.1140/epjst/e2013-01817-0.
61. Chaplin, M. *Water structure and science*, http://www.lsbu.ac.uk/water/ accessed on 8 Aug 2013.

XVI
On the surface tension of electrolyte solutions

Vincent S. J. Craig, Jian Cui, Thomas G. Brazier

Department of Applied Mathematics,
Research School of Physical Sciences,
Australian National University, Canberra ACT 0200, Australia.
Vince.Craig@anu.edu.au

The surface tensions of aqueous electrolyte solutions in general increase linearly with concentration over a wide range of concentration. Here we analyse this data using the Gibbs Adsorption Isotherm to evaluate the concentration of electrolytes in the surface layer and derive a simple expression whereby the concentration in the surface layer can easily be calculated.

Specific ion effects are manifest in all manner of measurements from the stability of proteins as originally investigated by Hofmeister[1] to the combining rules that govern the inhibition of bubble coalescence.[2,3] Here we wish to discuss a particular specific ion effect that is well known and appears at first glance to be very simple. If the surface tension of aqueous salt solutions are measured, a plot of surface tension versus concentration reveals a straight line over a wide concentration range (see Figure XVI.1).[4] Therefore the data can be described by the surface tension gradient $d\gamma/dC$, which is constant and positive indicating that the electrolyte ions are depleted from the interface. The same linear correlation between surface tension and concentration is found for other electrolytes with only a few exceptions.[4-6] The magnitude of the gradient varies but remains positive for most electrolytes other than acids. The surface tension is of particular interest because it reflects the interfacial

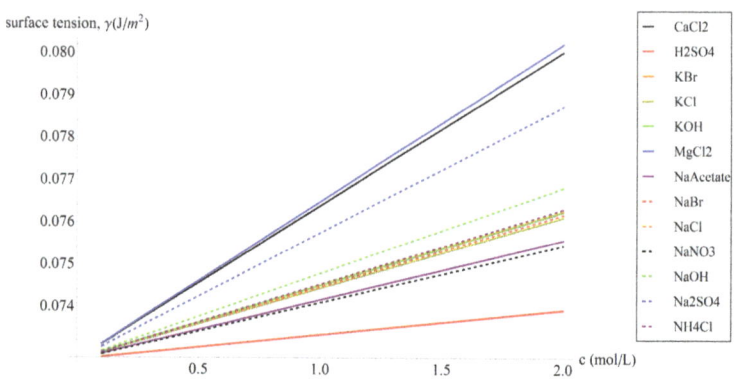

Figure XVI.1. Surface tension versus concentration for a range of electrolytes in water. Using slopes determined from experimental data.

concentration of ions, which is thought to be a significant contributor to the many observed specific ion effects. In the past decade surface sensitive spectroscopic techniques have provided a direct measure for investigating the surface propensity of individual ions,[7,8] however the data obtained by these techniques and advanced simulation methods have not been resolved with surface tension measurements. Resolution is not expected to be immediate or straightforward as the techniques may well probe a different surface depth, only some ions are revealed spectroscopically, the effect of individual ions cannot be measured by surface tension measurements and it is expected that the concentration profile of the ions in the interface is complex, being oscillatory.[9] We wish to investigate the form of the surface tension data in some more detail and attempt to quantitatively analyse the data using the Gibbs Adsorption Isotherm.

As surface tension is a thermodynamic quantity defined as the change in Gibbs Free Energy with area, one would expect that the concentration is best expressed as activity rather than molarity. The activity coefficient varies considerably from unity for electrolytes

On the surface tension of electrolyte solutions   343

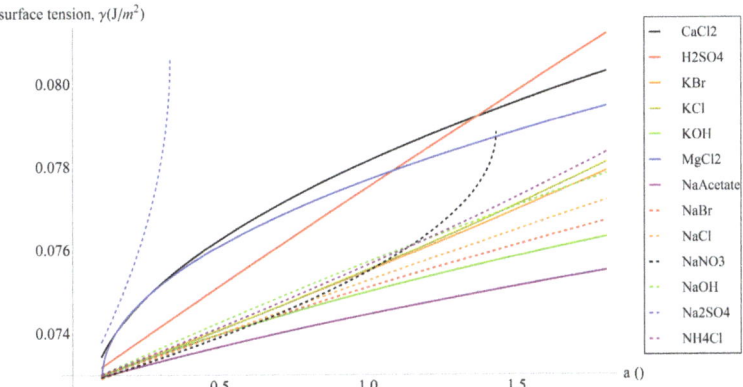

Figure XVI.2. Surface tension of aqueous electrolyte solutions as a function of the activity of solution.

and changes as a function of salt concentration. Therefore, one might expect that a plot of surface tension versus activity will not exhibit the linearity seen in plots of surface tension versus concentration. In Figure XVI.2 we have plotted the relationship between concentration and activity for a range of electrolytes. Within the accuracy of experimental measurements the relationship is linear for many of the electrolytes but not for all. What is not clear is why there should be a linear correlation. With reference to the Gibbs Adsorption Isotherm, the surface excess of electrolyte Γ can be calculated from the surface tension γ and the activity a, where R is the gas constant and T the temperature in Kelvin.

$$\Gamma_{electrolyte} = \frac{-1}{RT} \times \frac{d\gamma}{d\ln a} \qquad (1)$$

Note the surface excess is a relative concentration, that is, it is the concentration in the surface layer relative to the bulk. What is clear is that Γ is proportional to $-d\gamma/d\ln a$. As the surface tension increases with activity (and concentration) the surface layer is depleted with respect to the bulk by the magnitude of Γ. If the data is plotted as

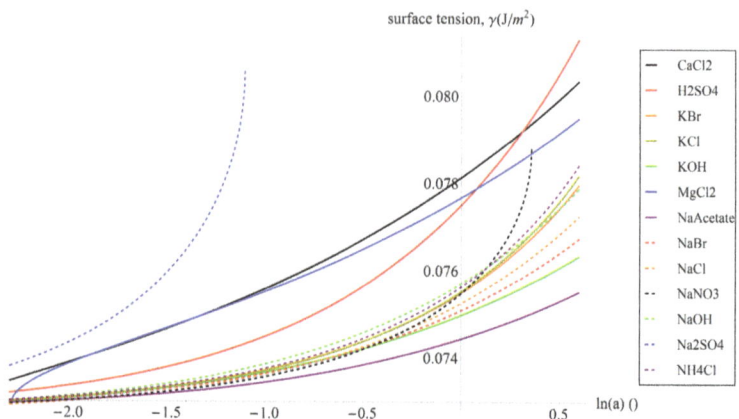

Figure XVI.3. Surface Tension versus the natural logarithm of the activity for aqueous electrolyte solutions. Within the paradigm of the Gibbs Adsorption Isotherm the slope of this plot is proportional to the surface excess.

surface tension versus the natural logarithm of the activity as in Figure XVI.3, the surface excess can be calculated using equation 1, where the $d\gamma/d\ln a$ term is the slope of the plot at a particular value of the activity.

Noting that the surface tension changes linearly with activity we can write $d\gamma/d\ln a = ma$, where $\dfrac{d\gamma}{da} = m$

combining with equation (1) we obtain a simple expression for the surface excess.

$$\Gamma_{electrolyte} = \frac{-ma}{RT} \qquad (2)$$

If R is expressed in J·mol K^{-1} then Γ is in units of mol m^{-2} and is therefore a surface layer concentration, that is a two dimensional concentration. In Figure XVI.4 we present the surface excess (two dimensional concentration) for electrolytes as a function of the bulk concentration. The Gibbs Adsorption isotherm chooses a particular dividing plane such that the surface excess of water is zero. It is

On the surface tension of electrolyte solutions 345

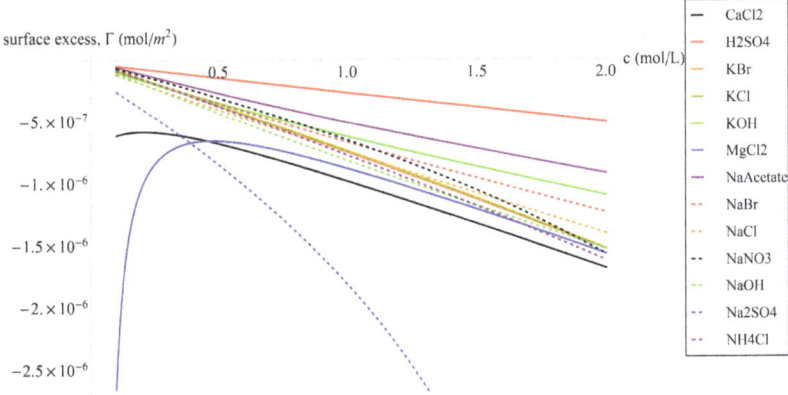

Figure XVI.4. Surface Excess (Γ) of electrolytes versus bulk concentration calculated using the Gibbs Adsorption Isotherm. The negative sign indicates that the surface is depleted of electrolyte.

unclear exactly where this dividing plane is located with respect to the interface. However, it is generally accepted that the interfacial depth is ~1 nm therefore by assuming a depth of the interface the surface excess can be converted to a three dimensional concentration. Moreover, if the slope of $d\gamma / d \ln a$ is positive, the surface excess will be negative. That is the concentration of electrolyte in the surface layer will in most instances be less than that in bulk, as is commonly understood. As such the actual concentration of the surface phase can be determined from the bulk concentration minus the surface excess when it is expressed as a 3D concentration. This concentration is of interest, as it will reflect the degree to which the surface layer is depleted of electrolyte.

As the depth of the interface is not actually known we have performed this calculation with a series of interfacial depths in an effort to discern the surface excess expressed as a three dimensional concentration. This enables the actual concentration of electrolyte in the surface layer to be determined.

The expression for this is

$$\Psi_{electrolyte} = c + \frac{ma}{RTl} \tag{3}$$

Where c is the bulk concentration and l is the thickness of the surface phase and will be less than 4 nm. This data is shown in Figures XVI.5, 6, 7. What is apparent is that for a surface layer of thickness 6Å, the thinnest layer that is suggested in the literature,[10] the calculated concentration of NaCl in the surface layer is negative. This is unphysical and implies that the surface layer thickness must be greater to account for the large depletion of NaCl in the surface. Surface layers of 8Å and above give positive concentrations of NaCl in the surface layer. If we make the assumption that the surface layer is the same for different types of electrolyte it is likely that the surface layer thickness has to be set to a value > 18Å to accommodate electrolytes that more strongly effect the surface tension, such as $MgCl_2$.

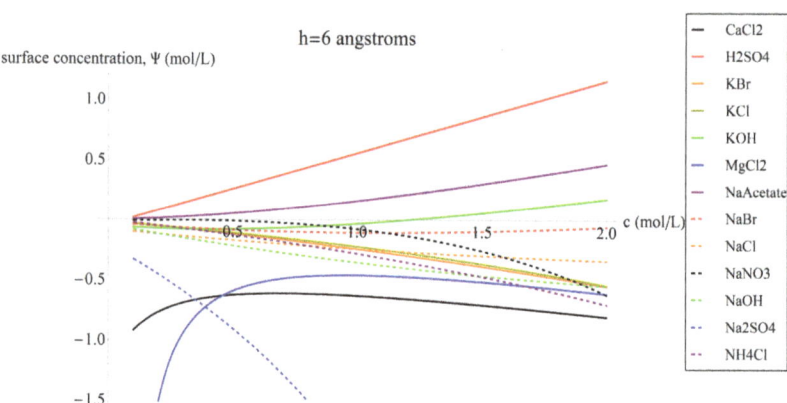

Figure XVI.5. Surface excess expressed as Molarity under the assumption that the surface layer is 6 Ångstroms thick versus the bulk concentration. Note that the surface concentration is determined by the sum of the bulk concentration and the surface excess (which is negative).

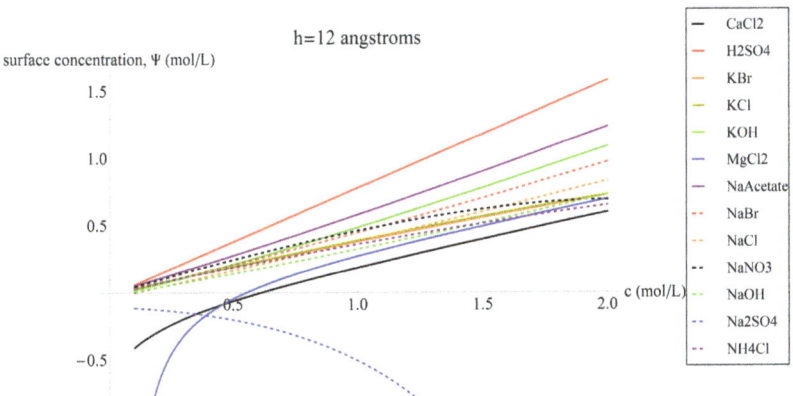

Figure XVI.6. Surface excess expressed as Molarity under the assumption that the surface layer is 12 Ångstroms thick versus the bulk concentration. Note that the surface concentration is determined by the sum of the bulk concentration and the surface excess (which is negative).

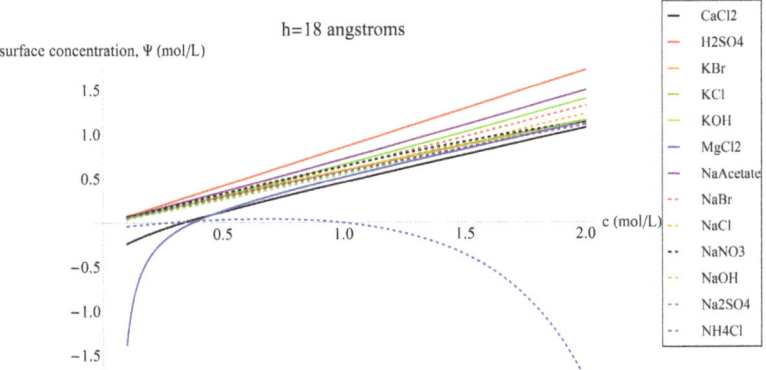

Figure XVI.7. Surface excess expressed as Molarity under the assumption that the surface layer is 12 Ångstroms thick versus the bulk concentration. Note that the surface concentration is determined by the sum of the bulk concentration and the surface excess (which is negative).

We have recently investigated the effect of electrolytes on the surface tension of non-aqueous electrolytes. The resulting data is very

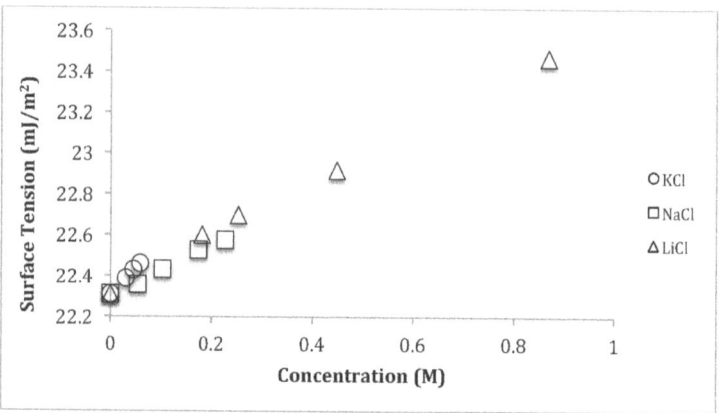

Figure XVI.8. Surface tension of electrolytes measured in methanol solutions as a function of concentration. Within the uncertainty in the measurements there is a linear correlation between surface tension and concentration for each electrolyte. The magnitude of the gradient is similar to what is measured for electrolytes in aqueous solution. The magnitude of the gradient in surface tension is 1.21×10^{-3} J m^{-2} /M for NaCl, 1.30×10^{-3} J m^{-2} /M for LiCl and 2.56×10^{-3} J m^{-2} /M for KCl.

similar in form to that of aqueous solutions in that the surface tension increases with electrolyte concentration and within error the surface tension versus concentration is linear. Data for NaCl, KCl and LiCl in methanol obtained using a KSV Cam 100 pendant drop surface tensiometer are presented in Figure XVI.8. Analysis using the Gibbs Adsorption isotherm could be conducted as above for electrolyte solutions, but we do not currently have the necessary information on the activity coefficients of electrolytes in these solutions. Regardless, the data suggests that the surface excess behaves in a similar manner to aqueous solutions, highlighting that the overall behaviour is determined primarily by the electrolyte, with the solvent playing a minor role.

Acknowledgments

We would like to thank Dr Drew Parsons for assistance with the differentiation leading to equation 2.

References

1. Kunz, W.; Henle, J.; Ninham, B. W. *Curr. Op. Coll. Interface Sci.* **2004**, *9* (1-2), 19-37.
2. Craig, V. S. J.; Ninham, B. W.; Pashley, R. M. *J. Phys. Chem.* **1993**, *97* (39), 10192-10197.
3. Craig, V. S. J.; Ninham, B. W.; Pashley, R. M. *Nature* **1993**, *364* (6435), 317-319.
4. Henry, C. L.; Dalton, C. N.; Scruton, L.; Craig, V. S. J. *J. Phys. Chem. C* **2007**, *111* (2), 1015-1023.
5. Weissenborn, P. K.; Pugh, R. J. *Langmuir* **1995**, *11*, 1422-1426.
6. Weissenborn, P. K.; Pugh, R. J. *J. Coll. Interface Sci.* **1996**, *184* (2), 550-563.
7. Petersen, P. B.; Saykally, R. J. *J. Phys. Chem. B* **2005**, *109* (16), 7976-7980.
8. Otten, D. E.; Petersen, P. B.; Saykally, R. J. *Chem. Phys. Lett.* **2007**, *449*, 261-265.
9. Jungwirth, P.; Tobias, D. J. *Chem. Rev.* **2006**, *106* (4), 1259-1281.
10. Pegram, L. M.; Record Jr., M. T. *Proc. Natl. Acad. Sci. USA* **2006**, *103* (39), 14278-14281.

XVII
The use of air bubbles to desalinate seawater without boiling

Muhammad Shahid, Richard M. Pashley

School of Physical, Environmental and Mathematical
Sciences, University of New South Wales, Canberra, Australia.
R.Pashley@adfa.edu.au

Summary

An unusual property of salt water, that is its ability to inhibit air bubble coalescence, has been used as the basis for a new method of seawater desalination. The coalescence, rise rate and vapour equilibration rate of bubbles within a simple bubble column is surprisingly complex and is still not fully understood. This is an important area because of its potential applications in desalination, solute concentration, evaporative cooling and a recently developed low temperature sterilization process. The thermal balance between the heat supplied to a column through warm gas bubbles and the heat required to fill the bubbles with water vapour can also be used as a fairly accurate method for determining the heat of vaporization of concentrated salt solutions.

Background

We need to drink about 2.5 litres of fairly pure, fairly salt-free water each day to survive. However, as is the case with all chemicals a small amount is fine but excess is dangerous (Paracelsus 1493-1541). In fact 4 litres of tap water taken quickly, or even worse purified distilled water, can cause hyponatremia, which has happened to marathon runners who have died from this rapid excess. Sports people now drink salty 'sports' drinks to prevent this loss of body salts. Our body relies on a fairly high level of salt (about 0.17M) in our body fluids and so drinking an excess of pure water rapidly dilutes this salt level and amongst other effects it damages the operation of our nerve cells.

Although we evolved from the salty sea several billion years ago, the sea has continued to dissolve rocks and collect salt and now has a higher salt level, of about 3% compared with the 1% in our body fluids. We cannot drink seawater because our bodies would lose more water in diluting the seawater than we could get back from drinking seawater. The recommended salt level limit for drinking water is <0.05%, so we need to dilute seawater by a factor of about 100. Of course, most of the water on the planet is in the sea (about 97%) and the remaining fresh water is mostly trapped in polar ice and glaciers. Only a very small proportion exists as potable water, readily accessible in fresh water rivers, lakes and reservoirs – that is only about 0.007% of the total water on Earth! It is not surprising, therefore, that the production of drinking water from seawater is one of the most important problems facing us in the near future.

Producing drinking water from seawater has a long history. Aristotle (384–322 BC) commented that pure water can be made by the evaporation of seawater. In ancient times, many civilizations used distillation to produce drinking water on their ships. Aristotle also carried out some experiments on removing salt from seawater by filtration and ion exchange, by flowing it through a high surface area porous material such as sand and clay. Simply by digging a hole near the seashore and allowing the seawater to percolate through the sand reduces the salt level. Modern ion exchange processes were developed following the discovery by Adams and Holmes in the 1930s that crushed phonograph records could be used as efficient ion exchange filters.[1] These were the first plastic or organic resins, which are now widely used by industry for water treatment and desalination.

The two main processes currently used for seawater desalination are based on the ancient methods of boiling and filtration. Ion exchange is normally used for the desalination of brackish water. In large scale, commercial thermal desalination processes, boiling is usually carried out under a reduced pressure to depress the boiling point. This process is very effective at producing clean water but it is costly in terms of energy requirements and is usually only cost effective when combined with an available source of waste industrial heat, for example from a power station. The second process of increasing importance is called reverse osmosis (RO) filtration. In this process an asymmetric filtration membrane is used which contains a thin surface layer of pores so fine that only water molecules can pass through. Unfortunately, high pressures (of about 70 atm) have to be used to force the water through the pores at a reasonable rate and the pores easily become clogged and so this process is also fairly costly as care has to be taken to pre-filter and clean the seawater prior to RO filtration.

It is possible to obtain high quality drinking water from seawater without boiling. This can be achieved using a remarkable but still unexplained property of salt in water, which was first discovered by Russian mineral flotation engineers in the 1930's, that adding salt to a flotation chamber significantly reduced bubble size and hence improved efficiency.[2] This reduction in bubble size occurs because the bubbles formed at a porous sinter, or frit, do not readily coalesce above a certain salt concentration. There is still no clear explanation for this phenomenon, although it has been well studied.[3-7] Some salts inhibit bubble coalescence and some have no effect and this can be described concisely by a combining table of cations and anions.[3,4] Common salt does cause bubble coalescence inhibition and this effect reaches a maximum at about 0.17M, which surprisingly, happens to be the salt level in the human body. It has been suggested that this effect is important because it is related to our evolution and that it protects our body from decompression sickness, even at atmospheric pressure.[3,4] Further increases in the salt level do not further increase the effect. The foaming of waves on the seashore is also largely due to this effect of salt, especially for clean seawater.

This inhibition of coalescence allows us to produce a high volume fraction of bubbles within a column of salt water. Warming the salt solution to about 50-60°C has only a slight effect and actually reduces bubble coalescence still further.[4] Hence, it is possible to continuously flow, initially dry, air bubbles through a column of heated seawater to produce a flow of saturated air from which pure water can be condensed.[8] This method has significant potential advantages over the other commonly used processes. For example, the seawater feed does not need pre-cleaning as it does in RO filtration and the problems associated with the control of boiling, in producing scale, are also removed. In addition, bubble column desalination is well suited to the direct application of sustainable sources of energy, such as wind power and solar heating.

If the bubble column is not heated, the high value of the latent heat of vaporization of water causes the column to cool. As an example, inlet dry air at about 25°C causes the column to cool to a steady state temperature of about 8°C.[9] It has, therefore, been suggested that this process offers a new technique for evaporative air conditioning.[9] The rate of cooling of the column depends upon the gas temperature, gas flow rate, initial column temperature and the volume of water in the column.[9] However, the final steady state temperature of the column only depends on the inlet temperature of the gas. The steady state equilibrium temperature is reached when the heat from the gas entering the system is precisely balanced by the heat required to vaporise water to completely saturate the incoming dry gas.[9]

Hence, the temperature attained by a thermally insulated bubble column, cooled by this evaporative cooling affect, can be calculated directly from the temperature of the inlet gas, its heat capacity and the heat of vaporization (ΔH_{vap}) of the water or salt solution in the column, at that steady state temperature, and the saturated water vapour density in air at that temperature (ρ_v in mol/m³). The heat capacity (C_p) of a gas such as nitrogen (or air) is quite constant over a fairly wide temperature range and has a value of about 1 J/gK, at a constant pressure of 1 atm.[9] The saturated water vapour density in the equilibrated bubbles can be obtained using the known or measured vapour pressure of the column solution, at the steady state temperature, via the ideal gas equation.

When the bubble column has reached the steady state temperature (T_e), each new bubble entering the column must, on average, supply precisely the amount of thermal energy (and contraction work) required to evaporate water to saturate that bubble, at that temperature.[9] The amount of heat (and work) supplied by each gas bubble, as it cools by ΔT, is given by the difference between the inlet gas temperature (T_i) and the temperature of the gas as it exits the system (T_o) multiplied by the specific heat of the gas, at constant pressure, in units of J/m³K. The exit gas temperature (T_o) may be equal to the steady state temperature of the column (T_e) or can be slightly higher.

Another work term, which has to be included in the energy balance, is the work supplied to the column by the decrease in pressure of the gas as it passes through the column and sinter. This work is given by the pressure difference (ΔP) between the gas entering and leaving the column and this can be readily measured using a differential manometer. The energy balance, once the bubble column has reached a steady state temperature, is therefore given by the relation:

$$\left[\Delta T \times C_p(T_e)\right] + \Delta P = \rho_v(T_e) \times \Delta H_{vap}(T_e) \qquad (1)$$

(in units of J/m³)

Typical results for seawater and concentrated NaCl solution, calculated using equation (1), for a column bubbled with nitrogen gas at different inlet temperatures are given in Figure XVII.1. These results clearly demonstrate the strong evaporative cooling effect of a bubble column and show that very hot gases are required to maintain a high operating temperature. This is simply because of the relatively

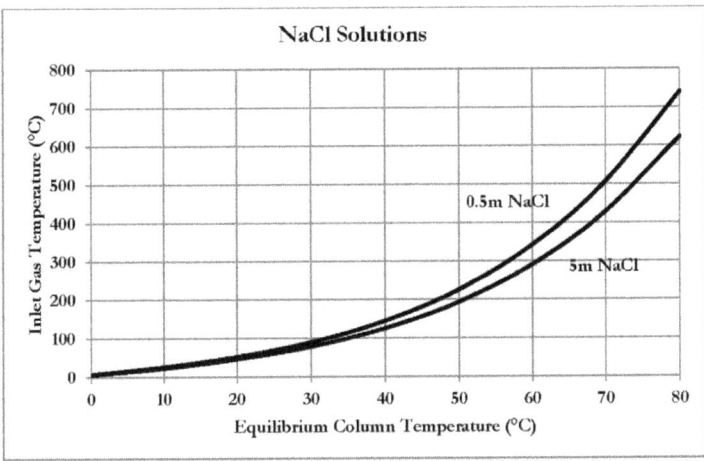

Figure XVII.1. Bubble column equilibrium temperatures calculated using equation (1) for nitrogen gas at 1 atm in concentrated NaCl solutions over a range of inlet gas temperatures.

low heat capacity of gases and the high latent heat of vaporization of water. Equation (1) has recently been used as a method for measuring the latent heat of vaporization of concentrated salt solutions.[9]

The higher the column temperature, the higher the water vapour density collected by the bubbles and hence the higher the rate of desalination. Separate, supplementary heating of the solution would probably be required for large-scale desalination applications. Current commercial thermal desalination units operate at a reduced pressure to give a lower boiling point, of about 80°C. The water vapour density produced in these processes is the same as that produced within a bubble column operating at the same temperature, i.e. 80°C.

A suitable high density bubble column can be produced by pumping air continuously through a 40-100 microns pore size glass sinter to produce a continuous stream of bubbles within a column filled with water or salt solution. When using an aqueous NaCl solution of about

0.15M, or more, finer bubbles are produced (of about 1-3mm diameter) giving an opaque column, because of the salt inhibition effect.[3,4] These bubbles rise at a limited rate of between about 15 and 35 cm/sec in quiescent water because they undergo oscillations in shape and rise trajectory which dampen their rise rate.[10,11] These oscillations also increase the rate of transfer of water vapour into the bubbles and enhance the rate of water vapour collection. Equilibrium vapour pressure within the bubbles is therefore attained quite quickly, within a few tenths of a second and the bubbles will therefore reach saturated vapour pressure within a travel distance of about 5-10 cm.[10] This means that a simple bubble column can be used to efficiently collect water vapour over a modest distance and time period. Unfortunately, however, only very limited data is currently available on the rise rate for different bubbles sizes and for higher column temperatures. Although there is a lack of detailed information on the fundamental processes involved in the bubble column evaporator, the technique has recently been used in the development of several new applications. These include sub-boiling desalination,[8] measurement of the latent heat of vaporization of concentrated salt solutions,[9] evaporative cooling[9] and, most recently, low temperature sterilization.[12]

Methods and techniques

A high surface area air/water interface can be continuously produced by pumping air through a porous glass sinter into a water-filled glass column. This is the basis of the bubble column evaporator. Bubbling at a modest rate through a 40-100mm glass sinter into a column filled with salt water at about 0.15M NaCl, or more, readily produces fine bubbles, in the approximate size range of 1-3mm diameter, and an opaque column. A typical apparatus used to control the temperature of the air flowing into a glass sinter column is shown in Figure XVII.2. In this system the (dry) air temperature can be varied using an electrical air heater with a thermocouple temperature monitor and an AC Variac

The use of air bubbles to desalinate seawater without boiling   357

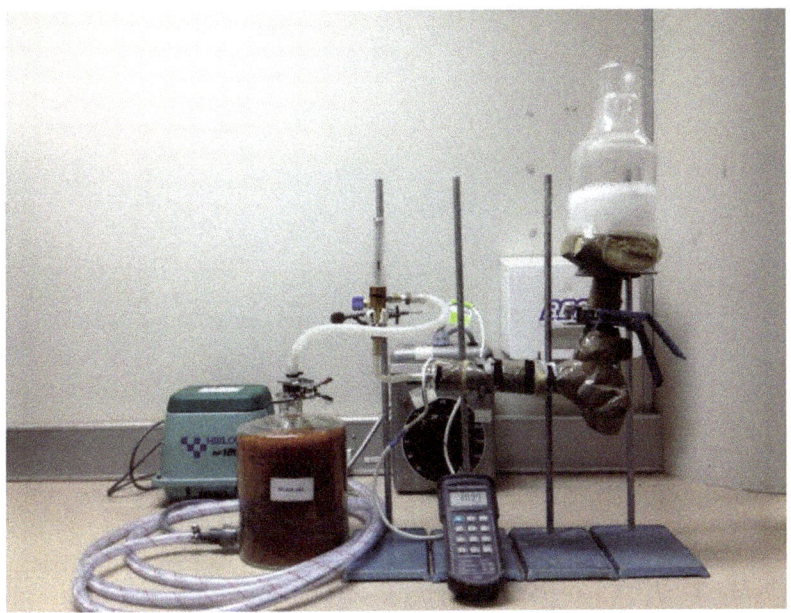

Figure XVII.2. Photograph of the bubble column desalination apparatus.

electrical supply controller. The actual temperature of the air flowing into the solution can also be measured using a thermocouple, with no solution present. The temperature of the column solution can be continuously monitored using a thermocouple positioned within the column solution. This type of system allows inlet gas temperatures up to 250°C. Further technical details are available in Refs. 9 and 12.

Some recent results on the effect of high temperature gases on bubble coalescence

The air bubbles formed within columns of pure water with an air inlet temperature of up to 250°C are quite large, up to a cm in diameter (see Figure XVII.3). However, addition of 0.15M NaCl (or higher concentrations) produces much finer bubbles (1-3mm diameter), due

Figure XVII.3. Photograph of a bubble column containing pure water with an inlet gas flow rate (RT) of 23 L/min at an inlet temperature of 250°C. The column solution temperature was about 53°C.

to coalescence inhibition, even at these high air inlet temperatures, as shown in Figure XVII.4. The only difference between the photographs in Figure XVII.3 & 4 is the addition of 0.5M NaCl to the bubble column. The remarkable observations shown in Figures 3 and 4 clearly demonstrate that using high temperature inlet gases has little or no effect on the inhibition of bubble coalescence caused by the addition of salt. The solution temperature of these columns quickly approaches the steady state value of about 53°C, as expected for this inlet gas temperature (see Figure XVII.1) and earlier studies have shown that bubble columns heated to about 60°C actually show a slight enhancement of the inhibition effect.[4] The effect of salt in inhibiting bubble coalescence is clearly dominant even though the gas inlet temperature is significantly above the boiling point of water. The film drainage time between two colliding bubbles will depend on many factors including bubble deformation but solution viscosity should also be a factor.[13] Hence it might be expected that coalescence would be enhanced at higher solution and gas temperatures, since the viscosity of the solution is reduced but this seems not to be the case.

Unfortunately, there is still no proper explanation for the effects of salt on bubble coalescence inhibition. Indeed, at first sight, adding salt would be expected to enhance coalescence because this increases the surface tension of water and hence the bubble's surface energy. Surface effects can arise because even though ions are repelled from the surface of water, due to an image-charge repulsion, once the bulk salt concentration is sufficiently high, ions will be forced to reside at

Figure XVII.4. Photograph of a bubble column containing of 0.5M NaCl solution with an inlet gas flow rate (RT) of 23 L/min at an inlet temperature of 250^0 C. The column solution temperature was about 53^0C.

the surface. To reduce energy, they will most likely adsorb as ion pairs and this could set up a local electrostatic field at the surface, which could immobilise adjacent water layers. If this happens it will induce a 'zero slip' boundary condition at the surface, which will reduce the rate of film drainage between two approaching bubbles, and so increase the likelihood of the bubbles separating before they can coalesce, especially within a turbulent bubble chamber. In addition, the observation that some salts inhibit coalescence and others do not,[3,4] could then possibly be explained in terms of their ability to create this local electrostatic field, as ions are forced into the surface region at high concentrations. Some salts will produce this immobilisation, whereas others will not. This important problem remains the subject of investigation.[6,7]

Increasing the salt concentration should effectively screen out any repulsive electrostatic force between charged bubbles.[14] Attractive van der Waals forces between bubbles may also be a factor,[14] as will any long-range hydrophobic forces.[14-16] However, the simple observation that forcing two bubbles slowly together in water or salt solution always produces coalescence suggests that the observations in bubble columns is related to dynamic processes. These will determine whether the film between two colliding bubbles can drain fast enough to allow coalescence before they are forced apart in the turbulence of the bubble column. It is generally accepted that bubbles will coalesce once the intervening film reaches about 50nm.[7]

As discussed earlier, it might be expected that as two air bubbles collide coalescence should be influenced by the rate of thinning of the intervening water film and this rate should depend on the viscosity of the intervening solution as well as on the rate of approach and the size of the bubbles, since these factors will also determine the degree of deformation of the bubbles.[13] The bubble sizes considered in this section, of 1-3mm, will undergo some degree of dimpling during collisions. Within a chaotic bubble column collisions occur continuously and within fractions of a second. If the film can drain during this collision time the bubbles will coalesce. Increasing the temperature of the column solution will significantly reduce its viscosity and this will act to increase the drainage rate of the film between approaching bubbles. This will also allow the film to drain before the bubbles are bounced apart within the turbulence of the column. Thus the combined effect of an increased column solution temperature and the very hot gas bubbles which must, at least close to the sinter, produce significant local heating in the solution surrounding the bubbles, should significantly reduce film viscosity and hence enhance bubble coalescence. However, this effect was not observed in these experiments.

The use of high temperature gases for efficient water vapour transfer

Since the use of high temperture inlet gases does not appear to affect the coalescence inhibition phenomena in concentrated NaCl solutions, studies have been carried out on the desalination efficiency of the bubble column process under these conditions. That is, high density bubble columns were used to continuously produce pure water (vapour) using high temperature inlet air in the range 150–250°C. In these studies the total weight loss of the solution was measured for a controlled total airflow, at different temperatures. For comparison, the expected vapour transfer rate for the bubble column operating under steady state conditions, discussed earlier, was also calculated.

At an inlet air temperature of 150°C, the measured total weight transferred was consistently greater, by about 6%, than the weight expected for a bubble column operating at thermal equilibrium. It was also found that the columns reached a steady-state operating temperature slightly above that expected from the energy balance in equation (1). Similar results were obtained for bubble columns operating with air inlet temperatures of 250°C, but with an 8% higher vapour transfer than expected for a similar column operating at thermal equilibrium. These initial results suggest that the water transfer rate obtained from a bubble column can be increased by using hot inlet gases.

The results obtained in these studies suggest that not all of the heat transferred to the column from the high temperature inlet gas was used to evaporate water and this produces a slightly higher column temperature than that expected for a column operating under thermal equilibrium. It is also interesting that although the expected thermal equilibrium, obtained from the balance of heat supplied and vaporisation energy, was not observed using high inlet gas temperatures, the column solutions still attained a steady state, but with a slightly higher temperature. Whatever the cause of this steady state, a higher rate of water transfer was produced by operating the bubble column under these conditions. These results suggest that the use of high gas temperatures might improve the thermal desalination process. Such gases are available from many industrial processes, which produce waste heat in the form of hot gases, often emitted directly to the atmosphere.

The effect of gas bubble humidity and vapour equilibrium on bubble coalescence

The cooling effect in a gas bubble column described by equation (1) and illustrated in Figure XVII.1 for NaCl solutions is caused by the thermal energy required to vaporize water into the continuous flow of dry gas bubbles. All of the earlier studies on the effect of salt on bubble coalescence within bubble columns have used dry inlet gases, which must cause the evaporation of water vapour into the bubbles formed at the sinter. However, this evaporation can be completely prevented by the use of inlet gas (air) pre-equilibrated with water vapour to produce a 100% relative humidity at the temperature of the bubble column.

Experiments have been carried out on bubble columns using both dry air and fully saturated air, comparing the degree of bubble coalesce by simple visual observation. These studies clearly demonstrate that the humidity and vapour evaporation process has no observable effect on the degree of bubble coalescence inhibition produced by added salt. Bubble column studies using high temperature gases also support the conclusion that the addition of salt is the main cause of coalescence prevention, since high temperature inlet gases must produce an even greater evaporation rate. Typical results are shown in the photograph in Figure XVII.5, which was for a bubble column containing 0.5m

Figure XVII.5. This is a photograph of a 0.5m NaCl salt solution within the bubble column with room temperature inlet gas pre-humidified to 100% relative humidity. The observed inhibition effect on bubble coalescence was found to be unchanged by using pre-humidified inlet gases.

NaCl solution with room temperature, saturated air inlet. The bubble size distribution and bubble density was found to be very similar to that observed for dry air inlet gas within the same column and salt solution at the same temperature.

These results clearly show that the evaporation of water vapour, which occurs soon after dry bubbles enter the column, has no effect on the bubble coalescence inhibition due to added salt. This result is surprising because it has been established that bubbles of these sizes reach equilibrium water vapour saturation within about a few tenths of a second or a travel distance of only about 5cm from the sinter. The water vapour must be transferred rapidly across the bubble surface and hence should produce a local salt concentration enhancement around the bubble. This effect should be further enhanced between two colliding bubbles close to the sinter. This local increase in salt concentration around the bubbles and between colliding bubbles would generate a repulsive osmotic pressure, which would act to force the bubbles apart, and hence may explain the coalescence inhibition effect. However, since there was no difference in the effect of salt on coalescence for 100% humid bubbles (with no net vapour transfer into the bubbles) and dry bubbles, clearly this evaporation process, and any resultant local osmotic pressure build up, appears to play no significant role in the phenomenon.

There are two possible explanations for this. Bubbles tend to coalesce when they reach approach distances of about 50nm.[7] However, any enhanced salt concentration may develop only in a thinner region of solution next to the bubbles and hence will have little influence on the film thinning process. It is also possible that the rapid motion of bubbles in the column and the local flow of solution past the bubble's surface acts to dissipate any significant build up in solute concentration.

Conclusion

The simple bubble column evaporator offers an interesting challenge to our understanding of the detailed processes involved in bubble rise rate, water vapour evaporation and the variable effects of different solutes on bubble coalescence inhibition. Fortunately, the most important and common salt, sodium chloride, acts in solution to inhibit bubble coalescence and that behaviour has been applied to the development of a wide range of useful techniques based on the bubble column evaporator. This complex system has in recent years been used to develop new methods for sub-boiling, thermal desalination, evaporative cooling, controlled solute concentration, low temperature sterilization and as a technique for the direct measurement of the latent heat of vaporization of concentrated salt solutions. These important applications suggest that the bubble column evaporator should be studied in greater detail in order to properly understand the fundamental processes involved.

Acknowledgements

The authors would like to thank Mr. Chao Fan for assistance with the high humidity inlet gas experiments.

References

1. Adams, B.A.; Holmes, E.L. *J. Soc. Chem. Ind.* **1935**, *54*, 1935.
2. Klassen, V.I.; Mokrousov, V.A. *"An Introduction to the Theory of Flotation"*. Butterworths, London. **1963**.
3. Craig, V.S.J.; Ninham, B.W.; Pashley, R.M. *Nature.* **1993**, *364*, 317-319.
4. Craig, V.S.J.; Ninham, B.W.; Pashley, R.M. *J. Phys. Chem.* **1993**, *97*, 10192-10197.
5. Craig, V.S.J. *Curr. Op. Coll. Interface Sci.* **2004**, *9 (1-2)*, 178-184.
6. Marcelja, S. *Curr. Op. Coll. Interface Sci.* **2004**, *9 (1-2)*, 165-167.

7. Horn, R.G.; Del Castillo, L.A.; Ohnishi, S. *Adv. Coll. Interface Sci.* **2011**, *168(1)*, 85-92.

8. Francis, M.; Pashley, R. *Desalination and Water Treatment.* **2009**, *12(1-3)*, 155-161.

9. Francis, M.J.; Pashley, R.M. *J. Phys. Chem.* **2009**, *113*, 9311-9315.

10. Leifer, I.; Patro, R.K.; Bowyer, P. *J. Atmos. Ocean Tech.* **2000**, *17*, 1392–1402.

11. Clift, R.; Grace. J.R.; Weber, M.E. *"Bubbles Drops and Particles"*. Academic Press, New York, 1978.

12. Shahid, M.; Pashley, R.M.; Rahman Mohklesur. A.F.M. *Deasalination and water treatment.* **2013**, 1-9.

13. Frankel, S.P.; Mysels, K.J. *J. Phys. Chem.* **1962** *66*, 190.

14. Israelachvili, J.N. *"Intermolecular and Surface Forces"*. Academic Press. 1992.

15. Israelachvili, J.N.; Pashley, R.M. *Nature.* **1982** *300*, 341-343.

16. Christenson, H.K.; Claesson, P.M. *Adv. Coll. Interface Sci.* **2001**, *91(3)*, 391-436.

XVIII
Water as a probe for the fractal structure evolution of cement: the effect of organic additives

Francesca Ridi, Emiliano Fratini, Piero Baglioni

Department of Chemistry "Ugo Schiff" and CSGI, University of Florence, 50019 Sesto Fiorentino (Firenze), Italy.
ridi@csgi.unifi.it

Overview

Cement is the binder at the base of concrete, the most used synthetic building material in the whole history. Since Roman times, its peculiar characteristics, developed because of the hydration reaction with water, have been widely exploited for the building of stunning and resistant edifices, memories of the Roman Empire. In spite of its widespread use and the abundant literature accumulated during a century of systematic scientific research on this material, the full understanding of its properties is far from complete. Several issues are still open, mostly related to the understanding of the hydration mechanism and to the comprehensive description of the microstructure. The hydration process is a complex series of chemical reactions between water and the anhydrous calcium silicates and aluminates constituting cement, which forms the hydrated phases responsible for the strengthening of the matrices.

Apart from being one of the leading actors in the curing process, water can act as probe to investigate the structural characteristics of the hydrating pastes. This contribution describes that the freezing-melting process of the water confined in the cement pastes can be followed by calorimetry and used to investigate the microstructure, providing information on the porosity and on the fractality of the evolving cement matrices.

Introduction

> There is [...] a kind of powder which from natural causes produces astonishing results. It is found in the neighborhood of Baiae and in the country belonging to the towns round about Mount Vesuvius. This substance, when mixed with lime and rubble, not only lends strength to buildings of other kinds, but even when piers of it are constructed in the sea, they set hard under water" (Marcus Vitruvius Pollio, Liber II, *De Architectura*, ~25 BC).

In these words the Roman engineer Vitruvius firstly described the properties of a material with surprising properties: a mixture of lime and crushed volcanic ashes was able to set under water, the resistance being increased along the time, in a way completely different to any other material. This "magic" material was termed *pozzolanic* from Pozzuoli, the city near Vesuvio where the ashes were taken from.

The reason of the success of the Roman concrete was the substitution of the usual crushed stones with volcanic ashes. According to the modern knowledge, a siliceous and aluminous material in itself does not possess any cementitious value, unless it is calcined and converted in an amorphous reactive form. The volcano naturally performed this calcination process at the Roman times. Nowadays, cement is artificially produced by high temperature calcination (1450 °C) of limestone and clay. Once cooled, the so obtained *clinker* is ground to give a mixture of calcium silicates and aluminates able to harden when mixed with water.

After more than a century of systematic studies, basic questions are still unsolved regarding the structural properties of cement pastes, on their effects on concrete behavior, and on the chemical and the physico-chemical mechanisms involved in the hydration reaction, especially in the presence of organic polymers. Most of these questions directly relate to the primary hydration product which is also the main binding phase of Portland cement pastes: the calcium silicate hydrate gel (C-S-H).[1]

The hydration reaction

When anhydrous cement is mixed with water, several exothermic chemical reactions take place both simultaneously and consecutively. These reactions are commonly denoted with the term *hydration* and schematically shown in Figure XVIII.1. The heat evolution versus time allows the identification of five main contributions to the overall hydration process. In the very first minutes after the adsorption of water on the surface of the dry powder, a small fraction of the silicate and aluminate phases dissolves producing a $Ca(OH)_2$ supersaturated solution with a final pH higher than 12. Moreover, on the same time scale some very fast conversions of the aluminate phases occur, that are responsible of a strong heat evolution usually modulated by the addition of gypsum. The second period, lasting few hours, is the induction (or dormant) period, characterized by a very low heat evolution and by a decrease in the workability of the paste. The dormant period ends

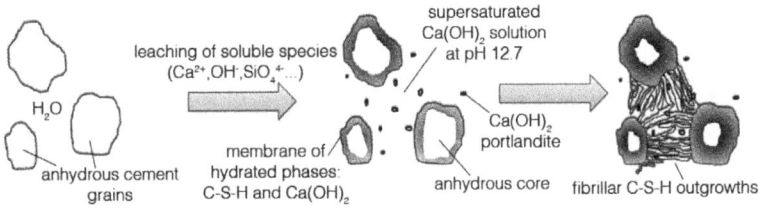

$$Ca_3SiO_5 + 3.1\ H_2O \rightarrow (CaO)_{1.7}\ SiO_2\ (H_2O)_{1.8} + 1.3\ Ca(OH)_2$$

Figure XVIII.1. Schematic representation of the early stages of C_3S hydration.

because new hydrated phases, mainly C-S-H, begin to precipitate from the solution, nucleating on the existing grains. These reactions are accompanied by a strong heat evolution, and are usually referred to as the *acceleration period*. The precipitation processes favor the further dissolution of the anhydrous phases, through an incongruent process. The hydrated phase responsible for the binding characteristics of the cement, is an amorphous calcium silicate hydrate, called C-S-H, having a high surface area and the properties of a rigid gel. During the subsequent period the hydration rate strongly decreases (*deceleration period*) and finally a slow, continued reaction occurs. This last part of the hydration is called *diffusional period*, because its kinetic rate is mainly determined by the diffusion rate of water molecules from the solution through the hydrated phases, to reach the anhydrous grains.

The microstructure of C-S-H

Many macroscopic properties of cement specimens (e.g. elasticity, compressive strength, resistance to degradation, transport phenomena) depend on the microstructure of C-S-H. For this reason, an important part of the literature is devoted to the non-trivial task of providing a comprehensive picture of C-S-H structure from nanometers (C-S-H unit globules, interlamellar water filled spaces) to tens of micrometers (largest capillary pores).[1-12] In particular many efforts have been oriented to assess the intrinsic relationship between the evolving microstructure and the final macroscopic properties.[13-15] For example, one of the most relevant industrial challenges in the cement chemistry field is the reduction (or at least the control) of the degradation mechanisms due to the permeation of salt-rich water because of atmospheric agents. This event is strictly related to transport phenomena taking place in the specimen. Moreover, the increasing use of organic polymers in the industrial practice, with the aim of modifying the hydration kinetics and tune the final properties of the material, compels to account for their effect on the developing C-S-H structure. In a recent work it was pointed out that the addition of last generation superplasticizers (SPs) to tricalcium silicate pastes modifies both the structure of the C-S-H basic globules and the overall microstructural arrangement.[16]

Many structural models have been proposed over the past years with the aim of detailing the microstructure of C-S-H in a systematic picture able to reconcile all the experimental evidences. The first models were based on the structural similarity with naturally occurring layered silicate minerals, such as tobermorite and jennite.[17] Because of the layered structure of these minerals, the existence of interlayer spaces containing strongly absorbed water was postulated. These models have been fairly successful in qualitatively explaining the shrinkage behavior and gas sorption properties of cement pastes. However, they are inadequate to fully address the C-S-H properties, especially regarding the viscoelastic response of C–S–H gel to mechanical loading (creep) and the relative humidity changes (drying shrinkage). More recently, models based on molecular modeling have been developed based on a bottom-up atomistic simulation metodology.[9] These theoretical approaches, validated by experimental measurements proved that the C-S-H gel structure includes both glass-like short-range order and crystalline features of the tobermorite.

The recognition of the colloidal and gel-like nanostructure of C-S-H has progressively gained importance. The first studies date back to the work of Powers and Brownyard, in the 1950s.[18] They described the broad structure of the material based on the experimental evidences coming from total and non-evaporable water contents and water vapor sorption isotherms. The colloidal description of C-S-H has been later included in other models[19] and nowadays it is definitely the most accepted assumption, being the most appropriate to address the complex properties of calcium silicate hydrate. A milestone in the development of the modern colloidal model for the C-S-H microstructure is the work published by Allen and coworkers in 1987, in which the microstructure of a hydrating cement paste was monitored by small angle neutron scattering.[20] In this work, the authors assessed the presence of a growing population of 5 nm gel globules just after the induction time. These globules successively aggregate in structures with correlation lengths around 40 nm. In recent years, based on the huge amount of data present in the literature, H.M. Jennings in a series of papers formulated a clear and coherent model for the calcium silicate hydrate microstructure.[21–23] The basic idea behind this model is that the bulk microstructure of the calcium silicate gel is formed as a consequence of the packing of the basic globules having peculiar shape and internal structure. Clusters of these particles group together in two packing densities, known as high density (HD) and low density (LD) C-S-H.[24] The first version of the model (*Colloidal Model-I*, CM-I) primarily focused on explaining how the properties of the material depend on the packing behavior of these basic globules.[21] In the description of the CM-I, the Jennings' effort was mainly directed to correlate the results obtained for the measurement of cement specific surface area by means of different techniques. In this version, the influence of the globules' internal structure on the bulk properties was not explored in detail. Recently, the CM-I has been improved and extended to explicitly take into account the smallest porosity of the C-S-H phase associated to the internal structure of the basic globules.[23] The *Colloidal Model-II* (CM-II) represents a significant advancement in the description of cement microstructure, since it reconciles many controversial data reported in the literature. In particular, this model recombines the "layer-like" into the "colloidal-like" scheme giving a rather exhaustive interpretation of the sorption isotherm experiments. According to CM-II, the microstructure of a cement paste can be schematically described as reported in Figure XVIII.2: the basic globule is a disk-like object, whose thickness is around 4 nm, having a layered internal structure similar to tobermorite and jennite. The water inside the globule is located both in the interlamellar spaces and in very small cavities (intraglobular pores, IGP), with dimensions around

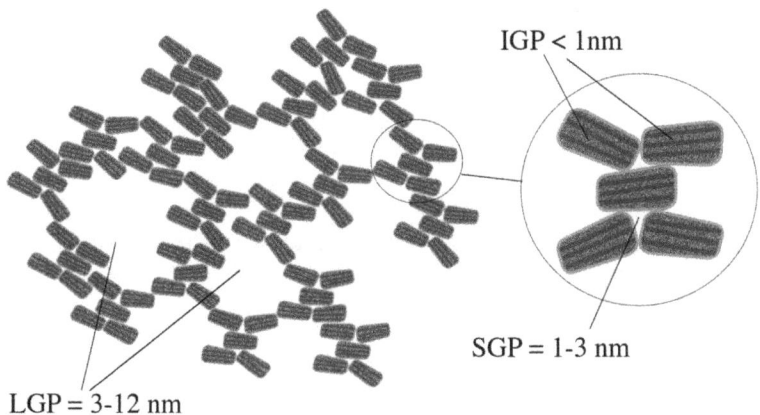

Figure XVIII.2. Schematic representation of the C-S-H microstructure according to the Jennings' *Colloidal Model II* (CM-II).

1 nm. The packing of these globules produces a porous structure, where two other main populations of pores can be identified: the small gel pores (SGP), with dimensions of 1-3 nm; and the large gel pores (LGP), 3-12 nm in size. The inclusion of the sub-nanometric porosity in the model justifies most of the experimental evidences, representing a crucial step for the understanding and the control of the relationships between structure and properties.

Using water to probe the porosity

The conventional methods to assess the porosity of cement samples are Mercury Intrusion Porosity (MIP) or analysis of images from Backscattering Scanning Electron Microscopy.

Alternatively, the thermal behavior of the confined water can be used to monitor the evolution of the cement paste porosity throughout the hydration process. The use of freezing/thawing cycles of samples whose porosity is saturated with water to assess their porosity is known as *thermoporometry*[25] and has been applied to the investigation of many solid systems. The application of this technique to cement samples can provide information on the evolving microstructure throughout the whole hydration process. The correct interpretation of the data, however, is not trivial and requires the assumption of a proper microstructural model.

By cooling a sample from room temperature to -80 °C and heating it again to room temperature (at slow rate, 0.5 °C/min to ensure the quasi-equilibrium conditions to be preserved), a typical thermogram of a cement sample obtained by Differential Scanning Calorimetry (DSC) appears as shown in Figure XVIII.3. The heating scan (from -80 °C to room temperature) shows a single hump over the whole temperature range, while the cooling scan presents some definite peaks. As documented in the literature,[26–32] being the distribution of

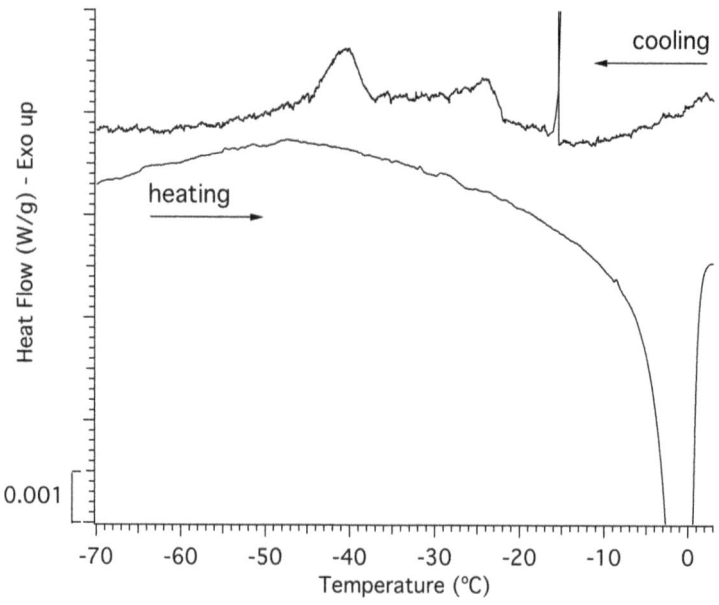

Figure XVIII.3. Heating and cooling of a typical LT-DSC thermogram recorded on cement samples.

the pore sizes in cement almost continuous, the hump in the heating curve is due to the melting of the water confined in cavities of progressively larger dimension. On the contrary, the appearance of peaks in the cooling scan has to be ascribed to the combination of the two water-freezing mechanisms: homogeneous and heterogeneous. The homogeneous nucleation is an activated process, because a free energy barrier must be surmounted to form a critical nucleus. The heterogeneous nucleation occurs at preferential sites and requires less energy than the homogeneous process. According to a very recent paper by Sanz et al.,[33] at low degree of supercooling (-20 °C < T < 0 °C), due to the high free energy required for the homogeneous formation of a critical cluster, only the heterogeneous nucleation is possible. For this reason, in a saturated system, the water confined in cavities whose dimension is large enough to host critical nuclei of at least ~8000 water molecules (size ≈ 8 nm, that is the size of the stable critical cluster at T≈ -15 °C) freezes *via* heterogeneous nucleation, even if it is isolated from the surface. Once the water in the capillary pores is frozen, the formed ice remains in contact with the liquid water present in the smallest cavities. By lowering the temperature below T< -20 °C, the size of the critical cluster sensibly decreases (at T≈ -35 °C the size reduces to about 3.5 nm corresponding to ~600 water molecules).[33] Moreover, the nucleation free energy barrier decreases, making the homogeneous process possible to occur. In these conditions both the homogeneous and the heterogeneous nucleation mechanisms become accessible, with comparable rates. Then water will freeze in the pores whose dimension is large enough to host the critical cluster stable at that temperature. The appearance of peaks in the cooling curve indicates, however, that the nucleation occurs at a preferential site, most likely at the pore entrance, where the liquid water is in contact with the surrounding ice. As previously reported in line with the *CM-II*, in the case of cement pastes, the hydration process originates two classes of nanometric porosities, SGP and LGP,[23] The dimensional range of SGP (namely, 1-3 nm) is compatible with the peak around -40

°C in the thermogram,[25,34,35] while the freezing of the water confined in LGP pores (3-12 nm) occurs in the -20/-35 °C temperature range.[25,34,35]

By following the changes of the DSC thermograms during the hydration it is possible to monitor the evolution of the microstructure. Furthermore we evaluated the effect of some common chemical additives on the microstructure of C-S-H, by following the evolution of the porosity during the hydration of C_3S pastes containing polymethacrylic acid chains partially esterified with polyethylene oxide lateral chains (PCEs).[36,37]

After the acceleration period the microstructure of the samples has formed, due to the growth of the hydrated phases. At this stage, the pastes have been soaked with water in order to saturate the whole porosity. This procedure allows extracting quantitative information on the total pore volume and on its evolution during the hydration reaction. The DSC thermograms registered afterward show three main features: an intense sharp peak in the -10/-20 °C range, due to the crystallization of bulk water contained in capillary pores; a peak in the -20/-35 °C region, corresponding to the crystallization of water in LGP pores; a peak at -40 °C due to the solidification of the water constrained in the SGP porosity.

According to a procedure reported elsewhere,[35] each peak was integrated, and the areas were used to calculate the amount of water involved. As the standard enthalpy of fusion varies with the temperature, to quantify the water we used a value of ΔH_0[38] estimated at the mean temperature corresponding to each integration range. Figure XVIII.4 shows the results of this calculation. The histograms display the pore volume of capillary pores, LGP and SGP as a function of the hydration time.

In all the samples the capillary pore volume evolves according to a depercolation process, showing a decreasing behavior during the hydration process as a result of the increase of the solid volume fraction (i.e. the hydrated products have molar volume higher than the anhydrous ones). In the hydrating C_3S/water paste (Figure XVIII.4A) the volume is almost constant during the first 14 days, and starts to decrease at 28 days showing a drastic drop only after 6 months. In this sample the depercolation threshold can be then estimated to occur after 28 days from the mixing time. The addition of PCEs sensibly decreases the capillary volume in respect to the C_3S/water sample and, in some cases, alters the percolation threshold. The sample with PCE102-2 (having among these polymers the lowest adsorption propensity, plasticizing efficiency[39] and retarding power[37]) shows a behavior very similar to the paste without additives, the capillary pore space remaining percolated throughout the first 28 days of hydration, and showing a reduction only after 6 months. When C_3S is hydrated in presence of PCE102-6 (Figure XVIII.4C) the depercolation of the capillary porosity is shorten (with respect to C_3S/water paste) at 7 days, while PCE23-2 and PCE23-6 (Figures XVIII.4D and XVIII.4E) maintain the depercolation threshold at 28 days after mixing, despite their very effective retarding action on the reaction kinetics, which cause the initial dormant period to stop after 42 and 260 hours respectively (as reported previously[37]).

The LT-DSC technique also provides the evolution of the finest microstructure in the samples. It is evident that the SGP volume is always higher than LGP, meaning that, apart from the capillary pores (acting as internal water reservoirs), the microstructure due to the growth of C-S-H gel mainly consists of a network of SGP pores. The analysis of the LT-DSC data shows that the volume of the nanometric pores (both SGP and LGP) in all the saturated samples does not change much, maintaining almost constant values throughout the hydration process.

The fractal nature of C-S-H

The hydration reaction of cement pastes consists in a "gelation" process, where the growing C-S-H gel links together the solid grains

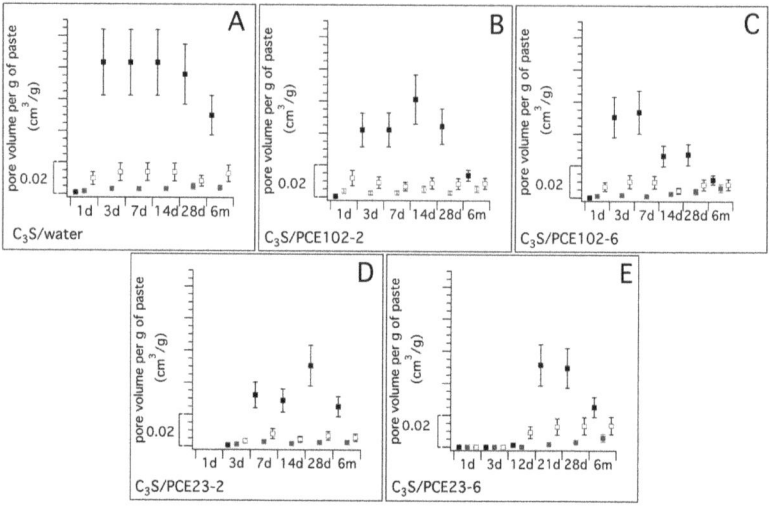

Figure XVIII.4. Histograms showing the evolution of the pore volume (cm³ per g of paste) during hydration: capillary porosity (solid grey), LGP (grey lines) and SGP (black dots).

and generates a disordered porous system. This kind of process is efficiently described by the *percolation theory*.[40,41] By a physical point of view the term percolation refers to the formation of a long-range connectivity in stochastic systems and its occurrence produces random fractal structures. Describing evolving porous systems through the percolation theory allows the evaluation of the *percolation threshold* and of the *fractal dimension*. Both these parameters are directly linked to the macroscopic characteristics of the material. For example the transport properties inside a cement specimen (and, as a consequence its resistance to the weathering) are greatly influenced by the *capillary porosity depercolation threshold*: at this time the capillary porosity starts to decrease, because of the growth of the hydrated phases, becoming "closed" towards the external surface. This implies that the *"in situ"* determination of the depercolation threshold could be important to understand and control the transport properties that, in most degradation mechanisms, govern the rate of damage of the specimen and hence the long term durability of the material.[32,42–44]

The fractality of cement pastes provides a quantitative measurement of the packing of the structure. For this reason, it has been extensively studied by means of small angle scattering (SAS) techniques[8,16,20,45–48] because the length scale of the structure (1-1000 nm) matches the dimension of the x-ray or neutron probe (i.e. the inverse of the scattering vector). These techniques do not require drying procedures and can be directly performed on wet samples. Hence, SAS techniques are particularly suitable to investigate the evolution of the fractal arrangement during the microstructure development.[49]

Using water to probe the fractality

As the heating scan directly depends on the distribution of the pore sizes, its analysis provides information on the fractality of the samples. Several studies in the literature evidence the fractal nature of C-S-H and extract the mass fractal dimension, D_m, as a function of the degree of saturation,[46] the hydration time,[45,49] and the degradation degree.[50] As already mentioned, SAS techniques are generally the methods of choice to investigate these properties.

$$\Delta T = T_m^0 - T_m$$

Previous works report the possibility to extract these quantities in the case of porous systems saturated with water from the DSC incremental volume distribution, in very good agreement with SAXS patterns measured on the same system.[51] The rationale behind the extrapolation of fractal properties from DSC relies on the fact that the melting temperature of an ice crystal confined in a pore is depressed of a quantity related to the radius R, being $R=r-l$, where r is the radius of the pore and l is the thickness of the non-freezable layer of water at the solid interface (previous estimations of l from NMR measurements report a value of 0.5±0.1 nm for porous glasses).[52] The

Gibbs-Thomson equation states that the melting temperature T_m and the radius R are inversely related as follows:

$$T_m = T_m^0 \left(1 - \frac{2\gamma V_s}{\Delta H R}\right) \tag{1}$$

where T_m^0 is the melting temperature of an ice crystal of infinite dimension, γ is the solid-liquid interfacial tension, ΔH is the specific melting heat, V_s is the specific volume of the solid. For the water case (assuming T_m^0=273.15 K, γ=40·10^{-3} Nm^{-1}, ΔH=334 Jg^{-1}, V_s=1.02 cm^{3}g^{-1}), equation 1 becomes simply:

$$\Delta T = \frac{68.29}{R} \tag{2}$$

where R is given in *nanometers*.

Heating from -80°C to room temperature a porous sample with pore size distribution $P(r)$ and saturated with water produces the melting of the liquid confined in pores of progressively increasing dimension. In other words, the registered heat flux is proportional to the incremental volume dV of the ice that melts at each T_m temperature. To provide a quantitative estimation, the detected heat flow must be independent on the heating rate. According to the literature[53] a DSC thermogram registered at rates lower than 2°C/min maintains the equilibrium conditions, yielding real information on the accessible porosity.

From the definition given above, the incremental pore volume per solid mass can be written as $dV = P(r)dr$, where $P(r)$ is the pore size distribution. To obtain dV, the heating DSC signal has been normalized with the total pore volume V_p, obtained by integration of the whole melting peak, scaled by the bulk water density value at 0 °C (0.9998 g/cm^3).

Consistently with the fractal nature of the systems it can be assumed[53] that the heat flow, J_q, measured by DSC on porous glasses is related to ΔT by a scaling law. Furthermore, in some papers[51,54] the mass fractal of wet silica gels were probed from DSC data and compared with the results obtained from SAXS. To compare the DSC and the

SAXS approaches, Vollet et al.[55] used a method originally proposed to link SAXS and nitrogen adsorption data on porous matrices. In their approach, the system was regarded as a homogeneous solid of density ρ_S where an incremental pore volume per solid unit mass $dV = P(r)dr$ was used to account for the change of the bulk density of the porous sample, $\rho(r)$, as a function of the pore-filling steps. The resulting process can be mathematically described as:

$$\frac{1}{\rho(r)} = \frac{1}{\rho_S} + \int_0^r P(r)dr \qquad (3)$$

As a matter of fact, $\rho(r)$ will then scale with r, in the fractal range $a \leq r \leq x$, as:

$$\rho(r) = \rho_S (r/a)^{D_m - 3} \qquad (4)$$

where a is the characteristic dimension of the smallest repeating unit generating the fractal and x is the maximum correlation length of the fractal aggregate. Combining equation 2 and equation 4 the following relation holds:

$$dV = A(\Delta T)^{D_m - 3} \qquad (5)$$

The plots dV vs ΔT on a log-log scale for the samples C_3S/water and C_3S/PCEs/water at different sampling times are reported in Figure XVIII.5.

Table 1 reports the D_m coefficients extracted from the fitting of the log-log plots in Figure XVIII.5 from $\Delta T_x \cong 1$ K to $\Delta T_a \cong 10$ K, corresponding to pores with radius between $\xi \cong 70$ nm and $a \cong 7$ nm, as calculated by means of equation 2. This range is directly comparable with that investigated by SAS techniques.[45,56]

In accordance with the compacting effect due to the growing hydrated phases, D_m is found to increase in all the samples from values around 2.0 (typical of poorly packed systems), after three days, to values around 2.6, after 28 days of curing. This data are in very good agreement with previous SAXS investigation.[45] In the early stage of the hydration the pastes containing superplasticizers exhibit D_m values lower than C_3S/water. This means that a less packed nanoscale structure is formed in PCE-containing pastes with high w/c values (w/c=0.4). The need of reducing the water content in real applications involving superplasticizers is well-known, as the excess water is known to increase the capillary porosity. The present investigation shows that high w/c values also induce the formation of a nanostructure less dense than that of the C_3S/water sample, especially in the first part of the hydration process. However, after 6 months D_m reaches values around 2.4-2.6 even in the C_3S/PCEs cases.

Table 1. Mass fractal dimension D_m (±0.1) for the analyzed pastes during the hydration.

	3 d	7 d	14 d	21 d	28 d	6 m
Water	2.0	2.3	2.5	---	2.6	2.5
PCE102-2	2.0	2.2	2.1	---	2.1	2.5
PCE102-6	2.1	2.2	2.2	---	2.4	2.7
PCE23-2	---	1.9	2.4	---	2.2	2.3
PCE23-6	---	---	---	2.6	2.5	2.5

Conclusions

It is known that some macroscopic properties of cement (degradation phenomena, elasticity, compressive strength, etc.) are influenced by porosity. For this reason the elaboration of methods able to easily access the characteristics of the pore structure during the hydration process is a task of primary importance. In this paper we showed that water can be used as an effective probe of the cement microstructure, by taking advantage of its peculiar freezing/thawing behaviour when confined in nanometric porosity. By means of DSC, it is possible to monitor the development of C-S-H microstructure during the hydration of C_3S in presence of PCEs estimating the volume of pores (capillary, large gel pores and small gel pores),

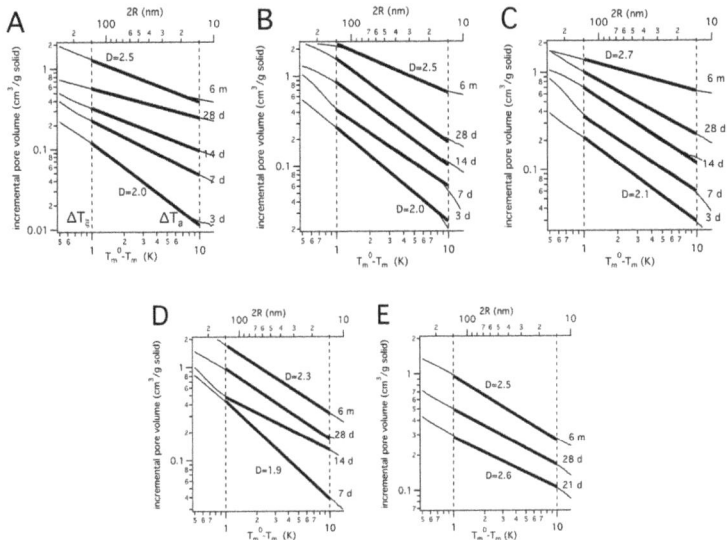

Figure XVIII.5. Incremental pore volume per solid mass as a function of the melting depression, $\Delta T = T \, \Delta T_m$ for the pastes investigated in this study: A) C_3S/H_2O, B) $C_3S/PCE102$-2, C) $C_3S/PCE102$-6, D) $C_3S/PCE23$-2, E) $C_3S/PCE23$-6.

and their evolution throughout the hydration. We also estimated the depercolation threshold of the capillary porosity and how this value is influenced by commonly used cement additives (polymethacrylic acid chains partially esterified with polyethylene oxide lateral chains). Furthermore, we showed that an accurate analysis of low temperature calorimetric measurements performed in equilibrium conditions could disclose the fractal dimension of the pastes, and these values are in good agreement with those obtained from small angle scattering measurements. These results, apart the significant information on the effect of additive on the curing process of cement pastes, show that the analysis of simple DSC measurements can provide a wealth of information on the fractal properties of this fundamental building material.

References

1. Allen, A. J.; Thomas, J. J.; Jennings, H. M. *Nat. Mater.* **2007**, *6*, 311–316.
2. Richardson, I. G. *Cem. Concr. Res.* **1999**, *29*, 1131–1147.
3. Tennis, P.; Jennings, H. *Cem. Concr. Res.* **2000**, *30*, 855–863.
4. Plassard, C.; Lesniewska, E.; Pochard, I.; Nonat, A. *Ultramicroscopy* **2004**, *100*, 331–338.
5. Nonat, A. *Cem. Concr. Res.* **2004**, *34*, 1521–1528.
6. Fratini, E.; Ridi, F.; Chen, S.-H.; Baglioni, P. *J. Phys.: Condens. Matter* **2006**, *18*, S2467–S2483.
7. Garrault, S.; Behr, T.; Nonat, A. *J. Phys. Chem. B* **2006**, *110*, 270–275.
8. Allen, A. J.; Thomas, J. J. *Cem. Concr. Res.* **2007**, *37*, 319–324.
9. Pellenq, R. J.-M.; Kushima, A.; Shahsavari, R.; Vliet, K. J. V.; Buehler, M. J.; Yip, S.; Ulm, F.-J. *Proc. Natl. Acad. Sci. USA* **2009**, *106*, 16102.
10. Alizadeh, R.; Beaudoin, J. J.; Raki, L. *Mater Struct.* **2011**, *44*, 13–28.
11. Chiang, W.; Fratini, E.; Baglioni, P.; Liu, D.; Chen, S. H. *J. Phys. Chem. C* **2012**, 1–8.
12. Ridi, F.; Fratini, E.; Milani, S.; Baglioni, P. *J. Phys. Chem. B* **2006**, *110*, 16326–16331.
13. Jennings, H. *Mater Struct* **2004**, *37*, 59–70.
14. Thomas, J. J.; Jennings, H. M.; Allen, A. J. *J. Phys. Chem. C* **2010**, *114*, 7594–7601.
15. Jones, C. A.; Grasley, Z. C.; Ohlhausen, J. A. *Cem. Concr. Comp.* **2012**, *34*, 468–477.
16. Chiang, W. C.; Fratini, E.; Ridi, F.; Lim, S. H.; Yeh, Y. Q.; Baglioni, P.; Choi, S. M.; Jeng, U. S.; Chen, S. H. *J. Coll. Interface Sci.* **2013**, *398*, 67–73.
17. Taylor, H. *Cement chemistry*; Thomas Telford Publishing: London, 1997.
18. Powers, T.; Brownyard, T. *ACI J. Proc.* **1947**.
19. Wittmann, F. H. In *Hydraulic cement pastes: their Structure a properties, Proceedings of a Conference held at University of Sheffield 8.-9.4.1976*; Cement and Concrete Association, 1976; pp. 96–117.
20. Allen, A.; Oberthur, R.; Pearson, D. *Philos. Mag. B* **1987**, *56*, 263–288.

21. Jennings, H. M. *Cem. Concr. Res.* **2000**, *30*, 101–116.
22. Thomas, J. J.; Jennings, H. M. *Cem. Concr. Res.* **2006**, *36*, 30–38.
23. Jennings, H. M. *Cem. Concr. Res.* **2008**, *38*, 275–289.
24. Thomas, J.; Jennings, H.; Allen, A. *Cem. Concr. Res.* **1998**, *28*, 897–905.
25. Brun M.; Lallemand A.; Quinson J.F.; Eyraud C. *Thermochim. Acta* **1977**, *21*, 59–88.
26. Sellevold E.J.; Bager D.H. In *7th International Congress on the Chemistry of Cement*; Paris, 1981; Vol. 4, pp. 394–399.
27. Bager, D. H.; Sellevold, E. J. *Cem. Concr. Res.* **1986**, *16*, 709–720.
28. Bager, D. H.; Sellevold, E. J. *Cem. Concr. Res.* **1986**, *16*, 835–844.
29. Bager, D. H.; Sellevold, E. J. *Cem. Concr. Res.* **1987**, *17*, 1–11.
30. Snyder, K. A.; Bentz, D. P. *Cem. Concr. Res.* **2004**, *34*, 2045–2056.
31. Bentz D.P.; Stutzman P.E. *ACI Mater. J. 103*, 348.
32. Bentz D.P. *J. Am. Ceram. Soc.* **2006**, *89*, 2606–2611.
33. Sanz, E.; Vega De Las Heras, C.; Espinosa, J. R.; Caballero-Bernal, R.; Abascal, J. L. F.; Valeriani, C. *J. Am. Chem. Soc.* **2013**.
34. Morishige, K.; Yasunaga, H.; Denoyel, R.; Wernert, V. *J. Phys. Chem. C* **2007**, *111*, 9488–9495.
35. Ridi, F.; Luciani, P.; Fratini, E.; Baglioni, P. *J. Phys. Chem. B* **2009**, *113*, 3080–3087.
36. Winnefeld, F.; Becker, S.; Pakusch, J.; Götz, T. *Cem. Concr. Comp.* **2007**, *29*, 251–262.
37. Ridi, F.; Fratini, E.; Luciani, P.; Winnefeld, F.; Baglioni, P. *J. Phys. Chem. C* **2012**, *116*, 10887–10895.
38. Hansen, E. W.; Gran, H. C.; Sellevold, E. J. *J. Phys. Chem. B* **1997**, *101*, 7027–7032.
39. Zingg, A.; Winnefeld, F.; Holzer, L.; Pakusch, J.; Becker, S.; Gauckler, L. *J. Coll. Interface Sci.* **2008**, *323*, 301–312.
40. Stauffer, D.; Aharony, A. *Introduction to Percolation Theory 2nd ed.*; Taylor & Francis: London, 1994.
41. Scherer, G. W. *Cem. Concr. Res.* **1999**, *29*, 1149–1157.
42. Bentz, D.; Garboczi, E. J. *Cem. Concr. Res.* **1991**, *21*, 325–344.
43. Garboczi, E. J.; Bentz, D. *Cem. Concr. Res.* **2001**, *31*, 1501–1514.

44. Bentz, D. P. *Cem. Concr. Comp.* **2006**, *28*, 427–431.
45. Kriechbaum, M.; Degovics, G.; Tritthart, J.; Laggner, P. *Progr. Colloid Polym. Sci.* **1989**, *79*, 101–105.
46. Winslow, D.; Bukowski, J. M.; Young, J. F. *Cem. Concr. Res.* **1995**, *25*, 147–156.
47. Heinemann, A.; Hermann, H.; Häußler, F. *Phys. B Condens. Matter* **2000**, *276*, 892–893.
48. Livingston, R. A. *Cem. Concr. Res.* **2000**, *30*, 1853–1860.
49. Degovics, G.; Laggner, P.; Tritthart, J. *Progr. Coll. Interface Sci.* **1992**, *89*, 335.
50. Thomas, J. J.; Chen, J. J.; Allen, A. J.; Jennings, H. M. *Cem. Concr. Res.* **2004**, *34*, 2297–2307.
51. Vollet, D.; Donatti, D.; Ruiz, A. I.; Gatto, F. *Phys. Rev. B* **2006**, *74*, 024208.
52. Rault, J.; Neffati, R.; Judeinstein, P. *Eur. Phys. J. B* **2003**, *36*, 627–637.
53. Neffati, R.; Rault, J. *Eur Phys J B* **2001**, *21*, 205–210.
54. Vollet, D. R.; Scalari, J. P.; Donatti, D. A.; Ruiz, A. I. *J. Phys.: Condens. Matter* **2007**, *20*, 025225.
55. Vollet, D.; Donatti, D.; Ruiz, A. I. *Phys. Rev. B* **2004**, *69*, 064202.
56. Teixeira, J. *J. Appl. Crystall.* **1988**, *21*, 781–785.

XIX
The emergence of structure from randomness in aqueous aggregation equilibria

Blake M. Rankin, Dor Ben-Amotz

Department of Chemistry, Purdue University,
560 Oval Drive, West Lafayette, IN 47907, USA.
bendor@purdue.edu

Abstract

Many aqueous aggregates of biological importance, such as those held together by hydrophobic and specific-ion interactions, are sufficiently weakly bound that the resulting structures may be viewed as teetering on the edge of randomness. In order to characterize such aggregation processes it is thus useful to compare the emergent structures with those predicted to exist in idealized random mixtures. Previously, we have shown that the binomial distribution may be used to both establish such a random mixing benchmark and use it to analyze processes ranging from the interfacial breaking of water hydrogen bonds to hydrophobic interactions between alcohols in water. Here we extend this random mixing analysis strategy by considering the influence of solute-ligand interaction energy on aggregate size distributions and thermodynamics. The results point to a shape invariant symmetry linking the solute-ligand interaction energy to the local ligand concentration in the first coordination shell of the solute. We further consider two types of aggregation equilibrium constants; one pertaining to the formation of a solute-ligand dimer and the other to the partitioning of ligand molecules between the bulk and solute coordination shell. Applications of our results are illustrated using both molecular dynamics (MD) simulations and experimental

Raman multivariate curve resolution (Raman-MCR) measurements pertaining to an ion-molecule interaction process in which the solute is *tert*-butyl alcohol (TBA) and the ligands are iodide (I$^-$) ions. Although our simulation and experimental results agree in implying that I$^-$ is expelled from the first coordination shell of TBA, the Raman-MCR results indicate a greater expulsion than that implied by the MD simulations.

1. Introduction

The vitality of biological structures necessitates the involvement of interactions whose strength does not greatly exceed ambient thermal fluctuations. Such interactions include those between hydrophobic groups in water, and those between hydrated counter-ions, as well as various specific and non-specific cross-interactions between ions and molecules. The weakness of such interactions is attested by the fact that the corresponding binary association equilibrium constants K_A often have magnitudes of the order of 1 M^{-1} – thus implying a degree of association that is not far from that expected in a random mixture consisting of nominally non-aggregating components.

In several recent publications we have applied random mixing statistics to aid in the interpretation of experimental results pertaining to the breaking of water hydrogen bonds at molecular hydrophobic interfaces,[1] hydrophobic interactions between alcohol molecules in water,[2] and interactions between aqueous ions and hydrophobic groups,[3] as well as molecular dynamics (MD) simulation results pertaining to both gas and liquid phase aggregation processes.[4] The latter study further suggested a means of combining random mixing predictions with simulation results in order to quantify the equilibrium constant and Gibbs energy pertaining to the partitioning of ions (and other ligands) between a bulk solution and the first coordination shell of a solute. Here we provide further theoretical justification for the latter suggestion, by showing that it is consistent with predictions obtained when solute-ligand interactions are explicitly incorporated into the random mixing analysis. In addition, we show how such a generalized random mixing analysis may be used to relate ligand partitioning equilibrium constants to those pertaining to the corresponding solute-ligand dimer formation process. Applications of this theoretical analysis strategy are illustrated using both experimental and simulation results pertaining to interactions between iodide (I^-) ions and the methyl groups of *tert*-butyl alcohol (TBA) dissolved in water. We have chosen to focus on this application here, in the spirit of the original aqua incognita conference, as there is currently an ongoing debate regarding the degree of affinity that large polarizable anions (such as I^-) have for both macroscopic air-water and oil-water interfaces,[5-10] as well as molecular hydrophobic hydration shells.[3, 11-14]

2. Random Mixing Statistics

The random mixing benchmark is formulated by considering an idealized mixture in which the concentration of each component is strictly uniform. In other words, a random mixture is one in which the average number of solute molecules within any volume V is exactly $V[c]$, where $[c]$ is the bulk concentration of the solute, as illustrated in Figure XIX.1. Thus, when the volume is sufficiently small that $V < 1/[c]$ then $V[c]$ is equivalent to the random mixing probability of finding the solute within the volume V. When applied to a solute-ligand aggregation process, we may equate V with the volume of the solute's first coordination shell. Moreover, if n is the maximum number of ligands that can simultaneously occupy the first coordination shell, then $p = (1/n)V[L]$ is the probability that a ligand will occupy any one of the n binding sites within the solute's coordination shell, where $[L]$ is the equilibrium concentration of the unbound ligand in the system of interest.

As a simple example, if the solute is assumed to be spherical and its first coordination shell is assumed to extend from r_1 to r_2, then the

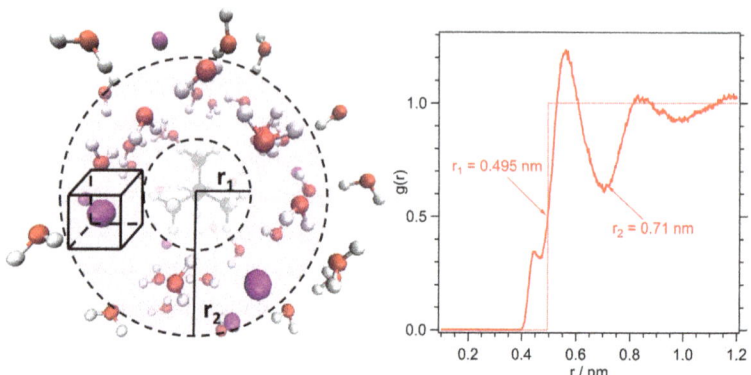

Figure XIX.1. MD simulation snapshot showing the first coordination–shell of a hydrophobic solute molecule containing both water and I⁻ ions. The dashed circles schematically represent the inner (r_1) and outer (r_2) extent of the solute's first coordination–shell. The cube illustrates that the first coordination–shell may be divided into n sub–cells, each of which may or may not include an I⁻ (ligand) molecule. The values of r_1 and r_2 may be obtained from the pair distribution function g(r), such as that shown in the right hand panel, which corresponds to the TBA (central carbon)–I⁻ g(r) obtained from MD simulations of a solution with [I⁻]=4.8 M. The faint curve in the right figure corresponds to a random mixing g(r) for a hard-sphere with an r_1 value of 0.495 nm.

first coordination shell volume is given by the following expression.[4]

$$V = \frac{4\pi}{3}\left(r_2^3 - r_1^3\right) \tag{0}$$

The latter estimate may be applied to non-spherical solutes by equating r_1 to r_2 with the leading edge and first minimum of the solute-ligand pair distribution function g(r) (as illustrated in Figure XIX.1). The random mixing probability $P(k)$ that k ligands will occupy the solute's coordination shell may be predicted using the following binomial distribution expression.

$$P_0(k) = \left[\frac{n!}{(n-k)!k!}\right] p^k (1-p)^{n-k} \tag{1}$$

Since n is equated with the maximum number of ligands that can occupy the solute's coordination shell, we may obtain a physically reasonable estimate of n from the solute-ligand accessible surface area divided by the surface area occupied by a single ligand, expressed in terms of the effective hard sphere diameters of the solute σ_1 and ligand σ_2 (as previously described).[4]

$$n \approx \pi\left(\frac{\sigma_1}{\sigma_2} + 1\right)^2 \qquad (2)$$

We now consider how the predicted aggregate size distributions would change if we introduce an additional interaction energy ε_1 between the solute and each bound ligand molecule. In other words, we consider an aggregation process for which the binding energy per ligand ε_1 is a k-independent constant. Thus, the total binding energy in an aggregate with k bound ligands is simply k times ε_1.

$$\varepsilon(k) = k\varepsilon_1 \qquad (3)$$

For such an aggregation process we obtain the following expression for the normalized probability that k ligands will occupy the solute's coordination shell.

$$P(k) = \frac{P_0(k)e^{-\beta k\varepsilon_1}}{\sum_{k=0}^{n} P_0(k)e^{-\beta k\varepsilon_1}} \qquad (4)$$

It is also interesting to note that the aggregate distribution functions $P(k)$ are linked to those pertaining to a random mixing process $P_0(k)$ through the following shape-invariant transformation. Physically, this invariance implies that the interaction energy ε_1 has the same effect on $P(k)$ as changing the ligand concentration in a non-interacting random mixture. Mathematically, this invariance stems from the fact that the functional dependence of P on k may be expressed as follows (to within a k-independent constant).

$$P(k) \propto \frac{1}{(n-k)!k!}\left[\left(\frac{p}{1-p}\right)e^{-\beta\varepsilon_1}\right]^k \qquad (5)$$

Thus, the functional form of $P(k)$ is invariant with respect to any changes in both ε_1 and p that leave the quantity $[p/(1-p)]e^{-\beta\varepsilon_1}$ unchanged. In other words, the aggregate size distribution obtained with some particular pair of ε_1 and p values is necessarily identical to that obtained in an idealized non-aggregating random mixture (for which $\varepsilon_1 = 0$) whose ligand binding site occupancy probability p_0 is related as follows to ε_1 and p.

$$p_0 = \left[\left(\frac{1}{p}-1\right)e^{\beta\varepsilon_1}+1\right]^{-1} \qquad (6)$$

The value of p_0 is thus also equivalent to the average site occupancy probability in the aggregate characterized by ε_1 and p. Stated in another way, the above invariant property implies that $[L]_0 = np_0/V$ is the local ligand concentration in the aggregate when the solute-ligand binding energy is ε_1 and the surrounding unbound ligand concentration is $[L] = np/V$. Thus, $K_P = p_0/p = [L]_0/[L]$ is equivalent to the partition coefficient of ligand molecules between the bulk and the aggregate (as further discussed in Section 3).

The above analysis implies that introducing the solute-ligand interaction energy ε_1 to the random mixing model is predicted to have a mathematically identical effect on the aggregate size distribution $P(k)$ as changing the bulk ligand concentration in a random mixture, as we had previously anticipated.[4] This also implies that the average number of bound ligands in a system with a solute-ligand interaction energy ε_1 is predicted to be $<k> = np_0$ (while $<k> = np$ is the corresponding number of ligands in the first coordination shell of a solution in an idealized random mixture with $\varepsilon_1 = 0$). However, it is important to stress that the above expressions for $P(k)$, p_0, and $<k>$ only pertain to systems for which the excess energy of an aggregate is given by Eq. 3 (as further discussed in Section 5).

3. Aggregation Equilibria

The binding of a single ligand to a solute molecule may be viewed as a simple dimerization reaction between the solute and ligand, $S+L \leftrightarrow SL$ (where S, L, and SL represent the solute, ligand, and solute-ligand dimer, respectively). The equilibrium constant for such an association reaction may be expressed as follows, in terms of the probabilities $P(0)$ and $P(1)$ that exactly 0 or 1 ligands occupy the first coordination shell of the solute, respectively.

$$K_A = \frac{[SL]}{[S][L]} = \frac{P(1)[S]_0}{P(0)[S]_0[L]} = \frac{P(1)}{P(0)}\frac{1}{[L]} \qquad (7)$$

The concentration $[S]_0$ corresponds to the total solute concentration, and thus $[SL]=P(1)[S]_0$, $[S]=P(0)[S]_0$, and $[L]$ are the equilibrium concentrations of the solute-ligand dimer, free solute, and free ligand, respectively. The standard Gibbs energy of such a dimerization reaction is related to the above equilibrium constant in the usual way.

$$\Delta G_A = -RT \ln K_A \qquad (8)$$

Thus, the value of ΔG_A dictates whether a solution which initially has a composition of $[SL] = [S] = [L] = 1$ M will be driven to further dimerize (if $\Delta G_A < 0$) or to further dissociate (if $\Delta G_A > 0$).

The partitioning of a ligand molecule between the bulk solution to the solute's coordination shell is determined by the following partitioning equilibrium constant.

$$K_p = \frac{[L]_{shell}}{[L]} = \frac{<k>}{V}\frac{1}{[L]} = \frac{np_0}{V[L]} \qquad (9)$$

Thus, the standard Gibbs energy pertaining to such a solute partitioning process may be expressed as follows.

$$\Delta G_p = -RT \ln K_p \qquad (10)$$

In other words, a value of $\Delta G_p = 0$ is equivalent to that pertaining to a random mixture (with $\varepsilon_I = 0$), while $\Delta G_p < 0$ implies that ligands are attracted to the solute ($\varepsilon_I < 0$), and $\Delta G_p > 0$ implies that ligands are repelled from the solute ($\varepsilon_I > 0$). Such a partitioning process is also closely related to that described by Pegram and Record, who performed an impressive thermodynamic analysis of a large amount of experimental thermodynamic data in order to obtain experimentally derived estimates of the partitioning of ions in aqueous salt solution to the coordination shells of small hydrocarbon and model peptide solutes.[15]

4. Comparisons of theoretical and simulation results

Figure XIX.2 demonstrates how a random mixing analysis may be applied to molecular dynamics (MD) simulation results pertaining to a TBA-I$^-$ aggregation process. These MD simulation results were obtained from an aqueous solution containing one TBA molecule and 1190 TIP4P water molecules, and thus $[S]_0 \sim 0.05$ M and $[L] \sim 4.8$

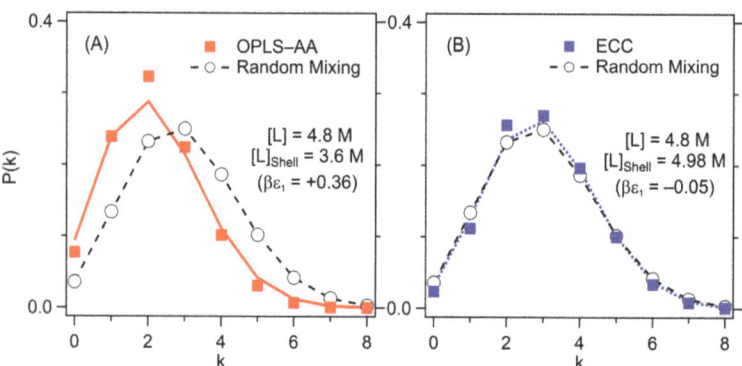

Figure XIX.2. $P(k)$ is the probability that k I$^-$ ions will be found in the first coordination shell of TBA. Panel (A) compares aggregate size distributions obtained from MD simulations (solid points) and random mixing predictions (open points). The solid curve represents predictions using Eq. 4, obtained from a best fit to the simulation points (while the open points connected by the dashed curve are predictions obtained using Eq. 1). Panel (B) shows the corresponding MD results performed using the ECC method (dotted curve).

M. The simulations were performed using OPLS-AA potentials for TBA and TIP4P water, and implemented in GROMACS (as further described previously).[4] The solid points were obtained by counting the number of times that exactly k iodide ions are found within the first coordination shell of TBA, where the first coordination shell is defined as a sphere centered on the central carbon of TBA with a maximum radius equal to the location of the second minimum in the TBA-I⁻ pair distribution function (shown in Figure XIX.1). Note that the first minimum in the TBA-I⁻ pair distribution function shown in Figure XIX.1 corresponds to I⁻ ions next to the -OH head group of TBA.

The open points and dashed lines in Figure XIX.2(A) pertain to random mixing predictions (with $\varepsilon_I = 0$), which implies that the I⁻ concentration in the first coordination shell of TBA is exactly the same as that in the surrounding bulk solution. These predictions are obtained assuming that $n=11$ is the maximum number of I⁻ ions that could occupy the first coordination shell of TBA (as inferred using Eq. 2 using $\sigma_1 = 0.45$ nm and $\sigma_2 = 0.54$ nm as the effective diameters of TBA and I⁻, respectively).[16, 17] The first coordination shell volume $V = 0.991$ nm³ is obtained using Eq. 0, with $r_1 = 0.495$ nm and $r_2 = 0.71$ nm (obtained from the leading edge and second minimum in the TBA-I⁻ pair distribution function shown in Figure XIX.1).

Comparisons of the random mixing benchmark predictions with simulation results shown in Figure XIX.2(A) clearly reveal that the simulated distributions are shifted to smaller aggregate sizes, and thus indicates that the local I⁻ concentration is lower than the bulk concentration, and so implies that I⁻ ions are expelled from the first coordination shell of TBA. More specifically, the local average I⁻ concentration in the first coordination shell of TBA is $[I^-]_{Shell} \sim 3.6$ M, which is significantly lower than the bulk I⁻ concentration $[I^-] \sim 4.8$ M.

Further quantitative information may be obtained from the MD simulation results shown in Figure XIX.2(A) by using Eq. 4 as a fitting function with ε_I as a fit parameter. Equivalently, we may equate the simulation results for the average number of I⁻ in the first coordination shell of TBA $<k> = 2.16$ with np_0 and then use Eq. 6 to obtain $\beta\varepsilon_1 = +0.36$ at 293K, since $p=(1/n)V[L] = 0.260$. Note that the positive value of $\beta\varepsilon_1$ is again consistent with the fact that I⁻ is expelled from the first coordination shell of TBA. The resulting predictions of Eq. 4, obtained from a best fit to the simulation results, are indicated by the solid curve in Figure XIX.2(A).

Although the agreement between simulated (points) and predicted (solid curve) aggregate size distributions $P(k)$ shown in Figure XIX.2(A) is reasonably good, it is certainly not perfect. This implies that that one or more of the assumptions made in deriving Eq. 4 are inconsistent with molecular interactions that gave rise to the simulation results. This is perhaps not surprising given that a number of simplifying assumptions were made both in deriving and applying Eq. 4. However, we have additional evidence suggesting that the most significant reason for the observed deviations may be traced to the influence of repulsion between two or more I⁻ ions in the first coordination shell of TBA. That evidence includes the fact that better agreement between simulated and predicted aggregate size distributions is obtained when the charge of I⁻ is decreased (as described below).

Previous studies have shown that the interactions of ions with hydrophobic interfaces can be quite sensitive to polarizability.[5-9] Moreover, recent studies have found that the influence of polarizability may be approximated using non-polarizable force fields with an electronic continuum correction (ECC), which accounts for electronic screening by reducing the effective charge of ions by a scaling factor of $(1/\varepsilon_{el})^{1/2} \sim 0.75$ (where ε_{el} is the high frequency dielectric constant of water).[18, 19]

Figure XIX.2(B) shows the results of our simulations of TBA-I⁻ aggregation obtained using the ECC method. The good agreement between these simulation results and random mixing predictions (with $\varepsilon_1=0$) clearly indicates that I⁻ is less strongly expelled from the coordination shell of TBA when the charge of I⁻ is reduced. Notice that the shapes of the simulated and predicted $P(k)$ distributions are now also in much better agreement with each other. Thus, the approximations made in deriving Eq. 4 are apparently consistent with simulation results for systems with reduced ligand-ligand electrostatic repulsion. Moreover, as will be further described in a forthcoming publication, predictions obtained using Eq. 4 are also in good agreement with simulations for other aggregations processes involving neutral solutes and ligands. Thus, it seems that the deviations between the points and solid curve in Figure XIX.2(A) are largely due to I⁻-I⁻ electrostatic repulsion.

5. Comparisons of experimental, theoretical, and simulation results

Figure XIX.3 compares experimental and theoretical results pertaining to the same aqueous TBA/NaI system described in Section 4. The experimental TBA concentration is 0.5 M and the I⁻ (NaI) concentration is varied from 0 M to 3 M. The points in Figure XIX.3 show upper and lower bounds obtained from the same set of experimental Raman-MCR measurements.[11] Thus, the shaded region between the points represents the physically reasonable range of the $<k>$ values obtained from the experimental data. For example, at a bulk I⁻ concentration of 1M, our Raman-MCR measurements imply that the average number of I⁻ ions in the first coordination shell of TBA is $0.10 \leq <k> \leq 0.23$. The latter lower and upper experimental bounds were obtained by assuming that the CH frequency shift induced by each I⁻ ion in the first coordination shell of TBA ranges from 4 cm⁻¹ to 9 cm⁻¹ (as further discussed previously).[11]

The dashed line in Figure XIX.3 shows random mixing predictions obtained using Eq. 1. The fact that the experimental points lie below this dashed line again implies that there are fewer I⁻ ions in the first coordination shell of TBA than predicted using the random

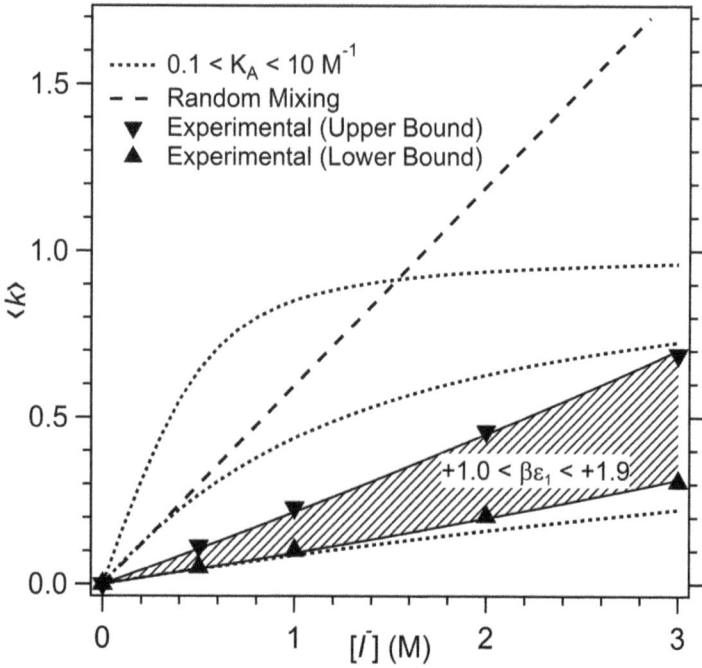

Figure XIX.3. Comparison of the average number of TBA–I⁻ contacts, $<k>$, obtained from Raman–MCR experiments (triangles), along with random mixing predictions (dashed line), and predictions obtained for a simple dimerization process with three different K_A values (dotted curves). The shaded region represents the physically reasonable range of experimentally derived $<k>$ values. The solid lines through the experimental points are predictions obtained using Eq. 4 with $\beta\varepsilon_I = +1.0$ (upper curve) and $\beta\varepsilon_I = +1.9$ (lower curve).

mixing model (with $\varepsilon_I=0$). The solid lines in Figure XIX.3 represent predictions obtained using Eq. 4 with non-zero values of ε_I ranging from $\beta\varepsilon_I = +1.0$ (upper solid line) to $\beta\varepsilon_I = +1.9$ (lower solid line). The fact that these best-fit values of ε_I are invariably positive, confirms that I⁻ is expelled from the first coordination shell of TBA.

The experimental results may further be compared to <k> values predicted assuming that the aggregation may be described as a simple dimerization of one TBA molecule and one I⁻ ion. The three dotted curves in Figure XIX.3 pertain to dimerization equilibrium constants of K_A = 0.1 M⁻¹, 1 M⁻¹, or 10 M⁻¹. The fact that the curved shapes of the dimerization predictions are not the same as the nearly linear experimental and random mixing predictions indicates that the aggregation of I⁻ around TBA is not limited to the adsorption of one I⁻ ion. Although the experimental results clearly do not have the same concentration dependence as dimerization predictions, the fact that the range of the <k> values obtained from experiment is less than expected from a dimerization model with K_A = 1 M⁻¹ confirms that the interaction between TBA and I⁻ is exceedingly weak. Moreover, the experimentally derived values of $\beta\varepsilon_I$ = +1.0 and $\beta\varepsilon_I$ = +1.9, combined with Eqs. 7 and 8, imply that 0.23 M⁻¹$\leq K_A \leq$0.52 M⁻¹ and +1.5 kJ/mol <ΔG_A< +3.7 kJ/mol.

The experimental Raman-MCR results may also be used to estimate the partition coefficient K_p. One way of doing this is illustrated using the aggregate size distribution results shown in Figure XIX.4 (all of which pertain to a 1M I⁻ concentration). The solid points (and shaded region) in Figure XIX.4 were generated using Eq. 4 with $\beta\varepsilon_I$ = +1.0 and +1.9 (as obtained from fits to the experimental data in Figure XIX.3). The open points connected by dashed lines are random mixing predictions (with ε_I=0). The shape invariant symmetry described in Section 2 implies that, rather than fitting experimental data to Eq. 4 with $\beta\varepsilon_I$ as a fit parameter, we may obtain the same information by fitting the experimentally-derived distributions to Eq. 1, using [L] as a fit parameter, and thus to obtain [L]$_{shell}$. Doing so results in an estimated I⁻ concentration of 0.16 M < [I⁻]$_{shell}$ < 0.38 M which, when compared to the bulk concentration of [I⁻] = 1 M, again implies that I⁻ is expelled from the coordination shell of TBA. Moreover, we may use Eq. 9 to obtain K_p =[I⁻]$_{shell}$ /[I⁻] and thus +2.4 kJ/mol < ΔG_p < +4.5 kJ/

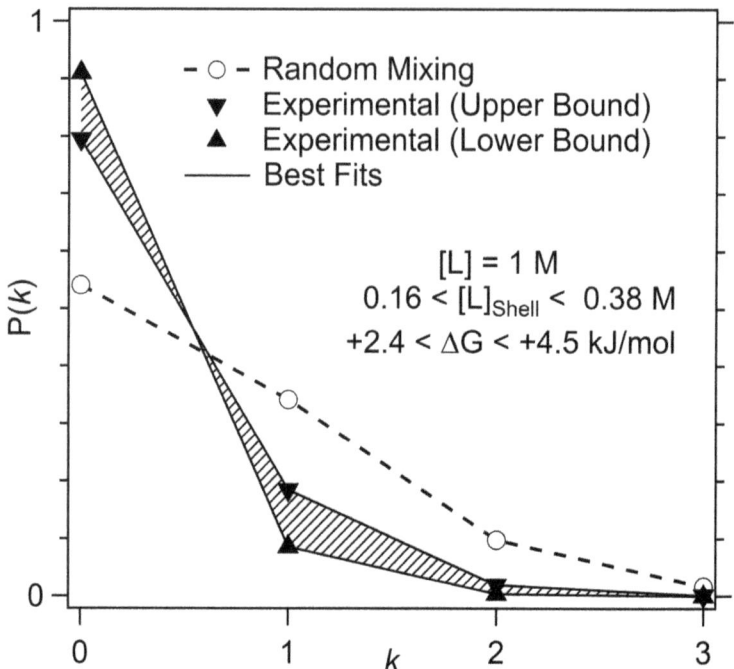

Figure XIX.4. The experimentally–derived aggregate size distributions (solid points and lines), assuming that $\beta\varepsilon_1 = +1.0$ or $+1.9$, and $[I^-] = 1$ M, are compared with the corresponding random mixing predictions (open points connected by a dashed line). These results indicate that there are fewer I^- ions in contact with TBA in the experiments than would be expected for a random mixing process. Note that the experiments imply that in a 1 M aqueous NaI solution only about 10% – 20% of the TBA molecules will be in contact with a single I^- ion (and virtually no TBA molecules will be in contact with two I^- ions), while in a random mixture one would expect nearly half of the TBA molecules to be in contact with one or more I^- ions.

mol. Although the experimentally derived uncertainty in ΔG_p is large, the sign is invariably positive, again confirming that I^- ions are driven away from TBA.

Figure XIX.5 contains results that provide a more direct quantitative comparison between the TBA-I^- aggregation experiments and simulations. Figure XIX.5(A) shows $g(r)$ simulation results pertaining to the distribution of I^- ions about the central carbon of TBA, excluding all those I^- ions that are in the first coordination shell of the OH head group of TBA (as further explained in the caption of Figure XIX.5). The solid g(r) curve is that obtained using OPLS-AA-TIP4P potentials and the dotted $g(r)$ curve is that obtained using the charge-screened ECC potentials. Note that these $g(r)$ curves differ

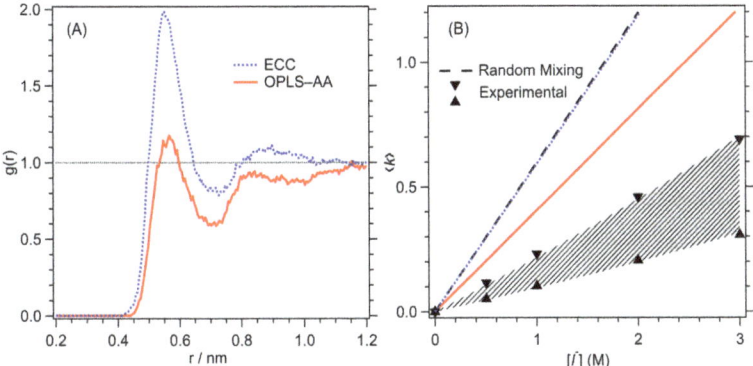

Figure XIX.5. (A) Pair distribution functions $g(r)$ pertaining to the interactions between the methyl groups of TBA and I^- obtained from OPLS-AA (solid) and ECC (dotted) MD simulations of a solution containing one TBA molecule and 4.8 M NaI. These $g(r)$ functions pertain to I^- ions about the central carbon atom of TBA, excluding those I^- ions that are within 0.42 nm of the TBA oxygen atom. The latter cutoff distance was determined from the first minimum in TBA (oxygen)–I^- $g(r)$. (B) The $<k>$ values obtained from these simulations are compared with those derived from the Raman–MCR experiments (triangles and shaded region, which are the same as those shown in Figure XIX.3). Our experimental, simulated, and theoretical results for the average number of TBA–I^- contacts indicate that at [L] = 1 M, $<k>_{\text{Raman-MCR}}$ = 0.10 – 0.23, $<k>_{\text{OPLS-AA}} \sim 0.41$, $<k>_{\text{ECC}} \sim 0.59$, and $<k>_{\text{Random Mixing}} \sim 0.60$.

from that shown in Figure XIX.1(B), which did include I⁻ ions near the TBA OH head group, as evidenced by the additional peak near r~0.45 nm in Figure XIX.1(B). We have used the simulation results shown in Figure XIX.5(A) to obtain the average number of I⁻ ions $<k>$ that are in contact with the hydrophobic groups of TBA. More specifically, we have obtained the solid and dotted lines in Figure XIX.5(B) assuming that the shape of g(r) is concentration independent (which is approximately correct over this concentration range), which is also consistent with the experimentally observed approximately linear dependence of $<k>$ on [I⁻]. The fact that all of the results in Figure XIX.5(B) fall below the idealized random mixing limit (dashed line) implies that both experimental and simulation results agree in predicting that I⁻ is expelled from the first coordination shell around the hydrophobic groups of TBA (although the ECC simulations are virtually indistinguishable from random mixing predictions). However, the different slopes of the experimental and MD-derived results means the experiments imply greater I⁻ expulsion than the simulations.

The deviation between the experimental and simulated results may be explained in several ways. First of all, it is important to note that the Raman-MCR results are subject to mathematical 'rotational ambiguity' inherent in MCR,[20] implying that the experimental $<k>$ estimates may possibly be smaller than the actual $<k>$ values. However, the range of our experimental estimates was generated in a way that should already reflect the possible extent of this rotational ambiguity, as our experimental $<k>$ values are consistent with the assumption that each I⁻ ion in the first coordination shell of TBA induces a TBA CH frequency shift of between 4 and 9 cm⁻¹. Thus, in order for the experimentally derived $<k>$ to increase, the shift per I⁻ ion would have to be smaller than the 4 cm⁻¹ predicted using hybrid quantum-classical calculations of aqueous TBA whose hydrophobic first hydration shell contains a single I⁻ ion.[3] Stated in another way, in order to obtain agreement between our Raman-MCR and non-polarizable (OPLS-AA-TIP4P) simulation results, would require assuming that a single I⁻ ion induces a CH shift of only ~2 cm⁻¹, which is probably too small to be physically reasonable. Moreover, as indicated by the dotted curve in Figure XIX.5(B), when the influence of polarizability is approximated using the ECC simulation the predicted <k> values are in essentially perfect agreement with random mixing predictions, which is even more inconsistent with experimentally derived estimates. Thus, it appears likely that I⁻ ions are more strongly expelled from the hydrophobic hydration shell of TBA than predicted by either the OPLS-AA or ECC simulations.

6. Conclusions

We have illustrated how delicate aqueous aggregation equilibria, such as those involved in the dynamic self assembly of biological structures, may be better understood and quantified by comparing experimental and simulation results with predictions pertaining to an idealized random mixture. More specifically, we have described the influence of both entropic random mixing statistics and energetic intermolecular interactions on aggregates formed when one or more ligands bind to a central molecule. Our theoretical results confirm that solute-ligand interactions should shift the mean number of bound ligands, but are predicted to leave the shape of the aggregates size distributions invariant (in the sense described in Section 2). We further show how the above entropic and energetic driving forces should influence both solute-ligand dimerization and the partitioning of ligand molecules between the coordination shell of a solute and the surrounding bulk solution.

Applications of our theoretical predictions are illustrated using both MD simulations and Raman-MCR measurements pertaining to the aggregation of aqueous I⁻ ions around the methyl groups of *tert*-butyl acohol (TBA). Both the MD and experimental results imply that I⁻ ions are weakly expelled from the hydrophobic hydration shell of TBA.

However, the shape of the corresponding aggregate size distribution differs from that implied by the shape invariant predictions described above. The agreement between the predicted and simulated shapes of the aggregate size distributions is found to improve as the charge on the I⁻ ion is reduced. This suggests the electrostatic repulsion between two or more I⁻ ions in the hydration shell of TBA has a significant influence on the shape of the aggregate size distributions. The latter conclusion is consistent with our additional theoretical and simulation results that will be described in a forthcoming publication. Although we have here focused primarily on aqueous ion-molecule interactions, the theoretical predictions and analysis strategies that we have used should readily be extendable to a wide range of other types of aggregation processes.

Acknowledgments and Meditations Regarding Galileo's Scientific Method

This work was funded by NSF (CHE-1213338) and a PRF Research Grant from Purdue University. The authors would also like to thank Ben Widom and Valeria Molinero for useful discussions and suggestions.

The following comments are contributed by Dor Ben-Amotz:

In 1613 – the same year as the debate that inspired this Aqua Incognita conference – Galileo wrote a letter to Father Benedetto Castelli that included the following remarkable explanation of his views regarding the relationship between scriptural and natural phenomena:[21]

> Holy Scripture has to be accommodated to the common understanding in many things which differ in reality from the terms used in speaking of them. But Nature being on the contrary inexorable and immutable, and caring not one jot whether her secret reasons and modes of operation be above or below the capacity of men's understanding:

it appears that, as she never transgresses her own laws, those natural effects which the Experience of the senses places before our eyes, or which we infer from adequate demonstration, are in no wise to be revoked because of certain passages of Scripture, which may be turned and twisted into a thousand different meanings.

Although Galileo's views are strikingly resonant with current scientific thinking and methodology, in his own time Galileo's ideas were vehemently opposed by a powerful inner-circle within the Jesuit community, whose preposterously irrational arguments eventually prevailed in precipitating Galileo's censure and imprisonment. I find it quite remarkable that, even today, there are those in the Jesuit community who retain an enthusiasm for undermining Galileo's credibility by seeking to uncover evidence of irrationality in his own thinking.[23] Thus, although it is acknowledged that "time proved Galileo right" his genius is nevertheless inexplicably attributed to his possession of "a way of seeing beyond what can be expressed by reasoned argument and experiment." The latter conclusion does not accurately reflect either Galileo's scientific method or his views, as expressed both in the above quotation and in his prescient devotion to "visual certainty" as the fountain of scientific discovery[22] – that is Galileo's genius!

References

1. Davis J. G.; Rankin, B. M.; Gierszal, K. P.; Ben-Amotz, D. *Nature Chemistry* **2013**, *5*, 796-802.
2. Wilcox, D. S.; Rankin, B. M.; Ben-Amotz, D. "Distinguishing Aggregation from Random Mixing in Aqueous t-Butyl Alcohol Solutions" *Faraday Discuss.* **2013**, *167*. DOI: 10.1039/C3FD00086A.
3. Rankin, B. M.; Hands, M. D.; Wilcox, D. S.; Fega, K. R.; Slipchenko, L. V.; Ben-Amotz, D. *Faraday Discuss.* **2013**, *160*, 255-270.
4. Rankin, B. M.; Ben-Amotz, D. *J. Phys. Chem. B*, **2013**, online ASAP. DOI:

10.1021/jp406413b.
5. Ben-Amotz, D. *J. Phys. Chem. Lett.* **2011**, *2*, 1216-1222.
6. Wick C. D.; Kuo, I-F. W.; Mundy, C. J.; Dang, L. X. *J. Chem. Theory Comput.* **2007**, *3*, 2002-2010.
7. Netz, R. R.; Horinek, D. *Ann. Rev. Phys. Chem.* **2012**, *63*, 401-418.
8. Stern, A. C.; Baer, M. D.; Mundy, C. J.; Tobias, D. J. *J. Chem. Phys.* **2013**, *138*, 114709.
9. Jungwirth, P.; Tobias, D. J. *Chem.Rev.* **2006**, *106*, 1259-1281.
10. Vazdar, M.; Pluhařová, E.; Mason, P. E.; Vácha, R.; Jungwirth, P. *J. Phys. Chem. Lett.* **2012**, *3*, 2087-2091.
11. Rankin, B. M.; Ben-Amotz, D. *J. Am. Chem. Soc.* **2013**, *135*, 8818-8821.
12. Paterova, J.; Rembert, K. B.; Heyda, J.; Kurra, Y.; Okur, H. I.; Liu, W. R.; Hilty, C.; Cremer, P. S.; Jungwirth, P. *J. Phys. Chem. B* **2013**, *117*, 8150-8158.
13. Rembert, K.B.; Paterová, P.; Heyda, J.; Hilty, C.; Jungwirth, P.; Cremer, P. S. *J. Am. Chem. Soc.* **2012**, *134*, 10039-10046.
14. Jungwirth, P. *Faraday Discuss.* **2009**, *141*, 9-30.
15. Pegram, L. M.; Record, M. T. *J. Phy. Chem. B* **2008**, *112*, 9428-9436.
16. Bondi, A. *J. Phys. Chem.* **1964**, *68*, 441-451.
17. Ben-Amotz, D.; Willis, K. G. *J. Phys. Chem.* **1993**, *97*, 7736-7742.
18. Leontyev, I.; Stuchebrukhov, A. *Phys. Chem. Chem. Phys.* **2011**, *13*, 2613-2626.
19. Vazdar, M.; Pluhařová, E.; Mason, P. E.; Vácha, R.; Jungwirth, P. *J. Phys. Chem. Lett.* **2012**, *3*, 2087-2091.
20. Fega, K. R.; Wilcox, D.S.; Ben-Amotz, D. *Appl. Spectrosc.* **2012**, *66*, 282-288.
21. Allan-Olney, M. *The Private Life of Galileo*. Nichols and Noyes, Boston, 1870, reproduced by BiblioLife LLC, see pages 83-84 footnote 1.
22. Galilei, G. *Sidereus Nuncius or The Sidereal Messenger Galileo Galilei*. The University of Chicago Press, 1989, page 62.
23. Caruana SJ, L. *Aqua Incognita*. Ballarat, **2014**, 1-17.

XX
Theory and modelling of ion specific hydration

Yu Shi, Travis Pollard, Thomas L. Beck

Departments of Chemistry and Physics, University
of Cincinnati, Cincinnati, OH 45221-0172 USA.
thomas.beck@uc.edu

Overview

Models of ions in water have progressed from continuum Born-type approaches to classical molecular-level treatments and recently to more accurate quantum mechanical studies. This chapter gives an overview of recent work aimed at a more thorough understanding of ion specificity in water, with a focus on fundamental aspects of hydration and interactions with the water surface. It is suggested that partitioning of interaction energies and the resulting free energies can lead to helpful insights into the hydration process. The quasi-chemical theory (QCT) provides a spatial partitioning of the hydration free energy that has led to a deeper understanding of a range of hydration problems. Links of the QCT to Lifshitz theory are outlined that may aid in obtaining quantitative results regarding the magnitude of the dispersion contribution. Length scales for specific ion hydration and the role of interfacial potentials are discussed. These issues are important for both fundamental and practical reasons, for example in setting an absolute single-ion free energy scale and in the development of ion-water force fields.

Introduction

In ionic solutions, there are intense electric fields on the ions and water atoms, with magnitudes up to several V/Å.[1] An example of a physical phenomenon resulting from these strong fields is crystalloluminescence in which visible light is emitted during ion crystallization,[1] a clear indication of significant alterations of the electronic states in the condensed phase. Models of alkali and halide ions as spherical balls with a charge embedded at the ion center, and with water molecules treated as a set of fixed partial charges distributed to mimic electrostatic potentials outside the molecule, thus appear oversimplified.

The theory of ions in water has a long and colorful history. The Born solvation model along with the Debye-Hückel theory constituted initial steps.[2] While often criticized and improved upon, these theories have been shown to possess sound physical underpinnings within limits.[3-5] Generally those limits involve detailed molecular interactions near a given ion, while distant interactions can be handled with simpler models (Gaussian, mean-field, or continuum). A second period of the theory of ions in solution addressed collective effects in ion transport, at the continuum level and led by Onsager.[2]

More recently, the advent of large-scale computing has led to extensive molecular-level simulation, initially at the classical level.[6] These simulations have yielded a better understanding of hydration structure, ion transport, and hydration thermodynamics. For example, there is a clear asymmetry between cation and anion hydration[4] that is not captured in dielectric models without seemingly ad hoc modifications of the effective ion size. In addition, work has been directed at understanding the behavior of ions near the water surface, including the effects of ion polarization on density profiles.[7,8]

In the last several years, there has been a strongly renewed interest in ion specific hydration and a wide range of Hofmeister effects.[9-11] These problems have been addressed from several angles that include classical[12,13] and quantum[14] simulation and Lifshitz theory.[15,16] While the classical simulations have given insights into a range of phenomena, an emerging conclusion is that ion hydration involves quantum mechanical effects in crucial ways. A quantum treatment is important for accurate models of induction,[14,17,18] charge transfer,[18] dispersion,[15,16] and water interfacial potentials.[19]

Recent theories of ion hydration can be categorized pictorially as top-down[9,15,16,20] and bottom-up.[3] In the top-down approach, ion-water interactions are viewed as emerging from the entirety of fluctuating charges on the ions and water; knowledge of the frequency-dependent response of the ion and the surrounding water bath leads to a full many-body treatment of the interactions through the Lifshitz theory. In the bottom-up approach, on the other hand, the ion and at least nearby waters are all treated at a molecularly detailed quantum mechanical level either through quantum chemistry methods or ab initio simulation techniques. Then appropriate statistical mechanical theory can be called into play for analysis of the bulk hydration thermodynamics.[3]

The present chapter focuses mainly on the bottom-up approach. First, a brief overview of interaction models is given. Then methods for partitioning interaction energies and free energies will be summarized. Part of the discussion will be aimed at suggesting a route towards linking the bottom-up and top-down approaches. Examples will be presented from our own work showing that classical models can lead to insights into the basic physics. Results will also be discussed that point out the limitations of classical models. The discussion will conclude with comments on the possible influence of water interfacial potentials on ion hydration.

Models

This section will give a brief overview of models that have been employed in studies of ion hydration. Ultimately, all of the interactions between an ion and water arise from the Coulomb potential and the smeared-out electron density obtained from quantum mechanics (the Schrödinger equation and the Pauli principle). The Hellman-Feynman theorem[21] illustrates this: given an accurate electron density for a particular nuclear configuration, all forces on the nuclei are due to classical electrostatics. This is not so helpful, however, since beyond a small number of atoms, nobody knows how to accurately solve for the exact charge density. [The Quantum Monte Carlo (QMC) method[22] is one exception, but it is not possible yet to model the motions of ions in bulk water with QMC.] Thus it has proven helpful to build models that divide up the interactions approximately into contributing parts.

At the classical level, the first step in the development of molecular models was the introduction of point charge models.[23] In these models, the water-water and ion-water interactions include point charges at the ion center and distributed on the water molecule, and some form of van der Waals potential with a repulsive wall and an attractive tail designed to mimic dispersion interactions. There has always been empiricism involved in developing point charge models, and in the end, it is surprising how well these models can represent a range of water and ion-water properties given their extreme simplicity (see Ref. 13 for an example). But of course they have drawbacks also: a poor representation of the direct ion-water interactions (no charge rearrangements, inaccurate repulsive wall designed to mimic exchange-repulsion, etc.), no polarization, only a very crude pairwise representation of dispersion interactions, and so on.

The next step involved the development of polarizable models[8] and the addition of distributed point multipoles on the atoms[24-26] (as discussed in Stone's book[27]). These more advanced models allow for a representation of collective polarization effects and a more accurate electrostatic potential outside of a given water molecule. Impressive results have been obtained regarding the solution structure and thermodynamics.[24-26] A byproduct of the increased accuracy is the introduction of additional parameters that are difficult to isolate and alter (without the simultaneous alteration of all parameters) when deviations from experiment are observed. In addition, it has been found that ion polarization can be significantly over-estimated in the classical models without empirical corrections that suppress the effect.[18,28]

The realization of the intensity of the interactions of ions with water, and their partial chemical character, has led to the development of more accurate quantum models. These models include the QM/MM[29] (quantum-mechanics/molecular-mechanics) and AIMD[14,30,31] (ab initio molecular dynamics) methods. In the QM/MM method, the spatial region that includes the ion and the nearby waters is modeled with quantum mechanics, while the more distant interactions are handled classically. This is a sensible approach, but there are always issues with how the QM/MM boundary is treated. The AIMD method utilizes Density Functional Theory[32] (DFT) to represent the electronic states and propagates the classical nuclei based on forces obtained from the quantum electronic distribution. The older DFT exchange-correlation functionals do not include dispersion interactions, however, and recent work has been directed at including dispersion at an approximate level.[33] The effects of dispersion are significant, leading to a water density increase of nearly 20% (to a value that agrees better with experiment).

It can be noted that there are two levels of quantum effects in aqueous solution. The first level, discussed above, concerns electronic effects. The second effect is that due to the motions of the water protons.[34,35] It has been shown that the zero-point effects of those protons are roughly equivalent to a 50 K temperature increase. It may be that those zero-point effects are not so crucial for modeling the direct ion-water interactions, but they may be important for the water structure in the surrounding shells.

The final level of modeling discussed here is a hybrid level that includes accurate quantum mechanics for the ion and the first shell water molecules, and then approximations for the more distant interactions. The quasi-chemical theory (QCT) developed by Pratt and coworkers[3,36-38] is an example of this level of theory in which the statistical mechanics of ion solvation is recast into a chemical equilibrium problem with the addition of interactions of the hydrated cluster with the surrounding water bath. This theory will be discussed below. An alternative approach is exemplified in two recent studies by Duignan, Parsons, and Ninham.[15,16] This work is an example of the top-down approach in which accurate quantum calculations are performed on the individual ions (the frequency-dependent polarizabilities), and then the bare ion is coupled to a continuum representation of water (using the experimental frequency-dependent dielectric constant). Both of these theoretical directions point towards the importance of a quantum mechanical treatment of ion hydration.

Partitioning

The physical situation of an ion that interacts strongly with the nearby waters but less strongly with waters outside the first shell begs for a theoretical treatment that partitions the problem spatially. The QCT is such a statistical mechanical theory.[3] The QCT is an exact partitioning of the hydration free energy, but it is ideally suited for judicious approximations for the distant interactions.

We start with the Potential Distribution Theorem (PDT) expression for the free energy:[3]

$$\mu_X^{ex} = -kT \ln\langle\exp(-\varepsilon_X/kT)\rangle_0 = kT \ln\langle\exp(\varepsilon_X/kT)\rangle \quad (1)$$

where

$$\varepsilon_X = U_{N+X} - U_N - U_X \quad (2)$$

is the interaction energy of the ion with the N waters. In the first expression for the free energy in Eq. 1, the sampling is performed with no ion in the system, while in the second form the sampling includes the ion. Both forms are exact, but neither can be employed directly in practical calculations involving ion solvation due to the strong interactions.

For illustration, consider a simple point charge model for an ion in water. Then consider the ion insertion as a two-step process in which an uncharged van der Waals particle is first inserted, followed by turning on the Coulomb potential. Then the electrostatic contribution to the free energy can be written as

$$\mu_{X,es}^{ex} = \langle \varepsilon_{X,es} \rangle_0 - kT \ln \langle \exp[-(\varepsilon_{X,es} - \langle \varepsilon_{X,es} \rangle_0)/kT] \rangle_0 \approx$$
$$\langle \varepsilon_{X,es} \rangle_0 - \frac{1}{2kT} \langle \delta\varepsilon_{X,es}^2 \rangle_0 = q \langle \phi_{np} \rangle_0 - \frac{q^2}{2kT} \langle \delta\phi_{np}^2 \rangle_0 \quad (3)$$

where here the uncoupled sampling (labelled with '0') includes the van der Waals particle. The approximation arises from a second-order cumulant expansion, and the final form expresses the electrostatic energies in terms of the net potential at the cavity center. That net potential arises in turn primarily from interactions near the cavity and from the surface potential of a distant interface:

$$\langle \phi_{np} \rangle_0 = \langle \phi_{lp} \rangle_0 + \langle \phi_{sp} \rangle_0 \quad (4)$$

where lp indicates the local potential and sp the distant surface potential. (There is no average potential arising from waters in the intermediate space since they exhibit random orientations.) Eq. 3 is the statistical mechanical basis of the Born model, but typically the Born model neglects the linear term. Also it can be seen that the ion radius in the Born model is a temperature-dependent quantity that depends on the electrostatic potential fluctuations.

If the interaction energies were truly Gaussian distributed (which in general they are not), then the potential fluctuations would be the same for uncoupled and coupled sampling. Previous work[4] has shown that the fluctuations differ between cations and anions, indicating an asymmetry to the hydration process. While there are deviations from Gaussian behavior, those deviations are not huge, and they arise from near-local interactions between the ions and water.[5] This helps explain why Born models can yield decent estimates of hydration free energies; if ion-size shifts are included for cations and anions separately (these shifts mimic the change in fluctuations), then remarkably consistent behavior is seen in the hydration free energies.[39] The results do suggest that local hydration asymmetry arising from water orientations near the ions (the water oxygen is of course closest for cations, and the water protons are hydrogen bonded to the anions) is important to consider.

The basic QCT formula results from insertion of unity (twice) into the relevant configurational integrals; the inserted integrals involve hard-particle interactions with a particle size λ. Here the hard-particle potential is labeled by $M(\lambda)$ or model potential, and allows for the insertion of a repulsive but continuous form amenable to simulation:

$$\mu_X^{ex} = kT \ln \langle \exp(-\varepsilon_M / kT) \rangle \\ - kT \ln \langle \exp(-\varepsilon_M / kT) \rangle_0 - kT \ln \langle \exp(-\varepsilon_X / kT) \rangle_M \quad (5)$$

The first term is the inner-shell term (zero for cavities smaller than the ion and then becoming negative), which is minus the work required to move the nearby waters out to the length scale specified by λ. The second term is the packing or cavity term (always positive), which is the work required to create a cavity of size λ in bulk water. The final long-range term involves all interactions of the ion with water, but the sampling is performed with the cavity included.

The QCT was developed over several years (summarized in Ref. 3), and has culminated in a beautiful exact formula that can be derived starting from Eq. 5:[38]

$$\mu_X^{ex} = -kT \ln K_{X,n}^{(0)} \rho_W^n + kT \ln p_X(n) + \mu_{XW_n}^{ex} - n\mu_W^{ex} \quad (6)$$

Here $K_{X,n}^{(0)}$ is the equilibrium constant for formation of the XW_n cluster in vacuum, ρ_W is the bulk equilibrium density of water, $p_X(n)$ is the probability of observing n waters in the λ-sized observation volume with the ion present (it is likely best to choose an observation volume that includes the first hydration shell and a value of n that reflects the most probable number of waters in the inner shell), $\mu_{XW_n}^{ex}$ is the free energy to insert the XW_n cluster into water, and μ_W^{ex} is the free energy to insert a water molecule into bulk water (-6.3 kcal/mol from experiment).

If all of the terms except $\mu_{XW_n}^{ex}$ are collected into one term, then the free energy can be written as

$$\mu_X^{ex} = -kT \ln\left[K_{X,n}^{(0)} \rho_W^n \exp\left(n\mu_W^{ex}/kT\right) / p_X(n) \right] + \mu_{XW_n}^{ex} = \mu_{X,n}^{ex} + \mu_{XW_n}^{ex} \quad (7)$$

which is the sum of a modified cluster term and the free energy to insert the cluster into bulk water. The cluster experiments used to determine the proton hydration free energy in Ref. 40 measure $K_{X,n}^{(0)}$ as a function of temperature. Except for the relatively small $kT \ln p_X(n)$ factor, the rest of the contributions to the first term cancel when looking at differences between cations and anions. It is relatively easy to show that, when looking at such differences, they can be computed as the difference between the free energies to insert the ions into the n-water cluster. Eq. 6 expresses the hydration in terms of the 'chemical equilibrium' for the 'reaction' of the ion with the inner-shell waters leading to the bound complex in solution. Then the key steps of the hydration process are primarily the formation of the XW_n cluster in the gas phase followed by insertion of the complex into water (with modifications related to the bulk properties of water).

Eq. 6 sets the stage for employing the strategy outlined above: an accurate quantum mechanical calculation of $K_{X,n}^{(0)}$ with the addition of an approximate estimation of $\mu_{XW_n}^{ex}$. The distant interactions are dominated by electrostatic effects, and several studies have shown that those long-ranged interactions tend to be Gaussian distributed.[5,41] Thus, it is not surprising that dielectric models for the hydration free energy of the cluster can yield accurate results when appended to the quantum cluster calculation.[36,37]

A variant of the QCT has been developed recently in which the interaction energies are partitioned based on the Local Molecular Field

Theory (LMFT) of Weeks and coworkers (see Ref. 5 and references therein). In this approach, the electrostatic interactions are divided into local and far-field components as in the Ewald formula for lattice sums. Ref. 5 shows how to employ this energetic partitioning in calculations of ion hydration free energies, and the approach has proven useful in analyzing the origins of observed specificity in ion hydration entropies and the behavior of ions near the water surface.[13,19,41]

Here we also suggest that Eq. 6 provides a link between the top-down and bottom-up approaches discussed above. Refs. 15 and 16 have provided an important step forward in modeling ions interacting with the full frequency response of water. But one possible concern might be raised in that the ion frequency-dependent polarizabilities are computed for the bare ions. Since the ion-water interactions are so strong and may even possess some chemical character, Eq. 6 provides a connection, namely through direct calculations of the cluster formation free energies using accurate quantum chemistry methods. Then the dispersion contribution to the XW_n hydration free energy $\mu_{XW_n}^{ex}$ could be estimated using Lifshitz theory (Figure XX.1). This work is currently in progress – it will be interesting to see to what extent the strong ion-water local interactions modify the frequency-dependent ion and water polarizabilities. In order to tease out the dispersion contribution to the cluster formation free energy, quantum chemical calculations using the SAPT (symmetry-adapted perturbation theory) method[27] should prove helpful. In the end, the SAPT dispersion energy results from a second-order perturbation expression similar to the one in Lifshitz theory, but the calculation is done with the ion interacting with the nearby waters.

Ion specific examples

Several examples from our own group are discussed in the context of classical and quantum models of ion specific hydration. The majority of studies employing Eq. 6 for obtaining the free energy have been

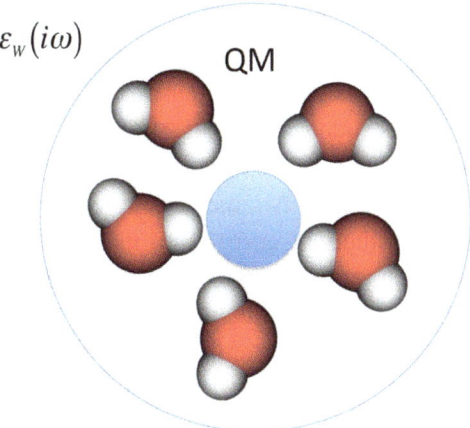

Figure XX.1. Representation of the QCT approach for computing dispersion contributions to ion hydration free energies. The inner-shell region around the ion is treated at an accurate quantum chemical level. The SAPT method can yield accurate estimates for the ion-water interactions in the first shell. The interactions of the cluster with the surrounding water bath can then be treated with Lifshitz theory.

directed at cations[36-38] (the strongly interacting OH⁻ ion is an exception[42]). The hydration shell around the smaller and less polarizable cations tends to be more symmetric than that for anions. Then a calculation of the cluster formation free energy can be accurately estimated as that for the cluster in a minimum energy structure and assuming a harmonic approximation for the vibrational contribution. The case of anions is more complex -- the larger more polarizable anions may actually prefer to reside at the cluster surface, and there are many structures close in energy.[43]

In order to deal with these issues, an approach was developed in Ref. 44 that allows for direct numerical evaluation of the inner-shell and packing terms using Bayesian methods. These methods were then used in a study of ion hydration employing the classical polarizable

AMOEBA force field[25,26] and the QCT.[45] The AMOEBA force field includes a van der Waals potential, fixed point multipoles on the atoms (through the quadrupole), and point induced dipoles distributed on the atoms. Several conclusions of that work are of interest in the present context. It was found that relatively small conditioning radii are sufficient to allow for a Gaussian treatment of the long-ranged contribution (as an average of two mean-field terms). This in turn allows for a division of the free energy into its various contributions: electrostatic, induction, and approximate dispersion. The ion hydration free energy depends more strongly on ion size than on polarizability, with partially cancelling first-order electrostatic and polarization changes. It was also concluded that electrostatic contributions are the main determinant of ion specificity (to some extent this result may be built into the classical model, however). The solvation structure around the ions depends more sensitively on the polarizability than on the free energies, with anisotropy increasing with polarizability. The average dipole moments of the anions (Cl^-, Br^-, and I^-) were found to be significantly larger than those observed in AIMD simulations and other quantum studies.[14,17,18] The possible role of over-polarization in the classical models of ions near the water surface was discussed.

Another paper[18] examined the hydration structure, polarization, and charge transfer for the Cl^- ion in water using classical AMOEBA simulations to generate the configurations and MP2 level quantum chemical calculations for the ion and nearby waters. Using Bader's Atoms in Molecules (AIM) approach,[46] a net charge transfer of 0.2 e was observed from the Cl^- ion to the surrounding waters. Recent spectroscopic experiments support this prediction.[47] The degree of over-polarization of the ion (classical vs. quantum) is by a factor of 2, a substantial discrepancy. The waters in the first solvation shell are slightly under-polarized compared with bulk water, and water-water interactions are more important than ion-water interactions in the (water) polarization.

Ref. 5 developed the LMFT approach for computing hydration

free energies and examined cation and anion hydration using the new method. A basic result of this work was the observation that, outside the first hydration shell, the interaction energies of the ion with more distant waters are strongly Gaussian distributed. This provides support for the use of dielectric models for the distant interactions as has been done in previous QCT calculations.

The alternative LMFT approach for modeling hydration free energies was then employed in a study of ion hydration entropies.[13] When referenced to the hydration entropy of water, a clear distinction between 'structure making' and 'structure breaking' ions emerges (kosmotropes and chaotropes, respectively: kosmotropes exhibit negative values for the difference, while chaotropes display positive differences).[48] Two interesting physical observations were made. First, it was found that the principal cause of the strongly ion specific behavior arises from the local electrostatic interactions (again this result depends to some extent on the simple models employed, and many-body dispersion provides another important contribution[15,16]). In fact, the local electrostatic part of the hydration entropy becomes *positive* for large chaotropic ions. The far-field electrostatic part of the entropy is small in magnitude and negative, however, consistent with a dielectric model. Thus, the behavior of waters in the first hydration shell is the major contributor to ion specificity in the entropy. The second observation is that the ion hydration entropies are remarkably close in magnitude to those for the isoelectronic rare gases, in seeming contradiction to intuition. The very simple point charge models employed in this study yielded quite good agreement with experiment for the entropies, except for the tough-to-handle F^- ion that exhibits some chemical character in its interactions with water (below).

A following paper[41] examined the Na^+ and I^- ions near the water surface, again modeled with simple classical point-charge models. The free energy was partitioned into cavity, dispersion, and local and far-field electrostatic contributions in a point charge model using the

LMFT approach. The two ions were moved from the center of a water slab to the dividing surface at the liquid-vapor interface. The relatively flat total free energy profile for the chaotropic I^- ion results from the near cancellation of several large contributions. There is a large attractive cavity contribution that tends to drive the ion towards the surface, but this contribution is largely cancelled by a repulsive local electrostatic term due to loss of hydrating waters near the dividing surface. There is a mildly attractive far-field electrostatic contribution that is due to the surface potential of the water surface (SPC/E water in these simulations). The dispersion contribution is weakly repulsive, again due to the loss of first-shell waters near the interface. For the kosmotropic Na^+ ion, the contributing free energy changes are smaller in magnitude than for the I^- ion. Both the local electrostatic and cavity contributions are weakly attractive for the interface. The dispersion contribution is very weakly repulsive, while the far-field electrostatic contribution (again due to the water surface potential) is repulsive, resulting in the overall repulsion from the interface. The division into the contributing components is instructive as to the origin of the overall driving forces, but it was noted that the individual pieces will differ considerably if the system is modeled quantum mechanically.

Ref. 19 then addressed more thoroughly the influence of water interfacial potentials on ion hydration in bulk and near the water surface. This issue has been addressed repeatedly from a variety of perspectives.[49] A statistical mechanical theory was developed[19] (summarized above) that shows that, rather than the separate local and far-field electrostatic contributions, the net potential at the center of an ion-sized cavity (the linear term in Eq. 3) provides a more direct physical picture of the influence of water interfacial potentials on hydration (Figure XX.2).[50,51] By comparison of the Marcus[52] and Tissandier et al.[40] experimental estimates of the proton hydration free energy, it was suggested that the net potential is -11.6 kcal/mol-e. This conclusion rests on the validity of the cluster-pair approximation (CPA) used in

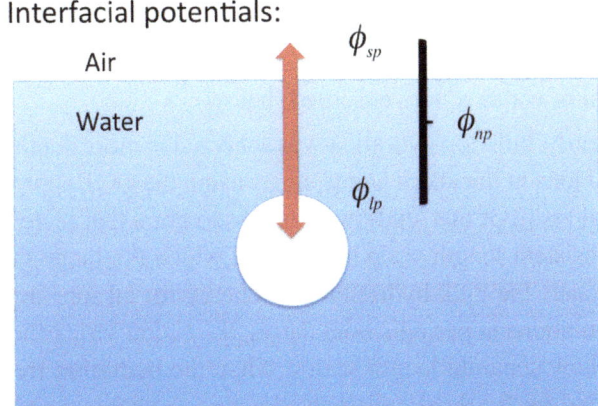

Figure XX.2. Schematic of the relevant water interfacial potentials that lead to the net electrochemical surface potential. The surface potential ϕ_{sp} is the potential change experienced by a non-interacting test charge crossing the liquid-vapor interface (from vapor to liquid). The local potential ϕ_{lp} is the potential change crossing the water-cavity boundary. The net potential ϕ_{np} is the sum of the two potentials. This potential shows some cavity size dependence but approaches a stable limit for cavities larger than roughly 6 Å in radius.[51] The values of the ϕ_{lp} and ϕ_{sp} potentials differ widely between classical and quantum models due to the very different traces of the water quadrupole moment tensor,[13,59] but the sum is of comparable magnitude due to a cancellation across the two boundaries.

the analysis of Ref. 40 (see below). Calculations were performed on clusters that included alkali halide ions and 105, 242, and 511 water molecules. The clusters were simulated with the SPC/E water model. The net potential was then obtained using classical and quantum calculations (with configurations drawn from the classical simulation). From the quantum calculations, the net potential for the n=105 cluster was estimated to lie between -7.4 and -10 kcal/mol-e depending on the level of quantum theory and the basis set. The result from the SPC/E model is -7.5 kcal/mol-e for the same cluster (n=105). Recent work by Mundy *et al.*[53] using DFT-AIMD simulations for water around the cavity suggests the value is closer to 0 kcal/mol-e. What is now clear is that the net potential is highly sensitive to the configurations of the waters near the cavity.[54] Thus the value of the electrochemical surface potential of water is not yet settled (below).

Recently, bulk ion hydration was analyzed in more depth for a wide range of ions in the alkali halide series using the QCT approach.[54] The principal result of that study was the observation that there appears to be a consistent length scale (6.15 Å) at which the inner-shell part of Eq. 5 equals the bulk hydration free energy for all ions in the series. For simulations in periodic boundaries, the cation series and the anion series show separate length scales; when the hydration free energies are shifted by 9.5 kcal/mol (one direction for cations and the other for anions), all ions possess a common length scale. Then the electrostatic potential at the center of the 6.15 Å cavity was computed (in periodic boundaries), with a result of +9.7 kcal/mol-e. The common length scale occurs when the cavity formation free energy accurately matches a simple Born model estimate for the long-ranged term. This result then strongly suggests that the cavity (or net or electrochemical surface) potential should be computed for a cavity of this size. The other interesting physical observation was that ion specificity occurs for length scales shorter than this cavity size; strongly ion specific behavior was observed for cavity sizes up to roughly 4.5 Å, followed by clear linear behavior in the inner-shell term with increasing cavity size (due to ion-water dipole interactions at larger distances).

This study shows that, even when a system is modeled in periodic boundaries, there is an 'interface' that needs to be considered, namely the boundary between the ion and the surrounding waters. The free energy shift associated with this boundary is of large magnitude (9.5 kcal/mol for SPC/E water), and thus should not be neglected. These points were first considered in Refs. 55 and 56, and developed further in Ref. 19, including the quantum case. An example of the importance of this factor is a recent simulation test of the TATB hypothesis[57] in which it was not included – the resulting free energy shift between the cation and anion is 19 kcal/mol (for the SPC/E water model), which implies a much larger deviation from equality of the free energies than actually exists for the given model. It is also important to note that, for the quantum case, the free energy shift would be of magnitude roughly 160 kcal/mol (for a periodic boundary setup), since the surface potential ϕ_{sp} of water has recently been determined to be roughly +3.5 V (or 80 kcal/mol-e)![19,58,59]

Interfacial potentials, the cluster-pair approximation, and single-ion free energies

In order to understand specific ion effects, single-ion hydration is a foundational issue that should be resolved in order to move on with confidence to more complex issues such as ions near the water surface, surface tensions, ion-ion interactions in bulk and near the surface, electrochemical interfaces, and ions interacting with proteins. The discussion above illustrates the potential importance of water interfacial potential effects in the analysis. A recent theoretical study of electrochemical interfaces provides another illustration.[60]

A first point to make is that the definition of a single-ion free energy scale is a delicate issue (see Ref. 61 and references therein). Once a water sample is placed in a beaker a range of phenomena take place. These include the auto-ionization of water and the inclusion of impurity ions (which is inevitable even with the most pure water samples). Ref. 61 shows that the resulting surface potential is a function of all of the ions in the system, and their effect does not go away even in the limit of infinite dilution. Also the makeup of the beaker itself could impact on the potential shift for a test charge moving into the water sample. For the equilibrium with the vapor (not a good conductor), the final result may even depend on the compositional path. A way out of this quandary is to define an idealized single-ion free energy as that for the ion at the center of a water droplet of mesoscopic size with no interference from other ions. This situation may not be attainable experimentally, but serves to target a situation in which the single-ion free energy has a definite meaning. Figure 16 of Ref. 7 shows, however, that the surface potential shifts with increasing ion concentration (measurable thermodynamically as opposed to the surface potential itself) are relatively small for monatomic ion pairs. Thus establishing the water interfacial contribution for an idealized droplet may still give helpful insights into ions near the water surface.

A second point to make (discussed in more detail in Ref. 19), is that there are two general kinds of thermodynamic single-ion tabulations. Examples of the first are the tabulations of Marcus,[52] Schmid, Miah, and Sapunov,[62] and values derived from the Latimer-Pitzer-Slansky (LPS) procedure.[39] These tabulations do not include a contribution from the water interfacial potentials (net potential of Ref. 19, or electrochemical surface potential). It is suggestive that the Marcus proton hydration free energy value and the LPS-derived value are exactly the same (-254.3 kcal/mol); neither of these values include the linear term in Eq. 3 in the analysis. This free energy can be considered a 'solid' estimate of the *bulk* hydration free energy of the proton (free of interfacial potential effects). The single-ion values of Tissandier et al.,[40] however, do in principle include a contribution from the net potential. Their method employs the clever cluster-pair approximation (CPA) that uses both bulk hydration and cluster free energy data in the analysis. Since the result is an estimate of the single-ion proton hydration free energy, any interfacial potential effects are included by definition (so long as the CPA analysis is valid).

We note that QCT results generally compare well with the Marcus values;[3,36,37] this is likely because there is little or no contribution in QCT from the net electrochemical surface potential. The small clusters in the inner shell do not include this collective effect, and a continuum dielectric model doesn't either; thus the QCT estimates the bulk (Marcus) hydration free energy. The QCT result in Ref. 37 for the hydration free energy of the proton using a dielectric continuum model for the distant interactions and the Zundel structure (2 waters) is -254.6 kcal/mol, within 0.3 kcal/mol of the Marcus value! We also note that when the long-ranged interactions are modeled at the molecular level (in periodic boundaries), and the results are shifted by the 9.5 kcal/mol result discussed above, the free energy values are also in excellent agreement with the Marcus result. The QCT thus yields a remarkably accurate result for this fundamental quantity when compared with experiment.

The conclusions of Ref. 19 rested on the assumption that the CPA estimate of the proton hydration free energy is accurate. Here we give a short outline of concerns recently raised concerning this issue,[63] and defer a more detailed analysis to a forthcoming paper. The observations in Ref. 40 and the re-analysis in Ref. 64 are that, when half the conventional anion-cation cluster hydration free energy difference is plotted as a function of the same bulk quantity (both available experimentally), the data for a wide range of ions display linear behavior with a common crossing point for all clusters in the size range $n=1$-6. Also the x and y values on the plot are the same at the crossing point.

These observations imply that the ion-n-water cluster formation free energy is the same for all of the $n=1$-6 clusters (for this hypothetical crossing point pair). Ref. 40 shows that the NaOH and NaF real ion pairs display behavior close to this idealized crossing point pair. The progression from $n=6$ to the infinite size limit is also discussed in detail. Two lines of evidence do *suggest* that the extrapolation procedure is valid. First, Figure 1 of Ref. 40 indicates convergence to the bulk value with increasing size. Second, it is apparent from Figure 1 of Ref. 64 that by $n=6$, the slope of the linear curve is already 90% of the way to the bulk (flat or zero-slope) value. But, as argued in Ref. 16, it is still possible that a transition could occur in the intermediate size range ($n=30$-100 or so). In order for the net potential discussed above to attain a value of close to zero, however, the shift would have to be substantial on the scale of the data -- the derived proton free energy would have to move from upward by 11.6 kcal/mol from -265.9 kcal/mol to -254.3 kcal/mol.

After careful analysis of the underlying assumptions of the CPA, we suggest here that this scenario is somewhat unlikely but possible. Therefore, we are initiating a study of water clusters that include the Na^+ and F^- ions in the intermediate size range using quantum methods. Preliminary classical results (TLB) using the SPC/E water model and the ion-water potentials from Ref. 12 also do not yield agreement with experiment for the $n=6$ cluster.

Resolving this issue is important for the development of ion-water force fields; several of the existing force fields have been parameterized to fit the Tissandier et al. single-ion free energies.[12] A common feature of these force fields is van der Waals well depths for anions that are smaller than those for cations. Since the dispersion interactions should definitely be larger for the anions, we can suggest two possible scenarios: 1) The Tissanadier et al. proton free energy is correct but the net potential for the chosen water model is incorrect (then the parameters are 'being asked' to make up for an error in the water model) or 2) The Tissandier et al. proton free energy is not correct and the parameters are being fit to data that does not reflect the correct physical situation.

What have we learned, and what's next?

This chapter has discussed theoretical and computational approaches for modeling the fundamental aspects of ion hydration. Two themes emerge from the discussion. First, classical molecular-level models can yield helpful insights into driving forces for ion hydration phenomena. Second, when we probe at a more fundamental level, it becomes apparent that, due to the intense fields that occur in the ionic solution, and the diffuse nature of the electron cloud in water and around the ion, a quantum-level treatment is necessary.

The top-down (Lifshitz) theoretical approach in principle includes contributions from all fluctuating charges on the ion and in water. Two recent papers by Duignan, Parsons, and Ninham[15,16] illustrate this view -- they provide a compelling indication that dispersion interactions are a significant contributor to ion specific hydration.

The bottom-up approach, on the other hand, seeks to represent the details of the quantum mechanical interactions with molecular reality for the first hydration shell. Then the more distant interactions can be handled with a Gaussian electrostatic model or even a closely related dielectric continuum theory. The quasi-chemical theory[3] (QCT) is a statistical mechanical approach that is ideally suited to provide a connection between these two views: an accurate molecular-level quantum mechanical treatment of the inner-shell, with approximations for the long-ranged interactions. In previous QCT work, the long-ranged interactions have been modeled as purely electrostatic ones employing the zero-frequency dielectric constant (or classical molecular simulations that produce an electrostatic potential). What will be pursued in the future is an assessment of many-body dispersion contributions with a combination of quantum chemical results for the inner shell followed by embedding the whole cluster into bulk continuum water (with the imaginary-frequency-dependent dielectric function taken from experiment, as done in Ref. 16).

As an indication of quantum effects on ion-water interactions, Figure XX.3 displays the electron density changes that occur with the binding of one water molecule to the F^- ion (obtained with high-level quantum chemical calculations); the distortion is intricate, with some charge transfer involved. The details of the charge rearrangement depend sensitively on the water orientation, here taken as the minimum energy hydrogen-bonded geometry. Previously the hydrogen bond in the F^-/water dimer has been shown to possess charge transfer and

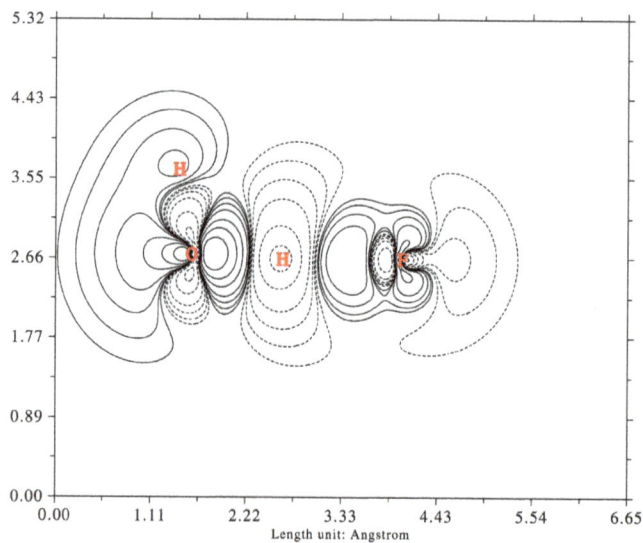

Figure XX.3. A contour map (in the plane of the complex) for the electron density change when a water molecule binds to the F⁻ ion, water on left, F⁻ on right. Distances are in Å. Solid lines indicate electron density buildup, while dashed lines imply electron density depletion. Note the large depletion region around the H-bonded water hydrogen, and the large region of electron accumulation on the far side of the water molecule. The polarization of the F⁻ ion is apparent but complicated.

chemical features;[65] the behavior for cations is quite different.[66] Figure XX.3 calls into question the ability of classical models to represent this level of complexity.

To close, we re-iterate the importance of an accurate quantum treatment of dispersion through another example. In Refs. 15 and 16, it was shown that a proper treatment of dispersion interactions is essential in understanding ion specific hydration. As part of that work, it was found that in the halide series, the F⁻ ion actually produces the largest dispersion contribution to the hydration free energy, a counter-intuitive result. In a forthcoming paper on the quantum mechanics of ion-water dimers,[66] we find the same trend using the SAPT method. The dispersion energy is roughly a third of the total binding energy. Since dispersion is a many-body correlated electron effect, it cannot be expected that a classical model would reproduce these results. The SAPT calculations also produce values for the exchange-repulsion, electrostatic, induction, and charge transfer energies. We observe a large charge transfer contribution for the F$^-$/water pair interaction energy (nearly as large as the dispersion energy), intimately related to the ion-water hydrogen bond. The remaining terms are all of magnitude as large or larger than the total binding energy and also display distinct ion specificity.

Looking to the future, a conclusive result concerning the extrapolation to the bulk limit for the proton hydration free energy using the CPA is an important goal. Without such a result, force fields developed based on the free energies of Ref. 40 can be questioned. On the other hand, given a firm result for the absolute proton free energy (that may include interfacial potential contributions), more accurate models can be developed that can be expected to produce reliable ion density profiles near interfaces. It is our belief that, while the goal of accurate single-ion hydration free energies is a challenging one, resolving these issues will aid in understanding ions near bulk interfaces, in small pores that occur for example in modern supercapacitors for energy storage, and near biological membranes and proteins.

Acknowledgements

We would like to acknowledge the support of this research by NSF grants CHE-1011746 and CHE-1266105 and a generous grant of computing time at the Ohio Supercomputer Center. We thank Lawrence Pratt, Dilip Asthagiri, Susan Rempe, Tim Duignan, Drew Parsons, Barry Ninham, Chris Mundy, Greg Schenter, Marcel Baer, Shawn Kathmann, and Kevin Leung for helpful discussions.

References

1. Sellner, B.; Valiev, M.; Kathmann, S. M. *J. Phys. Chem. B,* **2013**, *117*, 10869-10882.
2. Robinson, R. A.; Stokes, R. H. *Electrolyte Solutions*, Dover, New York, 1959.
3. Beck, T. L.; Paulaitis, M. E.; Pratt, L. R. *The Potential Distribution Theorem and Models of Molecular Solutions*, Cambridge, New York, 2006.
4. Hummer, G.; Pratt, L. R.; Garcia, A. E. *J. Phys. Chem.* **1996**, *100*, 1206-1215.
5. Beck, T. L. *J. Stat. Phys.* **2011**, *145*, 335-354.
6. Rasaiah, J. C.; Lynden-Bell, R. M. *J. Chem. Phys.* **1997**, *107*, 1981-1991.
7. Jungwirth, P.; Tobias, D. J. *Chem. Rev.* **2006**, *106*, 1259-1281.
8. Chang, T. M.; Dang, L. X. *Chem. Rev.* **2006**, *106*, 1305-1322.
9. Ninham, B. W.; Lo Nostro, P. *Molecular Forces and Self Assembly*, Cambridge, Cambridge, 2010.
10. Kunz, W.; Lo Nostro, P.; Ninham, B. W. *Curr. Opin. Colloid Interface Sci.* **2004**, *9*, 1-18.
11. Petersen, P. B.; Saykally, R. J. *Annu. Rev. Phys. Chem.* **2006**, *57*, 333-364.
12. Horinek, D.; Herz, A.; Vrbka, L.; Sedlmeier, F.; Mamatkulov, S.; Savkat, I.; Netz, R. R. *Chem. Phys. Lett.* **2009**, *479*, 173-183.
13. Beck, T.; *J. Phys. Chem. B*, **2011**, *115*, 9776-9781.
14. Baer, M. D.; Mundy, C. J.; *J. Phys. Chem. Lett.* **2011**, *2*, 1088-1093.
15. Duignan, T. T.; Parsons, D. F.; Ninham, B. W. *J. Phys. Chem. B* **2013**, *117*, 9412-9420.
16. Duignan, T. T.; Parsons, D. F.; Ninham, B. W. *J. Phys. Chem. B* **2013**, *117*, 9421-9429.
17. Guardia, E.; Skarmoutsos, I.; Masia, M. *J. Chem. Theory Comput.* **2009**, *5*, 1449-1453.
18. Zhao, Z.; Rogers, D. M.; Beck, T. L. *J. Chem. Phys.* **2010**, *132*, 014502.
19. Beck, T. L. *Chem. Phys. Lett.* **2013**, *561-562*, 1-13.
20. Parsegian, V. A. *Van der Waals Forces: A Handbook for Biologists, Chemists, Engineers, and Physicists*, Cambridge, Cambridge, 2006.
21. Levine, I. N. *Quantum Chemistry*, 7th ed., Prentice Hall, New York, 2013.

22. Foulkes, W. M. C.; Mitas, L.; Needs, R. J.; Rajagopal, G. *Rev. Mod. Phys.* **2001**, *73*, 33-83.

23. Ciccotti, G.; Frenkel, D.; McDonald, I. R. *Simulation of Liquids and Solids: Molecular Dynamics and Monte Carlo Methods in Statistical Mechanics,* North-Holland, Amsterdam, 1987.

24. Fanourgakis, G. S.; Xantheas, S. S. *J. Phys. Chem. A* **2006**, *110*, 4100-4106.

25. Grossfield, A.; Ren, P.; Ponder, J. W. *J. Am. Chem. Soc.* **2003**, *125*, 15671-15682.

26. Ren, P.; Ponder, J. W. *J. Phys. Chem. B* **2003**, *107*, 5933-5946.

27. Stone, A. J. *The Theory of Intermolecular Forces,* Oxford, Oxford, 1996.

28. Masia, M. *J. Phys. Chem. A* **2013**, *117*, 3221-3226.

29. Tongraar, A.; Rode, B. M. *Chem. Phys. Lett.* **2005**, *403*, 314-319.

30. Ho, M.-H.; Klein, M. L.; Kuo, I.-R. W. *J. Phys. Chem. B* **2009**, *113*, 2070-2074.

31. Heuft, J. M.; Meijer, E. J. *J. Chem. Phys.* **2005**, *122*, 094501.

32. Parr, R. G.; Yang, W. *Density Functional Theory of Atoms and Molecules,* Oxford, Oxford, 1989.

33. Schmidt, J.; VandeVondele, J.; Kuo, I.-F. W.; Sebastiani, D.; Siepmann, J. I.; Hutter, J.; Mundy, C. J. *J. Phys. Chem. B* **2009**, *113*, 11959-11964.

34. Beck, T. L. in *Free Energy Calculation: Theory and Applications in Chemistry and Biology*, ed. Pohorille, A., Chipot, C., Springer-Verlag, New York, 2007, 389-422.

35. Morrone, J. A.; Car, R. *Phys. Rev. Lett.* **2008**, *101*, 017801.

36. Asthagiri, D.; Pratt, L. R.; Paulaitis, M. E.; Rempe, S. B. *J. Am. Chem. Soc.* **2004**, *126*, 1285-1289.

37. Asthagiri, D.; Pratt, L. R.; Ashbaugh, H. S. *J. Chem. Phys.* **2003**, *119*, 2702-2708.

38. Asthagiri, D.; Dixit, P. D.; Merchant, S.; Paulaitis, M. E.; Pratt, L. R.; Rempe, S. B.; Varma, S. *Chem. Phys. Lett.* **2010**, *485*, 1-7.

39. Ashbaugh, H.; Asthagiri, D. *J. Chem. Phys.* **2008**, *129*, 204501.

40. Tissandier, M. D.; Cowen, K. A.; Feng, W. Y.; Gundlach, E.; Cohen, M. H.; Earhart, A. D.; Coe, J. V.; Tuttle, T. R. *J. Phys. Chem. A* **1998**, *102*, 7787-7794.

41. Arslanargin, A.; Beck, T. L. *J. Chem. Phys.* **2012**, *136*, 104503.
42. Asthagiri, D.; Pratt, L. R.; Kress, J. D.; Gomez, M. A. *Chem. Phys. Lett.* **2003**, *380*, 530-535.
43. Kim, J.; Lee, H. M.; Suh, S. B.; Majumdar, D.; Kim, K. S. *J. Chem. Phys.* **2000**, *113*, 5259-5272.
44. Rogers, D. M.; Beck, T. L. *J. Chem. Phys.* **2008**, *129*, 134505.
45. Rogers, D. M.; Beck, T. L. *J. Chem. Phys.* **2010**, *132*, 014505.
46. Bader, R. F. W. *Atoms in Molecules: a Quantum Theory*, Oxford, Oxford, 1994.
47. Xiong, K.; Asher, S. A. *J. Phys. Chem. A.* **2011**, *115*, 9345-9348.
48. Collins, K. D.; Neilson, G. W.; Enderby, J. E. *Biophys. Chem.* **2007**, *128*, 95-104.
49. Randles, J. E. B. *Phys. Chem. Liq.* **1977**, *7*, 107-179.
50. Baer, M. D.; Stern, A. C.; Levin, Y.; Tobias, D. J.; Mundy, C. J. *J. Phys. Chem. Lett.* **2012**, *3*, 1565-1570.
51. Ashbaugh, H. S. *J. Phys. Chem. B* **2000**, *104*, 7235-7238.
52. Marcus, Y. *Ion Solvation,* John Wiley, New York, 1985.
53. Baer, M. D.; Schenter, G. K.; Mundy, C. J.; Remsing, R. C.; Weeks, J. D., private communication, 2013.
54. Shi, Y. Beck, T. L.; *J. Chem. Phys.* **2013**, *139*, 044504.
55. Harder, E.; Roux, B. *J. Chem. Phys.* **2008**, *129*, 234706.
56. Vorobyov, I.; Allen, T. W. *J. Chem. Phys.* **2010**, *132*, 185101.
57. Schurhammer, R.; Engler, E.; Wipff, G. *J. Phys. Chem. B* **2001**, *105*, 10700-10708.
58. Kathmann, S. M.; Kuo, I. W.; Mundy, C. J.; Schenter, G. K. *J. Phys. Chem. B* **2011**, *115*, 4369-4377.
59. Leung, K. *J. Phys. Chem. Lett.* **2010**, *1*, 496-499.
60. Cheng, J.; Sprik, M. *Phys. Chem. Chem. Phys.* **2012**, *14*, 11245-11267.
61. Pratt, L. R. *J. Phys. Chem.* **1992**, *96*, 25-33.
62. Schmid, R.; Miah, A. M.; Sapunov, V. N. *Phys. Chem. Chem. Phys.* **2000**, *2*, 97-102.

63. Hünenberger, P.; Reif, M. *Single-Ion Solvation: Experimental and Theoretical Approaches to Elusive Thermodynamics Quantities,* RSC Publishing, Cambridge, 2011.
64. Kelly, C. P.; Cramer, C. J.; Truhlar, D. G. *J. Phys. Chem. B* **2006**, *110*, 16066-16081.
65. Thompson, W.H., Hynes, J.T., *J. Am. Chem. Soc.* 2000. 122. 6278-6286.
66. Pollard, T., Beck, T., in preparation, 2013.

XXI
Comparison of mechanical and thermodynamical evaluations of electrostatic potential differences between electrolyte solutions

Xinli You, Mangesh I. Chaudhari, Lawrence R. Pratt

Chemical & Biomolecular Engineering Department,
Tulane University New Orleans, LA 70118, USA.
lpratt@tulane.edu

Introduction

The mean electric field in a homogenous conducting material being zero, the electrostatic potential is spatially constant throughout a conducting phase. A value $\phi^{(\alpha)}$ can be thus assigned to conducting phase α as the potential of the phase. For an ion of type j and charge q_j, that potential of the phase augments the chemical potential

$$\mu_j^{(\alpha)} = q_j \phi^{(\alpha)} + \overline{\mu}_j^{(\alpha)} \qquad (1)$$

additively. Since the chemical potentials govern the chemistry and phase equilibrium of materials, the potential of the phase is a central concept for experiments which transfer charge at phase boundaries, and specifically for electrochemistry.[1]

The bulk compositions of conducting phases at equilibrium are electrically neutral

$$\sum_j q_j n_j^{(\alpha)} = 0 \qquad (2)$$

Therefore $\phi^{(\alpha)}$ is not represented in

$$G = \sum_j \mu_j^{(\alpha)} n_j^{(\alpha)} \qquad (3)$$

the Gibbs free energy of that phase. This suggests that $\phi^{(\alpha)}$ cannot be obtained by classic thermodynamic techniques that transfer neutral material combinations between two phases with neutral compositions. This suggestion has been closely analysed throughout the history of this subject, and is the standard view. In recent years, this standard view has been called the "Gibbs-Guggenheim Principle."[2]

Nevertheless, molecular simulation calculations now provide specific molecular-resolution descriptions of interfaces between solutions, which generally do support non-zero mean electric fields. Analysis of those interfacial electric fields leads to[3-5]

$$\phi^{(\alpha)} - \phi^{(\gamma)} = \int_{z^{(\gamma)}}^{z^{(\alpha)}} z \rho_q(z) dz / \varepsilon_0 \qquad (4)$$

with $\rho_q(z)$ the density of electric charge, and we take the interface to be planar and perpendicular to the z-axis. In contrast to the thermodynamic relation Eq. (1), Eq. (4) is a mechanical route to evaluate contact potentials. This does not necessarily alter the conclusions of the thermodynamic analyses, but provides another way to learn about contact potentials. The example (Figure XXI.1), which involves a capped ethylene-oxide oligomers in coexisting water and n-hexane phases,[6] shows that the electric potential change can be calculated straightforwardly with Eq. (4).

An important goal for comparisons of the thermodynamic and mechanical routes to contact potentials is to see how they are consistent with one another. That has been a challenge, and indeed has resulted in a several surprises that have not been fully internalised in current research in this area. Explanation of several of those surprises is the program for the discussion that follows.

The mechanical formula on the basis of molecular dipole and quadrupole moments

Eq. (4) relates the surface potential directly to the dipole moment of the charges in the interfacial region. Solvent contributions to the integral Eq. (4) arising from molecular multipole moments involve only molecular dipole and quadrupole moments because

$$\rho_q(z) = -\left(\frac{d}{dz}\right) P_z(z) + \left(\frac{d}{dz}\right)^2 Q'_{zz}(z) + \ldots \tag{5}$$

where terms higher-order in the gradients have not been written out because they will be irrelevant in the integral Eq. (4). Eq. (5) deliberately follows conventional textbook notation.[7] $P_z(z)$ is the density at elevation z of the z–component of molecular dipole moments, referenced to a specified molecular center. Similarly, $Q'_{zz}(z)$ is the density at height z of the zz –component of molecular quadrupole moments. Because of the gradients in Eq. (5), integration by parts of Eq. (4) gives

$$\phi^{(\alpha)} - \phi^{(\gamma)} = \int_{z^{(\gamma)}}^{z^{(\alpha)}} P_z(z) dz / \varepsilon_0 - \Delta Tr \mathbf{Q'} / 3\varepsilon_0 \tag{6}$$

The original papers should be consulted for the specific notation.[4,5] The right-most term of Eq. (6) appears not to involve the interfacial structure, but only the molecular-structure and the change indicated by Δ of the molecular quadrupole density between the two isotropic phases, the isotropy permitting the trace *"Tr"* notation. Each of the terms on the right of Eq. (6) depends on the molecular-center chosen to evaluate the molecular moments. The integral on the right of Eq. (4) obviously does not depend on a choice of a molecular center. Thus both terms of Eq. (6) are required to obtain a center-independent result. Change of the molecular center changes the molecular quadrupole contribution, presumably independent of interfacial structure, but also the molecular dipole contribution which does reflect interfacial structure.

This result has been surprising. It has been argued that the quadrupole contribution arises from considering of electric potentials

Comparison of mechanical and thermodynamical evaluations 437

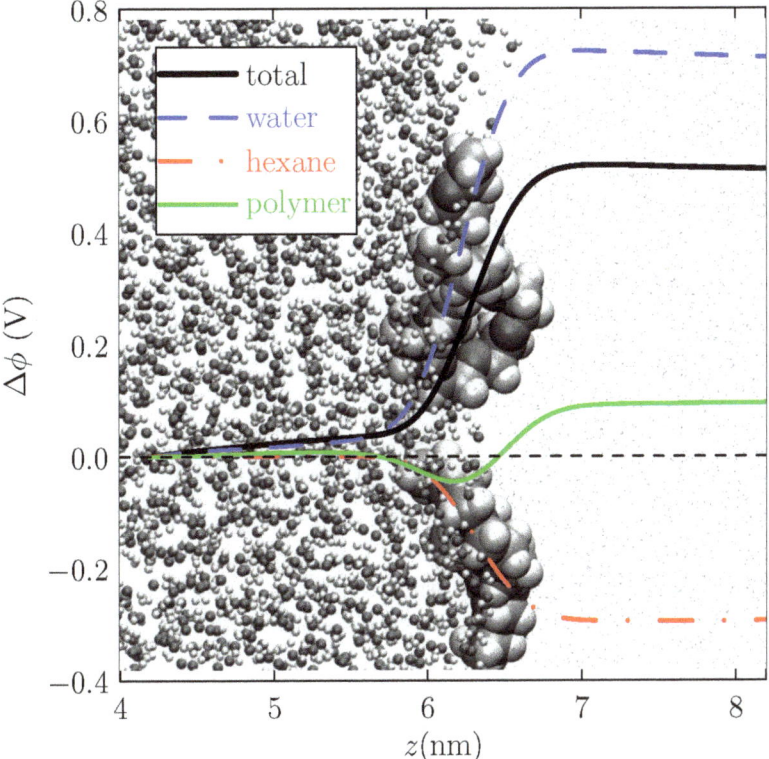

Figure XXI.1. Electrostatic potential change with progress from a liquid aqueous phase (left), through the planar solution interface, and into liquid n-hexane (right). This calculation included capped poly-ethylene-oxide oligomers described in Ref. 6; they are referred to here as "polymer". Consistent with previous calculations discussed above, the aqueous phase has a negative potential when considering the water alone, described by the SPC/E model.[6] This is due to the dominance of the quadrupole moment density exhibited in Eq. (6). When flexible and extended molecules are considered, Eq. (6) would be awkward though Eq. (2) applies directly.

inside molecules, but only the electric potential in intermolecular spaces should be considered, and therefore the quadrupole contribution should be dropped.[8] The difficulty with this argument is that the contact potential no longer has just one, center-independent, value.

This quadrupole contribution leads to another surprise. Standard molecular simulation models of liquid water utilise partial charges to describe intermolecular forces. Using those partial charges to evaluate the required molecular moments yields a value of $\phi^{(l)} - \phi^{(g)}$ through either Eq. (4) or Eq. (6). For the SPC/E model that value is about -0.55 V at T=298 K.[9] But the distribution of those partial charges is different from the distribution of charge in a water molecule based on evaluation of electron densities. The electron-density based quadrupole contribution changes in sign and order-of-magnitude to produce a net value of about 3.6 V.[10] How this huge change should fit into the picture of the statistical thermodynamics of electrolyte solutions is not yet settled.

This result has been surprising in another respect because it makes explicit how a substantial electric polarization can be induced by purely compressional interactions that change the density. This issue must be addressed also in evaluating polarization responses of a solution to solutes which inevitably alter local solvent densities.[11]

Contact potential from phase equilibrium of ions

We elaborate the chemical potential[12] of Eq. (1)

$$\mu_j^{(\alpha)} = q_j \phi^{(\alpha)} + kT \ln\left[\rho_j^{(\alpha)} \Lambda_j(T) / q_j^{(\text{int})}(T)\right] + \Delta \overline{\mu}_j^{(\alpha)} \qquad (7)$$

to indicate the contribution $\Delta \overline{\mu}_j^{(\alpha)}$ derived conventionally from intermolecular interactions, and an ideal contribution, the middle term. The ideal contribution depends the bulk density of the ion $\rho_j^{(\alpha)}$ and on factors that only depend on the temperature T. We consider two conducting phases in equilibrium, and for initial simplicity consider a 1-1 electrolyte which dissolves in both phases, e.g., tetraethylammonium tetrafluoro-borate TEABF$_4$. The conditions of equilibrium include matching of the chemical potentials Eq. (7) in both phases for all ions. Consequently $\mu_{TEA^+}^{(\alpha)} - \mu_{BF_4^-}^{(\alpha)}$, in which the ideal concentration factors cancel because of electro-neutrality, must be the same in both phases. This matching between the two phases implies

$$2e\left(\phi^{(\alpha)} - \phi^{(\gamma)}\right) = -\left[\left(\Delta\overline{\mu}_{TEA^+}^{(\alpha)} - \Delta\overline{\mu}_{BF_4^-}^{(\alpha)}\right) - \left(\Delta\overline{\mu}_{TEA^+}^{(\gamma)} - \Delta\overline{\mu}_{BF_4^-}^{(\gamma)}\right)\right] \quad (8)$$

In another setting,[13] this calculation has been called the *plasma approximation,* though it is not approximate here. The plasma physics lesson[13] is to prefer charge-neutrality as a basis for evaluation of potential changes instead of Poisson's equation, Eq. (4).

Eq. (8) has been surprising because the several factors on the right-side are ion specific and have non-trivial, non-zero infinite-dilution limits. Thus the infinite-dilution limiting value is ion-composition specific. To emphasise the surprise, this has been called a "phantom ion effect"[14] and discussed by remarking that the contact potential "depends on which ions are not there."[5] This inflammatory language has been unhelpful for understanding this basic phenomenon. One step in straightening-out our understanding is to insist in referring to the quantities $\phi^{(\alpha)}$ as potentials of the (conducting) phase, *not* a property of the interfaces of a dielectric solvent. Thus a contact potential between solutions of $TEABF_4$ depends on the characteristics of the electrolyte $TEABF_4$, and that is true of the limiting infinite-dilution value also.

A mathematical mechanism[15] for Eq. (4) and Eq. (8) to be consistent is the following: the Debye screening length κ^{-1}, through the combination

$$\kappa^2 = \sum_j \rho_j q_j^2 / kT\varepsilon \quad (9)$$

with ε the permittivity of the solvent, provides a length-scale for an electrolyte solution at low concentration. At low concentrations, this Debye length is large on a molecular-scale. In considering Eq. (4) we can identify such a Debye length for each phase, and consider each side of the integration range individually. But we will suppress that additional notation here for the sake of simplicity. Adopting this length scale for considering the integral Eq. (4) we see

$$\int_{z^{(\gamma)}}^{z^{(\alpha)}} z \rho_q(z) dz = \left(\frac{1}{\kappa^2}\right) \int_{\kappa z^{(\gamma)}}^{\kappa z^{(\alpha)}} (\kappa z) \rho_q(\kappa z) d(\kappa z) \qquad (10)$$

This assumes that the charge density will decay to zero at long distances from the interface on the length scale κ^{-1}. Because κ^2, Eq. (9), is linearly proportional to the solution concentrations and then to the magnitude of $\rho_q(\kappa z)$, it is natural to expect that the right-side of Eq. (10) has a non-zero infinite-dilution limit. Indeed this is how simple theories of electrolyte solution interfaces work: the requirement Eq. (8) becomes a condition of solvability and Eq. (4) is then consistent for the ionic contributions.

This mathematical mechanism requires adequate description of inter-ionic screening lengths, but does not depend on non-physical issues such as choice of statistical mechanical ensemble.[14]

As a physical mechanism, this picture is explained as follows: Consider again a two-phase system and a dissolved electrolyte such as $TEABF_4$. Suspend theoretical consideration of electrostatic interactions for the moment. Suppose that one phase, say γ, was prepared with $TEABF_4$ concentrations that satisfy electro-neutrality, Eq. (2). Would the concentration of TEA^+ and BF_4^- satisfy electro-neutrality in the opposite phase, α? The obvious answer is "no, these ions are different from one another." But now considering electrostatic interactions again, the physical requirement is that composition of phase α must be electro-neutral. How is this achieved? The answer is that the ion densities are layered in the interfacial region just to make-up the electrostatic potential change to be Eq. (8). Specific examples are available in the references.[5,15]

Concluding discussion

We propose several recommendations. Firstly, the mechanical route to contact potentials makes them well-defined on the basis of permitted, molecular-resolution interfacial information. That does not change the Gibbs-Guggenheim Principle that single-ion transfer free energies are not obtained from thermodynamic information. Secondly, though the contact potential can be regarded as dipole potential, it is not derived from a density of molecular dipoles even for such a compact solvent molecule as H_2O. Thirdly, resist regarding $\phi^{(\alpha)} - \phi^{(\gamma)}$ as a property of a dielectric solvent interface. This is because, fourthly, potentials between electrolyte solutions depend on ionic composition even at low electrolyte concentrations. Finally, be amazed and relieved that the contact potentials, e.g. Eq. (1), are not reflected in standard thermodynamic analyses.

References

1. Sommerfeld, A. *Thermodynamics and Statistical Mechanics, Lectures on Theoretical Physics*, Volume 5, Academic Press (New York), 1956.
2. Pethica, B. A. *Phys. Chem. Chem. Phys.* **2007**, *9*, 6253-6262.
3. Landau, L. D.; Lifshitz, E. M. *Electrodynamics of Continuous Media*, Pergamon Press, New York, 1980.
4. Wilson, M. A.; Pohorille, A.; Pratt, L. R. *J. Chem. Phys.* **1989**, *90*, 5211-5213.
5. Pratt, L. R. *J. Phys. Chem.* **1992**, *96*, 25-33.
6. Chaudhari, M. I.; Pratt, L. R. Microstructures of capped ethylene oxide oligomers in water and n-hexane. In *Oil spill remediation: colloid chemistry based principles and solutions*, 2012, ed. P. Somasundaran, R. Farinato, P. Patra, and K. Papadopoulos. See also: *http://arxiv.org/abs/1303.6597*
7. Jackson, J. D. *Classical Electrodynamics*, 1998, Wiley. See Chapter 6.
8. Matsumoto, M.; Kataoka, Y. *J. Chem. Phys.* **1989**, *90*, 2398-2407.

9. Sokhan, V. P.; Tildesley, D. J. *Mol. Phys.* **1997**, *92*, 625-640.
10. Leung, K. *J. Phys. Chem. Lett.* **2010**, *1*, 486-499.
11. Beck, T. L. *Chem. Phys. Lett.* **2013**, *561*, 1-13.
12. Beck, T. L.; Paulaitis, M. E.; Pratt, L. R. *The Potential Distribution Theorem and Models of Molecular Solutions,* Cambridge University Press, 2006.
13. Chen, F. F. *Plasma Physics and Controlled Fusion*, Plenum Press, New York, 1984.
14. Levin, Y. *J. Chem. Phys.* **2008**, *129*, 124712.
15. Nichols III, A. L.; Pratt, L. R. *J. Chem. Phys.* **1984**, *80*, 6225-6233.

XXII
Modelling water as a continuum solvent to understand ion-specific effects

Tim Duignan, Drew Parsons, Barry Ninham

Applied Mathematics Department, Australian National University, Canberra, ACT, 2602, Australia.
tim@duignan.net

1 Introduction

That water is an excellent solvent for ions is a precondition for life. Electrolyte solutions are as important as they are ubiquitous, and their properties depend sensitively on the specific constituent ions. An explanation of this specificity (Hofmeister effects) is a precondition for our quantitative understanding of a huge range of biological and industrial processes.[1,2]

In pursuit of this main aim we begin with the simplest properties of these solutions. The simplest properties will have the least number of "moving parts", and so their explanation should be the most straightforward. Thereafter models that explain these properties can be generalised to explain more complex systems with confidence. Reasonable choices for the simplest properties of electrolyte solutions are ionic solvation energies, osmotic coefficients, and surface tensions.

Currently our understanding of all of these properties is inadequate. We have some qualitative explanations, but complete quantitative agreement with experiment has been elusive unless parameters for each ion or salt are adjusted to reproduce experiment. Worse still different parameters for the same salt are required for each experiment. There is still even a degree of disagreement about the underlying physical mechanisms responsible for some observed behaviours.

The central challenge in calculating these properties is the calculation of free energies. For example the free energy of an ion in water can be compared directly with experimental solvation energies. The free energy change as an ion approaches another ion, and the free energy change as an ion approaches the air-water interface can be used to determine the osmotic/activity coefficients and the surface tension of electrolyte solutions respectively. This is made possible with the use of well-established statistical mechanical methods, such as the hypernetted chain approximation or the Poisson-Boltzmann equation,

1.1 Primitive models

The classical or primitive model of electrolyte solutions treats the ions as charged spheres, with adjustable radii. The solvent water is treated as continuum characterised by its dielectric constant only. In the case of solvation energies this "primitive" approach corresponds to the Born model.[3] It gives a reasonable estimate of solvation energies. However, the size parameters have to be adjusted to unphysical values to fit experiment precisely. For example, cation and anion crystal sizes have to be adjusted by different constants.[4]

In the case of osmotic coefficients and surface tensions this primitive model approach gives the Onsager-Samaris and Debye-Hückel theories.[2] These models are adequate at very low concentrations where ions often behave indistinguishably from each other. However, to get agreement at higher concentrations where there is significant ion-specificity, the size parameters again have to be adjusted to unphysical values to bring the theory into agreement with experiment. For instance, the model must use a size parameter for iodide that is smaller than fluoride's in order to reproduce the low surface tension of iodide electrolyte solutions, and the fitted size parameters are non-additive for the osmotic coefficients.[5]

1.2 Explanations

The earliest explanation of specific ion effects is the idea that ions change the structure of the surrounding water, i.e., some ions break the structure of water (chaotropes) and some increase the structure of water (kosmotropes). This idea has fallen out of favour recently[6,7] due to evidence that the water structure is not significantly affected beyond the first or second hydration layer.[8] This hypothesis does however highlight one of the reasons for the failure of primitive type models, which is their inability to account for changes in the water surrounding an ion.

A second hypothesis for explaining these effects is that the ions experience non-electrostatic interactions, i.e., dispersion interactions[2,9] which will depend on a balance of ion size and polarisability and hence will give ion-specificity. Omission of dispersion forces is tantamount to neglecting quantum mechanics. This hypothesis has mainly been applied to ion interactions with surfaces, where it has been established that it clearly makes a substantial contribution to ionic distribution profiles near interfaces, and hence to measurable macroscopic properties.[10,11,12]

It is unlikely however that simply supplementing primitive models with these potentials will be enough to build a quantitatively accurate and predictive model. This is because of the aforementioned failure of the primitive models to account for the role of water adequately. This line of reasoning leads to a third explanation of specific ion effects, namely that although the water structure is only altered in the first one or two water layers surrounding an ion, this water interacts with the ion in way that is unique to the ion, depending on its size, or more specifically, surface charge density,[13] as well as other specific properties of the ion such as the sign of its charge, and its polarisability. Other effects, such as chemical interactions may also play a role, particularly for apparently unusual ions such as fluoride. For ions to interact with other molecules in solution these short-range water molecules must be removed or rearranged resulting in highly ion-specific short-range interactions. This ion specificity in the interaction of ions with water is clear from the solvation energy of ions where it varies dramatically, firstly depending on size, but also on the nature of the ion. For instance, the solvation energy of fluoride and potassium are quite different despite the fact that they are similar in size.[14] This mechanism of ion-specificity is subsumed in the term hydration effect. Due to the difficulty of modelling this effect, it is used to justify the inclusion of additional fitted parameters in many models, which would otherwise not agree with experiment.

2 Modern Approach

2.1 Explicit Solvent Approach

This postulated hydration effect has lead to the view that it is necessary to know the specific orientation and position of the water molecules around ions. This belief has focussed research on developing accurate molecular dynamic simulations of electrolyte solutions using classical simulation. These have had mixed success. Additionally, the possibility of "chemical interactions" between an ion and the water molecules in the first layer has spurred attempts at *ab initio* calculations of ions in water, whether through full DFT molecular dynamic simulation,[15] hybrid QM/MM approaches[16] or static geometry optimized clusters.[17]

The notion that electrolyte solutions must be described with models that explicitly track the position and orientation of water molecules may be true. If it were so, it would be very unfortunate. This is because these models impose dramatic computational demands. Pure quantum mechanical approaches are impractical for anything but simple systems and simple properties. With significant approximations the range of applicability can be widened. Even so, scaling them up to the large systems relevant for biology, or using them to test large classes of molecules for particular properties is prohibitively expensive.

In addition, these approximate models involve many fitted parameters. As a result it is far from certain that they will work outside the range of experiment they have been adjusted to reproduce. It also means that when building a model there are many decisions to be made with regards to parameter choice and functional form, which may lead to substantially different predictions and conclusions. For example the solvation energies of individual ions is still a matter of debate due to a substantial variation in theoretical values that depend on the model used.[18] Thirdly, even if agreement with independent experiment is observed, the models do not necessarily provide a conclusive qualitative explanation or guidance as to the physical cause. For instance the surface excess of large anions at the air water interface has been observed in agreement with experiment, but the physical cause of this excess is still the matter of debate.[19]

2.2 Continuum Solvent Models

One key advantage of the primitive models of electrolyte solutions is that they treat the solvent as a continuum, including only the average effect of the solvent molecules. This approach is dramatically less computationally demanding, allowing them to be scaled up to situations of biological interest, and to be applied to large sets of molecules. They also generally provide direct and unambiguous physical insight due to their derivation. For these reasons these models are still widely used in applications. For example, the Born and Debye-Hückel models are very useful pedagogically and for informing intuition, more advanced continuum models are also useful in the simulation of biologically important systems.[20,21] This is despite the fact that until recently they have not proven totally satisfactory[18] at reproducing the simplest properties of electrolyte solutions.

The clear and obvious problem with these models is that they are not rigorously true in the microscopic limit. That is, where the solutes are of a size similar to water molecules, i.e., a few Ångstroms.

This is clear from the mathematical derivation of the approach, which requires fields to be averaged over volumes containing many molecules.[22] It is however possible that the models remain somewhat accurate outside the range they are strictly applicable to. If this were the case it would be greatly advantageous, for the reasons discussed above. Correspondingly, it would be more than unfortunate if these models were to go undiscovered because enough effort was not focussed on improving them.

We should then proceed by assuming the approach is valid and building the best models that we can, then comparing their predictions with experiment. This should allow us to determine at what size scale these approaches break down, and the degree of severity of the approximations. As mentioned above many continuum solvent models are focussed on describing complex properties that are difficult to address with explicit solvent models.[20] This is reasonable as it may seem that there is little point in modelling properties with these simple and approximate models, when they can be, or soon will be described by sophisticated and comprehensive explicit solvent simulations.

However, there are three important reasons to focus on explaining these simple properties with continuum solvent models. Firstly, as outlined in the introduction, any model of electrolyte solutions needs to reproduce the simplest properties first. Secondly, simple models are often more useful and provide more insight than complex ones even if they are less accurate. Einstein captured this insight well when he said, "the supreme goal of all theory is to make the irreducible basic elements as simple and as few as possible without having to surrender the adequate representation of a single datum of experience." [23] Thirdly, it will actually be very useful to be able to compare the implicit and explicit solvent models in order to test that they are physically realistic, i.e., the partitioning of energies should be similar with both approaches.

As mentioned above explicit solvent models can suffer significant problems due to the large number of fitted parameters. However, many continuum solvent models have been built which also rely on parameters adjusted to reproduce experiment and hence suffer all the same problems. As Feynman pointed out, "The first principle is that you must not fool yourself, and you are the easiest person to fool."[24] We can apply some strict criteria to our model which can help prevent the risk of this: The model needs to rely on a minimum of parameters, which have good physical justification from independent experiment or *ab initio* calculation, and which can be used to reproduce a wide range of experimental properties for a wide range of molecules.

To build a model that meets these criteria we need to account for both of the key mechanisms that cause ion specificity outlined above. Firstly, this means inclusion of dispersion interactions of the ions with water and other molecules in water. It turns out that the continuum model is particularly suited to this type of calculation due to the development of Lifshitz theory and its extensions to include electrolytes.[2,25] Secondly, the direct interaction of the ion with the water in the first solvation shell is needed, including the energy of cavity formation. This is the energy of breaking the water-water bonds to form the cavity the ion occupies. Once this is accurately modelled it also needs to be included in the interaction energy of ions in water. For instance as a lithium ion approaches a surface or another ion there will be a large cost of removing the strongly bound water from its first solvation layer.

There are a number of improvements which can still be made to continuum solvent models which may bring them into agreement with experiment while meeting the criteria laid out above. For example, ab initio calculations can be used to accurately determine dipole and higher order multipole moment polarisabilites, as well as the size parameters of ions in water. These must be attempted before any conclusions about the range of applicability of these models are justified. Here we outline some promising applications of continuum models to specific ion effects and outline future steps to build on them.

2.3 Solvation Energies

The most fundamental and hence important properties of electrolyte solutions are the solvation energies of individual ions. Continuum models have, in the past, proven inadequate at reproducing these energies adequately.[18] This has been used as an argument for the necessity of explicit solvent approaches. Particularly the inability of continuum models to explain the charge hydration asymmetry; the puzzling result that ions of the same size can have significantly different solvation energies.[19]

We have recently presented a model[26,27] that extends the Born model of solvation energies to include dispersion and cavity formation energies to reach good quantitative agreement with experiment. All the parameters used in the model have good external physical justification and the model explains charge hydration asymmetry. Normally this is attributed to the fact the water has a different orientation around cations and anions.[19] Although there is some evidence for this from simulation, the effect varies substantially depending on the water model chosen and some water models with physically accurate multipole moments do not see this effect at all.[28] Also, and crucially, simulation can not explain the fact that cations of the same size can have significantly different solvation energies, i.e., copper(I) and lithium, or silver(I) and sodium. This is accounted for in our model by the dispersion interaction that depends on the balance between the ion's polarisability and its size. This leads to a large dispersion contribution to the solvation energy of certain ions, i.e., fluoride, copper(I) and silver(I), satisfactorily explaining their anomalously large solvation energies. The model also includes a simple cavity formation contribution. This partially cancels the dispersion energy leaving the values similar in size to estimates based solely on the Born model. Promising extensions of this model to account for partial molar volumes and entropies, as well as *ab initio* calculations that qualitatively confirm the conclusions are currently underway.

The fact that this model shows agreement with the solvation energies of 11 cations and anions including the variation in energy among ions of the same size, without the use of fitted parameters is a strong indication that it is modelling the strength of the ion-water interaction quite well. It should therefore be capable of incorporating one of the key physical origins of ion-specificity discussed above: The change in the ion-water interaction energy as an ion approaches another ion or an interface. We therefore believe that the use of explicit solvent models may not be necessary for many applications. Interestingly, these results imply that the dispersion interaction explains a significant degree of specific ion hydration. Consequently, these interactions play a doubly important role for ionic interactions in water. Firstly, there is the direct interaction between the ion and another particle or surface in water. Secondly, there is a change in the dispersion interaction of the ion with the surrounding water. Whether these promising conclusions hold up as well for divalent and trivalent ions remains to be seen.

As mentioned above, models that can be considered satisfactory need to be applicable to a whole range of experiments without adjusting parameters from case to case, e.g, for mixed electrolytes. This approach is exemplified by Kunz et al. in Ref. 29, where one consistent model with one set of fitted parameters is used in an attempt to reproduce the osmotic coefficients and surface tensions simultaneously. The approach had limited success but serves as a framework for future work.

The osmotic coefficients and surface tensions are, after all, the simplest properties of electrolyte solutions beyond the solvation energies. It is a logical next step to extend our (so far promising) solvation model to explain them. Predicting these properties requires the calculation of the free energy of interaction of two ions and of an ion with the air (or oil) water interface. These energies are essentially equivalent to the solvation energy of the solutes as a function of separation. We should therefore be able to improve upon Ref. 29 by basing the calculation on a solid foundation, namely our satisfactory solvation model. New quantum chemical methods and physical insights have come to light in recent years that can also be used to improve agreement. Two additional publications in the literature present these new physical insights and methods.

2.4 Osmotic Coefficients

The first of these models is the model outlined by Lund *et al.*[30] It provides an explanation of the affinity of like sized ions for one another. The central advance of this model of ion-ion interactions is that it includes the cavity in the dielectric medium which changes shape as the two ions come together, this accounts for the removal of the water from between the ions, including the important observation that a large ion must remove a lot of water from the surface of a small ion which will cost a large amount of energy. Lund *et al.* refer this to as a "shadowing effect". It results in a clear and simple explanation for Collins' law of matching water affinity, with a semi-quantitative model to support it. They also demonstrate that the same calculation can be done more accurately by using *ab initio* ion-ion calculations with a polarisable continuum solvent model to describe the background water.

The problem with their approach is that it is not based on a model that accurately reproduces the solvation energies of ions, and hence ion-water interactions, including the dispersion interaction of the ion with the surrounding water. As a result it does not yet provide quantitative agreement with experimental values.

2.5 Surface Tensions

The other important application of a continuum solvent model to these simple properties of electrolyte solutions is the model of Levin et al.[31,32] that reproduces the surface tension of electrolyte solutions with reasonable accuracy. There are some improvements that can be made to this model. Firstly, it is not based on a satisfactory model of solvation energies. As a result, the cavity formation energy used is too small[27] and the model requires ad hoc parameters adjusted to reproduce experiment. For instance, fluoride feels a hard wall repulsion from the air-water interface at a distance where iodide feels almost no repulsion at all.[31] Thirdly, it neglects the important surface potential and dispersion contribution to the ion-water interaction, which explains why such a small cavity formation energy needs to be used.

3 Conclusion

We have earlier developed[26,27] a satisfactory continuum model of the solvation energies of monovalent ions. The success of this model implies that it should be possible to develop adequate models of more complex specific ion effects, while neglecting the explicit position and orientation of the solvent molecules. This is because continuum solvent models can incorporate the two key drivers of ion specificity: dispersion interactions and changes in ion hydration. This class of models are highly advantageous due to their low computational demand and the physical insight they provide.

Promising work is currently underway, building upon the foundation of our simple solvation model. The more complex properties of electrolyte solutions that we are currently expanding the model to embrace are: entropies of solvation, partial molar volumes, osmotic coefficients, and surface tensions. The models are hoped to be quantitative and predictive, and hence able to reproduce these properties without the use of additional parameters to fit experimental values. This work will provide a comprehensive understanding of electrolyte solutions, which can be built upon to explain and predict the properties of the many complex and important systems where electrolyte solutions play a central role.

References

1. Lo Nostro, P.; Ninham, B. W. *Chem. Rev.* **2012**, *112*, 2286–2322.
2. Ninham, B. W.; Lo Nostro, P. *Molecular Forces and Self Assembly*; Cambridge University Press: Cambridge, 2010.
3. Born, V. M. *Z. Phys.* **1919**, *1*, 45–48.
4. Rashin, A. A.; Honig, B. *J. Phys. Chem.* **1985**, *89*, 5588–5593.
5. Robinson, R. A.; Stokes, R. H. *Electrolyte Solutions*; Butterworth & Co.: Devon, 1959.
6. Zhang, Y.; Cremer, P. S. *Curr. Opin. Chem. Biol.* **2006**, *10*, 658–663.
7. Zhang, Y.; Cremer, P. S. *Annu. Rev. Phys. Chem.* **2010**, *61*, 63–83.
8. Omta, A. W.; Kropman, M. F.; Woutersen, S.; Bakker, H. J. *Science* **2003**, *301*, 347–349.
9. Ninham, B. W.; Duignan, T. T.; Parsons, D. F. *Curr. Opin. Colloid Interface Sci.* **2011**, *16*, 612–617.
10. Boström, M.; Williams, D. R. M.; Ninham, B. W. *Langmuir* **2001**, *17*, 4475–4478.
11. Boström, M.; Kunz, W.; Ninham, B. W. *Langmuir* **2005**, *21*, 2619–2623.
12. Parsons, D. F.; Ninham, B. W. *Colloids Surf. A: Physicochemical and Engineering Aspects* **2011**, *383*, 2–9.
13. Kunz, W. *Curr. Opin. Colloid Interface Sci.* **2010**, *15*, 34–39.
14. Tissandier, M. D.; Cowen, K. A.; Feng, W. Y.; Gundlach, E.; Cohen, M. H.; Earhart, A. D.; Coe, J. V.; Tuttle, T. R. *J. Phys. Chem. A* **1998**, *102*, 7787–7794.
15. Baer, M. D.; Mundy, C. J. *J. Phys. Chem. Letters* **2011**, *2*, 1088–1093.
16. Tongraar, A.; Michael Rode, B. *Phys. Chem. Chem. Phys.* **2003**, *5*, 357–362.
17. Zhan, C.-G.; Dixon, D. A. *J. Phys. Chem. A* **2004**, *108*, 2020–2029.
18. Hunenberger, P.; Reif, M. *Single-Ion Solvation: Experimental and Theoretical Approaches to Elusive Thermodynamic Quantities*; The Royal Society of Chemistry, 2011.
19. Netz, R. R.; Horinek, D. *Annu. Rev. Phys. Chem.* **2012**. *63*, 401-418.
20. Chen, J.; Brooks, C. L.; Khandogin, J. *Curr. Opin. Struc. Biol.* **2008**, *18*, 140–148.

21. Gallicchio, E.; Paris, K.; Levy, R. M. *J. Chem. Theory Comput.* **2009**, *5*, 2544–2564.
22. Jackson, J. D. *Classical Electrodynamics*, 3rd ed.; John Wiley & Sons, Inc.: New York, 1998.
23. Einstein, A. "On the Method of Theoretical Physics." *The Herbert Spencer Lecture,* Oxford, June 10, 1933.
24. Feynman, R. P. *Surely You're Joking Mr. Feynman*; W. W. Norton, 1985; p 350.
25. Mahanty, J.; Ninham, B. W. *Dispersion Forces*; Academic Press: London, 1976; p 230.
26. Duignan, T. T.; Parsons, D. F.; Ninham, B. W. *J. Phys. Chem. B* **2013**, *117*, 9412–9420.
27. Duignan, T. T.; Parsons, D. F.; Ninham, B. W. *J. Phys. Chem. B* **2013**, *117*, 9421–429.
28. Mukhopadhyay, A.; Fenley, A. T.; Tolokh, I. S.; Onufriev, A. V. *J. Phys. Chem. B* **2012**, *116*, 9776–9783.
29. Kunz, W.; Belloni, L.; Bernard, O.; Ninham, B. W. *J. Phys. Chem. B* **2004**, *108*, 2398–2404.
30. Lund, M.; Jagoda-Cwiklik, B.; Woodward, C. E.; Vácha, R.; Jungwirth, P. *J. Phys. Chem. Letters* **2010**, *1*, 300–303.
31. Levin, Y.; dos Santos, A. P.; Diehl, A. *Phys. Rev. Lett.* **2009**, *103*, 257802.
32. dos Santos, A. P.; Diehl, A.; Levin, Y. *Langmuir* **2010**, *26*, 10778–10783.

XXIII
The impact of ionic solvation energy and water structure on surface forces

Drew F. Parsons

Research School of Physical Sciences and Engineering,
Australian National University, Canberra, ACT 0200, Australia.
Drew.Parsons@anu.edu.au

Abstract

The structure of water in an interfacial region differs from bulk due to surface-induced water ordering. The difference is seen both in the local water density and in the water polarisability, particularly orientational polarisability as water dipoles are oriented towards the interface. The impact of this phenomenon in continuum models is to replace the dielectric constant of the medium with a position-dependent spatial dielectric function. The relationship that this kind of dielectric function bears with the electrostatic potential is known in the Poisson equation. But an additional effect of the spatially dependent dielectric function on ionic solvation energies (including a change in the Born energy of the ion) has not been widely recognised. The dielectric constant in an interfacial region has been reported to fall to values around 5, far from the bulk value of 78. The corresponding spatially-dependent ionic solvation energy therefore introduces a strongly repulsive ion-surface interaction which must be included as an additional "nonelectrostatic potential" in the Boltzmann factor, determining ion concentrations in a Poisson-Boltzmann model. Consequently a strongly repulsive surface force—a primary hydration force—is obtained with a range corresponding to the range of the surface-induced water ordering, usually the thickness of several water layers.

1. Introduction

The stability of a suspension of colloidal (microscopic) particles in an aqueous solution, or more generally the adhesion of one material to another, is important in a vast range of areas, among them food and cosmetic emulsions, the flotation process for extracting minerals, biofouling of marine vessels, adhesion of proteins or bacteria to biological membranes. Colloidal stability is determined by the forces experienced between the surfaces of two particles. The classic DLVO theory of Derjaguin and Landau,[1] Verwey and Overbeek[2] interpreted the total force between surfaces as the balance been two competing components. On the one hand, quantum mechanical van der Waals forces typically provide an attractive force between surfaces. On the other hand the formation of a layer of adsorbed counterions at a charged surface tends to drive a repulsive force. The latter adsorption repulsion tends to be diminished at higher salt concentrations, but can dominate at low concentrations. On balance, the overall force is typically repulsive at low salt concentrations, allowing the formation of a stable suspension of particles, and attractive at higher salt concentrations, leading to the coagulation of particles out of solution.

The DLVO model provides a useful picture for interpreting the behaviour of real systems under some conditions. But experimental observation has found that real systems frequently behave in a manner contradictory to DLVO theory.[3,4] In particular, short range repulsion is often observed under conditions where theory would expect attractive forces to dominate. The non-DLVO repulsive force commonly has a decay length below 1 nm. This was attributed to the presence of water layers near a surface, and the phenomenon has come to be known as the "hydration force". But the precise nature of the force has remained under debate.[5] Early interpretation[3] that the hydration force arises due to surface-induced water ordering was consistent with theoretical analysis of the impact of water structure.[6] But other theoretical and experimental work suggests that other mechanisms can play a role in the formation of short range repulsion. Short range repulsion observed in soft surfaces such as lipid membranes,[7-9] could be in part attributed to fluctuations in the position of the surface.[10] And experimental work hints at the role of adsorbed ions on the strength of the observed repulsion.[4,7,8] The effect of ions is sometimes described as a "secondary hydration force".[4] Indeed, short range repulsion between surfaces has been reported in theoretical models when a repulsive interaction between ions and the surface is included,[11-13] acting for instance against the normal attractive interaction between a counterion and a charged surface.

Primary (water structure) and secondary (ionic) hydration forces can be united by considering the effect that water structure has on ion-surface interactions. Water structure was indirectly included via the ion-surface interaction employed by Marčelja, who used a potential of mean force derived from molecular dynamics calculations with explicit water. Parsons employed a somewhat artificial repulsive barrier to exclude some ions from the surface regions,[12,13] invoking an analogy to Collins' Law of Matching Water Affinity[14] with the argument that chaotropic (weakly hydrated) ions would not penetrate the surface hydration layer of a hydrophilic (strongly hydrated) surface. Surface-induced water structure was therefore introduced in the form of the surface hydration. But the ad-hoc nature of the implementation of the idea, acting only on chaotropic ions, can be improved. We present such an improvement here.

2. Water structure and spatially-dependent dielectric constant

Continuum solvent models do not explicitly include solvent molecules, but instead include their cooperative effect via a dielectric function. In the simplest continuum models the properties of the solvent are captured in a single parameter, the dielectric constant ε, thereby ignoring all spatial structure of the solvent. But in general the dielectric constant depends on the density of the solvent. The Clausius-Mossotti relation, for instance, relates the dielectric constant to the density ρ and polarisability α of water molecules,

$$\frac{\varepsilon - 1}{\varepsilon + 2} = \frac{4\pi}{3}\rho\alpha \qquad (1)$$

There is evidence from computer simulation (with explicit water molecules), that the density of water falls in the vicinity of a surface,[15] whether hydrophobic or hydrophilic, leading to spatial water structure in the form of a distance-dependent density, $\rho(z)$ (taking z as the distance from the surface). From the Clausius-Mossotti relation above, it follows that this would result in a distance-dependent dielectric function $\varepsilon(z)$, which could be used in continuum models. For instance in Poisson-Boltzmann models, the Poisson equation takes the form

$$\varepsilon_0 \frac{d}{dz}\left(\varepsilon(z)\frac{d\psi(z)}{dz}\right) = -\sum_i q_i c_i(z) \qquad (2)$$

where ε_0 is the permittivity of the vacuum, $\psi(z)$ is the electrostatic potential, q_i is the charge on the ith ion in solution and $c_i(z)$ is its concentration profile.

2.1. Spatially-Dependent Born Energy

The Poisson equation with distance-dependent dielectric function does not alone introduce short range repulsion. But consider the Boltzmann component of the Poisson-Boltzmann model, which defines the ion concentration profiles via the relation

$$c_i(z) = c_{i0}\exp\left[-\mu_i(z)/kT\right] \qquad (3)$$

where c_{i0} is the concentration of the ith ion in bulk solution, and $\mu_i(z)$ is the interaction potential of the ion at distance z from the surface. In traditional Poisson-Boltzmann models, the interaction potential is simply taken as the electrostatic energy $\mu_i^{el}(z) = q_i\psi(z)$. But more generally, $\mu_i(z)$ should be understood as the excess chemical potential of the ion relative to bulk. The chemical potential in bulk contains internal energies such as the electrostatic Born energy (solvation energy)

$$\mu_i^{\text{Born}} = -(q_i^2/8\pi\varepsilon_0 a)(1 - 1/\varepsilon_s), \quad (4)$$

where ε_s is the value of the dielectric constant in bulk solvent and a is the cavity radius of the ion. The impact of these bulk internal energies are reflected in the bulk concentrations c_{i0}.

But if the dielectric constant varies with distance, it follows that the Born energy likewise varies with distance. This effectively gives rise to an additional interaction potential, a spatially varying solvation energy,

$$\Delta\mu_i^{\text{Born}}(z) = (q_i^2/8\pi\varepsilon_0 a)(1/\varepsilon(z) - 1/\varepsilon_s). \quad (5)$$

Note that if the dielectric constant falls in the surface region due to a fall in local water density, then the corresponding Born interaction energy $\Delta\mu_i^{\text{Born}}(z)$ will be repulsive, pushing ions away from the surface. In this way a repulsive ion-surface interaction is obtained, with no artificial distinction between cosmotropic and chaotropic ions (instead, the distinction between them enters via the cavity radius a). This interaction energy must be included alongside the classic electrostatic energy, such that the total interaction potential used in the Boltzmann relation (Eq. 3) becomes

$$\mu_i(z) = \mu_i^{el}(z) + \Delta\mu_i^{\text{Born}}(z). \quad (6)$$

Other nonelectrostatic contributions to the total interaction energy such as ionic dispersion energies may be considered.[16] For simplicity we here consider the impact of water structure on the ion interaction energy only through its effect on the Born term $\Delta\mu_i^{\text{Born}}$, which is the dominant component of ion solvation energies.[17] Note that water structure also has a direct impact through the appearance of $\varepsilon(z)$ in the Poisson equation itself, Eq. 2.

2.2. Water Structure: Density Profile

We represent surface-induced water ordering through the water density profile $\rho(z)$. We adopt a model for the water density profile adapted from the profile found in molecular dynamics simulations of Schwierz, Horinek and Netz.[15] The water density profile is handled as a monotonic function which decreases from the bulk value ρ_0 as the surface is approached. Specifically, the functional form

$$\rho(z)/\rho_0 = [1 + \tanh\{(z-z_0)/w\}]/2$$

is used. The parameters z_0 and w were adjusted to follow Schwierz' profile after dropping the oscillations between z = 1.3–5Å in order to obtain a monotonic profile. Under this model, the water density starts to fall away from its bulk value within 3 Å of the surface. However under the Clausius-Mossotti transformation, Eq. 1, the spatial reach of the corresponding dielectric function extends slightly further, with $\varepsilon(z)$ deviating from its bulk value within about 4 Å.

An alternative mechanism for generating a spatially dependent dielectric constant is dielectric saturation.[18] In this case the static polarisability α of water molecules is attenuated due to a loss of orientational freedom in the presence of an electric field, leading to a reduction in the dielectric constant. However in our experience we have found that the electric field next to typical surfaces is too low ($< 10^9$ V/m) for dielectric saturation to be greatly significant.

3. Calculated surface forces

We calculate the force curve resulting from the surface-perturbed water density for 1mM KCl between two mica surfaces, following experimental measurements.[4] The surface charge is taken to be -0.01 Cm^{-2} (constant charge). Force curves are calculated using the Derjaguin approximation, where force curves F/R for curved surfaces with radius R are estimated from the excess free energy G^* of flat surfaces (relative to their energy at large separation), such that $F(L)/R = 2\pi G^*(L)$. The calculated force curve is shown in Figure XXIII.1. We see that the theoretical curve including the spatially dependent electrostatic Born energy, Eq. 5 follows experimental data well, with repulsion increasing at small separations below 1 nm. The classic DLVO curve, by contrast, exhibits a repulsive peak at 1 nm but then becomes attractive at smaller separations.

The underlying origin of the short range repulsion can be determined by breaking the total force into components:[19]

$$F = F_{en} + F_{el} + F_{Born} + F_{Ham}. \qquad (7)$$

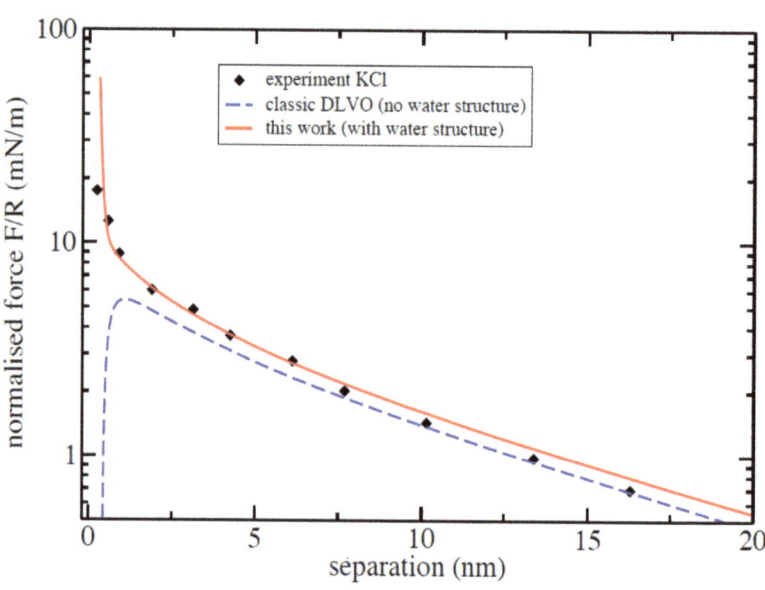

Figure XXIII.1. Force-distance curves for mica surfaces with -0.01 Cm^{-2} surface charge in 1 mM KCl. Symbols indicate experimental values at pH 6.[4] The solid red curve indicates calculation with water structure (including the spatial Born energy, Eq. 5). The dashed blue curve indicates the classic DLVO calculation without water structure.

F_{en} refers to the force due to the configuration entropy of ions, arising from their concentration profiles $c_i(z)$ relative to bulk concentrations c_{i0}. The corresponding free energy term is

$$G_{en} = kT \sum_i \int_0^L dz \left[c_i(z) \ln \frac{c_i(z)}{c_{i0}} - c_i(z) + c_{i0} \right]. \quad (8)$$

F_{el} refers to the force arising from the electrostatic energy G_{el} of the electric field,

$$G_{el} = \frac{1}{2} \int D \cdot E = \frac{\varepsilon_0}{2} \int_0^L \varepsilon(z) \left(\frac{d\psi}{dz} \right)^2 dz \quad (9)$$

F_{born} is the additional contribution due to the spatially varying Born solvation energy,

$$G_{Born} = \sum_i \int_0^L dz\, c_i(z) \Delta\mu_i^{Born}(z) \quad (10)$$

taking $\Delta\mu_i^{Born}(z)$ from Eq. 5. F_{Ham} is the force due to the Hamaker (Casimir-Lifshitz or van der Waals) energy,

$$G_{Ham} = -A/12\pi L^2 \quad (11)$$

whose magnitude is given by the Hamaker constant, taken here with a nonretarded value of 2.29 kT for the mica-water-mica interaction.

The decomposition of the total force into these terms is shown in Figure XXIII.2a. Repulsion arises from both the entropic and spatial Born contributions. But the entropic term is cancelled to some degree by the attractive electrostatic contribution. The importance of the entropic term can be understood by comparing with the decomposition of the classic DLVO force, Figure XXIII.2b. In the classic DLVO case, the entropic component is again repulsive, but after cancellation against the electrostatic component, is dominated by the Hamaker contribution at separations below 1nm, leading to short-range attraction. With water structure included, the entropic term

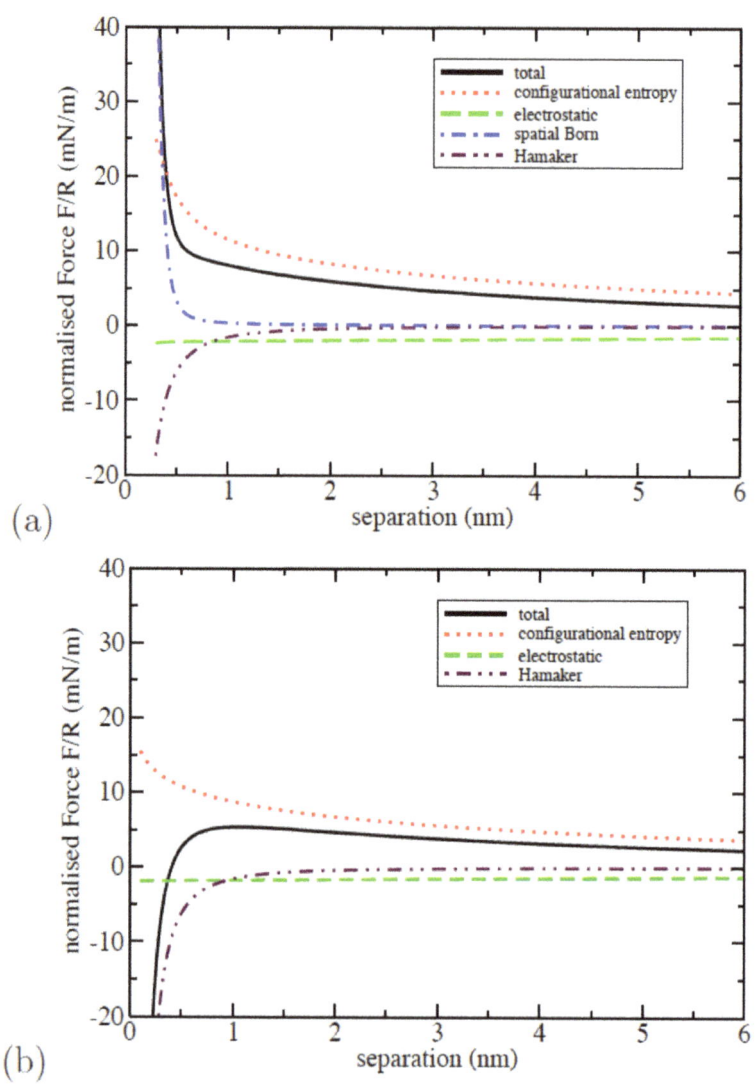

Figure XXIII.2. Force-distance curves for mica surfaces with -0.01 Cm^{-2} surface charge in 1mM KCl, decomposed into separate contributions (see Eq. 7). (a) This work, including water structure and the spatial Born solvation energy (Eq. 5). (b) Classic DLVO, without water structure.

becomes slightly larger in magnitude, allowing entropic repulsion to remain dominant over Hamaker attraction at separations below 1nm. On top of that, at the smallest separations below 0.6nm the repulsive spatial Born solvation term reaches very large magnitudes, boosting short range repulsion.

This increase in the configurational entropy component of the total force can be understood further from the ion concentration profiles, shown in Figure XXIII.3 for a 4 nm separation between surfaces. The addition of the repulsive spatial Born energy (Eq. 5) results in a depletion zone at the surface where the counterion concentration drops to zero. The coion Cl^- is similarly depleted at the surface, but in the absence of other nonelectrostatic interactions such as ionic dispersion, the coion does not play a significant role. The presence of the depletion zone means that the counterion concentration must rise in the middle region between the surface in order to maintain electroneutrality. It is this increase in counterion concentration outside the depletion zone which leads to the increase in the repulsive configurational entropy.

4. Discussion and conclusions

The appearance of the ion depletion zone along the surface could be described as a kind of surface hydration layer, the language used in Ref. 12, in the sense that it arises from surface-perturbed water structure. In Ref. 12 the depletion zone was constructed artificially, imposed as a distance of closest approach and conceptually applied only to chaotropic ions. A different mechanism is presented in this work, where the dielectric constant of water falls in the surface region due to a change in water structure, leading to the appearance of the depletion zone as a consequence of the change in the Born energy of the ions. This seems a more natural mechanism for building a depletion zone, and therefore obtaining short range repulsion. It requires no artificial distinction between chaotropic and cosmotropic ions.

In the model presented here we introduced water structure via

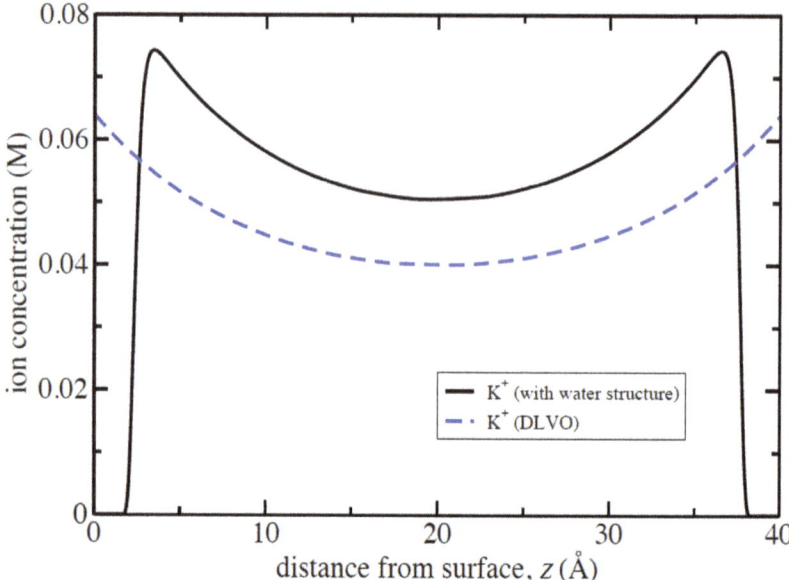

Figure XXIII.3. Counterion (K$^+$) concentration profiles for 1 mM KCl between mica surfaces at 4nm separation, calculated with (this work) and without (classic DLVO) water structure.

the idea of a surface perturbed water density $\rho(z)$, as suggested by computer simulations. But the calculation actually uses a surface perturbed dielectric function, $\varepsilon(z)$. The use of $\rho(z)$ is simply a means of obtaining $\varepsilon(z)$, with the help of the Clausius-Mossotti relation, Eq. 1. The same result of short range repulsion via a surface perturbed $\varepsilon(z)$ may be obtained by a physics other than a spatially varying density profile. Specifically, it seems likely that the rotational freedom of water molecules in the "surface hydration layer" would be impeded, reducing the orientational polarisability of water. This could be handled under the same methodology presented here, by applying a spatially varying polarisability, $\alpha(z)$ rather than (or in addition to) a varying density $\rho(z)$ in the Clausius-Mossotti relation.

We conclude that a salt contribution to short range repulsion can appears as an indirect effect of water structure via the configurational entropy of ions. A surface depletion zone with a fall in counterion concentrations is formed as a result of the spatially dependent repulsive Born solvation energy. This repulsive ion interaction energy appears due to a fall in the dielectric constant of water close to the surface (here, due to a drop in water density). This entropic repulsion is augmented by the direct contribution to the total force from the spatially dependent Born solvation energy of ions (again arising from the fall in the value of the dielectric function near the surface).

References

1. Derjaguin, B.; Landau, L. *Progr. Surf. Sci.* **1993**, *43*, 30–59.
2. Verwey, E. J. W.; Overbeek, J. T. G. *Theory of the Stability of Lyophobic Colloids*; Elsevier Publishing Company, 1948.
3. Israelachvili, J. N. *Faraday Disc.* **1978**, *65*, 20–24.
4. Pashley, R. M. *J. Coll. Interface Sci.* **1981**, *83*, 531–546.
5. Zemb, T.; Parsegian, V. A. "Editorial overview: Hydration forces." *Curr. Op. Coll. Interface Sci.* **2011**, *16*, 515–516.
6. Marčelja, S.; Radić, N. *Chem. Phys. Lett.* **1976**, *42*, 129-130.
7. Petrache, H. I.; Zemb, T.; Belloni, L.; Parsegian, V. A. *Proc. Natl. Acad. Sci. USA* **2006**, *103*, 7982–7987.
8. Aroti, A.; Leontidis, E.; Dubois, M.; Zemb, T.; Brezesinski, G. *Coll. Surf. A* **2007**, *303*, 144-158.
9. Demé, B.; Zemb, T. *Curr. Op. Coll. Interface Sci.* **2011**, *16*, 584-591.
10. Schneck, E.; Netz, R. R. *Curr. Op. Coll. Interface Sci.* **2011**, *16*, 607-611.
11. Marčelja, S. *Curr. Op. Coll. Interface Sci.* **2011**, *16*, 579–583.
12. Parsons, D. F.; Ninham, B. W. *Coll. Surf. A* **2011**, *383*, 2–9.
13. Ninham, B. W.; Duignan, T. T.; Parsons, D. F. *Curr. Op. Coll. Interface Sci.* **2011**, *16*, 612–617.
14. Collins, K. *Biophys. Chem.* **2006**, *119*, 271–281.
15. Schwierz, N.; Horinek, D.; Netz, R. R. *Langmuir* **2010**, *26*, 7370–7379.

16. Parsons, D. F.; Boström, M.; Lo Nostro, P.; Ninham, B. W. *Phys. Chem. Chem. Phys.* **2011**, *13*, 12352–12367.
17. Duignan, T. T.; Parsons, D. F.; Ninham, B. W. *J. Phys. Chem. B* **2013**, *117*, 9421–9429.
18. Paunov, V.; Dimova, R.; Kralchevsky, P.; Broze, G.; Mehreteab, A. *J. Coll. Interface Sci.* **1996**, *182*, 239-248.
19. Parsons, D. F.; Ninham, B. W. *J. Phys. Chem. C* **2012**, *116*, 7782–7792.

XXIV
Nanobubbles of dissolved gas in bulk aqueous solutions of electrolyte

Nikolai F. Bunkin[†,‡,*], Alexey V. Shkirin[‡], Valeriy A. Kozlov[†,‡], Artyom L. Sendrovitz[‡]

‡: A.M. Prokhorov General Physics Institute, Moscow, Vavilova 38, 119991 Russia.

†: Bauman Moscow State Technical University, Moscow, 2nd Baumanskaya 5, 105005 Russia.
nbunkin@kapella.gpi.ru

Abstract

Results of experiments with the dynamic light scattering, phase microscopy and polarimetric scatterometry allow us to claim that long-living gas nanobubbles and the clusters of such nanobubbles are generated spontaneously in an aqueous solution of salt, saturated with gas. The charactertistic sizes of nanobubbles and clusters are found. The inverse problem of the light scattering in ionic solutions has been solved, which provided means to determine the characteristic parameters of the nanobubble clusters. These experimental results can be treated as evidence for a special role of ions in generation and stabilization of gas nanobubbles.

I. Introduction

The present study is devoted to experimental proof of the existence of quasi-stable gas nanobubbles in ionic solutions under normal conditions, i.e. at room temperature and atmospheric pressure. We can formulate our goal as follows. There exists a problem with the interpretation of results of the molecular light scattering in water.[1] Basically, the so-termed transverse scattering coefficient, which is conventionally denoted in literature as R_{90}, is completely described by the known thermodynamic parameters of liquid.

$$R_{90} = \frac{\pi^2}{2\lambda^4}\left(\rho\frac{\partial\varepsilon}{\partial\rho}\right)_T^2 \beta_T kT \frac{6+6\Delta_u}{6-7\Delta_u} \quad (1)$$

Here λ is the wavelength, ρ is the density of liquid, ε is its dielectric permittivity, β_T is isothermal compressibility, and Δ_u is the depolarization coefficient for the scattering of unpolarized light. This coefficient can be measured in direct experiments. It turned out that in most liquids except water we deal with good agreement between the theory and the experimental results, while in water the disagreement is significant. Based on the literature, the experimental values of R_{90} for water generally exceed the theoretical ones by 1.5-2 times. It is clear that there exist some entities, which give an additive contribution to the molecular scattering from water.

Additionally, if we focus an ultrasonic wave of sufficient intensity in a liquid, we will see a track of vapor-gas bubbles in the focal volume of an ultrasonic lens; this phenomenon is termed cavitation. The rupture strength of liquid is expressed as $\sigma n_l^{1/3} \sim 10^4$ atmospheres, where σ is the surface tension coefficient, n_l is the number density of molecules of the liquid (all numerical estimates will be hereinafter made for water). At the same time, experimental data indicate that the cavitation can be induced at the amplitudes of sound wave ~ 1 atmosphere. Hence, for the cavitation to occur the long-lived centers of the cavitation should exist. The steady gas bubbles could serve as such centers.

The mechanical equilibrium condition for micron-sized bubble is given by the known Young–Laplace equation:

$$P_{in} = P_{atm} + \frac{2\sigma}{R} > P_{atm} \qquad (2)$$

R is the radius of the bubble, P_{in} is the pressure of gas inside the bubble, P_{atm} is the pressure of atmospheric gas above the surface of the liquid, i.e. the atmospheric pressure. Eqn. (2) implies that the pressure of gas inside the micron-sized bubble is always higher, than the pressure of the same gas above the liquid surface. Thus, the solution of the gas in the liquid, whose content, according to the Henry law, is controlled by the pressure P_{atm}, appears to be unsaturated with respect to the pressure P_{in} of the same gas inside the bubble. Therefore, such bubble is diffusively non-stable; the gas escapes from the bubble by diffusion kinetics, and the bubble collapses. This fact was analyzed in the literature.[2,3] As was shown, if we deal with a bubble with the radius 10^{-3} cm, the lifetime of this bubble does not exceed 10 s. This time drastically falls with decreasing the radius of the bubble. For example, if bubble radius is ~ 100 nm, the bubble lifetime does not exceed 10 ms. Hence, for the diffusion stability we should require that the surface tension to be somehow compensated; only in this case the pressure inside the bubble will be equal to the external pressure, and the bubble will be stable both mechanically and with respect to diffusion of gas. One of the mechanisms of such compensation is associated with solid impurities introduced from the outside; inasmuch as gas bubble, attached to solid interface, has a possibility to become stable both mechanically and diffusively. The model, where the stable heterogeneous centers of cavitation arise due to the presence of solid impurities, is widely accepted.[4-7] It is necessary to note that even fine filtration of liquid cannot completely suppress the cavitation effect. At the same time, it is well known that the cavitation ability in aqueous media increases at adding various salts, despite new solid particles are not introduced into the water sample together with the salt.[8] However, in our study we do not deal with nanobubbles attached to the hydrophobic substrate; this case was explored comprehensively.[9-21] We consider only the bulk nanobubbles; one of the possible mechanisms for compensating the surface tension forces is the selective adsorption of anions at the water-gas interface, see Ref. 22 and the references therein. It is necessary at once to note that the adsorbed ions (here we speak about the ions of inorganic salts) can give rise to a slight growth of the surface tension coefficient (see, e.g., the tables in Ref. 23 and also Refs. 24 and 25), which should be taken into account in numeric estimations. This *bubble-stabilized-by-ions* model (or *bubston* model) was first put forward in the study,[26] and then was developed in Refs. 27-30.

The mechanical stability condition for bubston is expressed as:

$$P_{in} + P_e = P_{atm} + 2\sigma/R. \tag{3}$$

Here $P_e = -\left(\dfrac{\partial \Phi_e}{\partial V}\right)_T$ is the pressure, caused by the presence of charge Q_0 at the spherical interface of the bubble, and Φ_e is the energy of electrostatic field of the system in question. Our system is a spherical area of radius $r \geq R$, surrounding the charged bubble of radius R. The electric field inside this area is non-zero; for that reason, the Helmholtz energy and the volume V are found as

$$\Phi_e = \dfrac{1}{2\varepsilon} \int_R^r \left[Q^2(x)/x^2\right] dx$$

and

$$V = \dfrac{4\pi}{3}\left(r^3 - R^3\right)$$

Thus, we arrive at

$$P_e(R) = -\frac{1}{4\pi R^2}\frac{\partial \Phi_e}{\partial R}$$
$$= \frac{1}{8\pi\varepsilon}\frac{Q^2(R)}{R^4} = \frac{1}{8\pi\varepsilon}\frac{Q_0^2}{R^4} = \frac{2\sigma}{R} \quad (4)$$

It is clear that the pressure P_e expands the bubble and thus is capable of compensating the squeezing surface tension forces at a certain radius R. The approach for calculating the charge Q_0 is described in Refs. 29 and 30. Our theory predicts that the bubston size is ~ 10 - 100 nm. Additionally, separate bubstons are capable of coagulation to one another with the formation of the bubston clusters at the micron scale. We distinguish two mechanisms of the bubston coagulation. The first mechanism is related to self-aggregation of bubstons; such clusters include several tens of bubstons. The second mechanism is related to the bubston aggregation on solid nanoimpurities, which cannot be removed from the liquid via filtration; the size of such cluster is ~ 1 micron. In Refs. 31-37 we describe experiments to study the bubstons in equilibrium aqueous solutions of salt, saturated by dissolved air. Furthermore, some evidence supporting this hypothesis was inferred from the observations of the low-threshold laser-induced breakdown in water and aqueous electrolytic solutions transparent for the laser beams employed.[38-45] The next section is devoted to direct experimental proof of the existence of bubstons and their clusters with the help of modulation interference microscopy, dynamic light scattering and polarimetric scatterometry.

II. Experimental part

Let us start with the technique of modulation interference microscopy, which was described in our previous studies.[32,35,36] This microscope allows us not only to find the sizes of the objects under study, but also to determine their refractive index n (the optical density) with respect

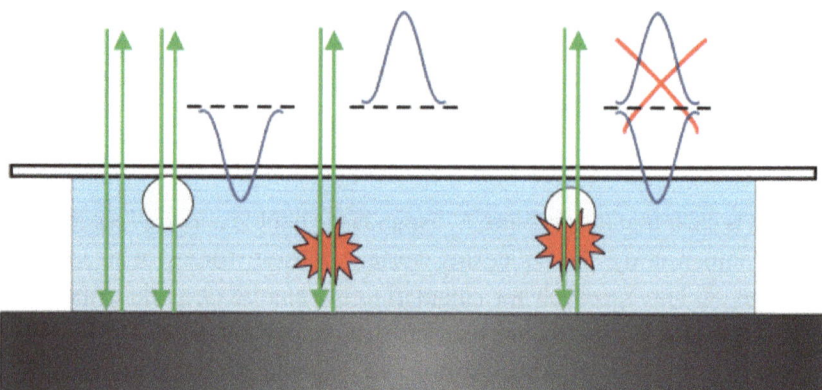

Figure XXIV.1. Schematic diagram, illustrating the principle of the phase microscopy. The cell with a thin liquid sample has a mirror substrate and is covered by transparent (made of glass) cover-plate. The phase shift δ between the object wave and the reference wave is measured. The refractive index n of that particle can be higher or lower than the refractive index of the liquid n_0. In the case, where $n < n_0$ (e.g., gas bubble) the value of $\delta < 0$, while at $n > n_0$ (solid particle) $\delta > 0$.

to ambient medium. In this experiment the phase shift δ between the object and the reference waves of the interferometer is measured. In this process the object wave passes through the cell with the liquid sample having the refractive index n_0 (see Figure XXIV.1, and Ref. 32), then the wave is reflected from the mirror substrate of the cell, again passes through the same liquid sample, and finally illuminates a CMOS-matrix.

The reference beam strikes the same matrix, so we measure the interference pattern intensity I. In the absence of any particles in the liquid the phase shift δ_0 can be set in the way that $\delta_0 = 2\pi m + 3\pi/2$, $m = 0, 1, 2,...$, and the intensity I is a uniform function across the matrix. The interference pattern intensity is $I \sim cos\,(\delta + \delta_0)$. Therefore, if δ_0 is adjusted as indicated above, then the measured intensity I is highly sensitive to changing the argument $(\delta + \delta_0)$ associated with the presence of particles. When the object wave passes through a spherical particle of radius R, the phase shift is

$$\delta = \frac{4\pi}{\lambda} R(n - n_0) \qquad (5)$$

where $\lambda = 532$ nm, n is the refractive index of the particle. By measuring δ we can subdivide colloid particles into those having high and low (with respect to water) optical density. In our microscopy experiments we have investigated the layers with 30 micron thickness of deionized water and aqueous solutions of various salts. The liquid samples have been filtered beforehand with the help of 200 nm porous membrane filter, which is a common practice. The specific feature of our experimental procedure is the following: we found that filtering results in a slight degassing of liquid. Thus the filtered samples were kept for some time in the cells open to atmospheric air in a dust-free area, i.e. we deliberately allowed the atmospheric air to penetrate inside the liquid. In some cases we purged dust-free nitrogen through liquid sample. We also studied various suspensions of colloid particles. Figure XXIV.2, obtained in the white light of the microscope, shows different particles in liquids.

In Figure XXIV.2 (*a*) we can see the particles of colloid silica

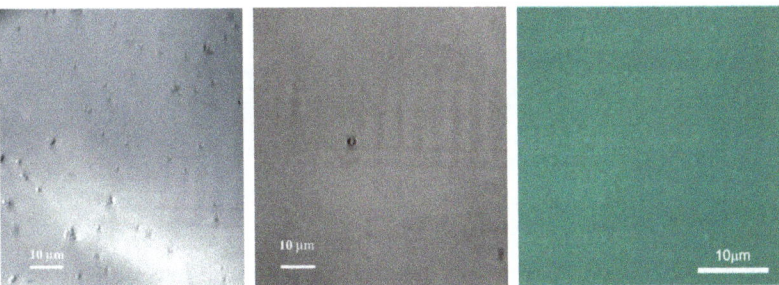

Figure XXIV.2. (*a – c*). Colloid silica particles ($R = 0.6$ μm) in the microscope white light (*a*). Monolithic gas bubble ($R = 1.3$ μm) in the microscope white light (*b*). Particles, spontaneously nucleated in aqueous NaCl solution (0.8 M) in the microscope white light (*c*).

in water; we see a sharp interphase boundary around them. Figure XXIV.2 (*b*) illustrates specially generated micron-sized gas bubble is shown; the interphase boundary is also clearly seen. Note that the silica particles and the micron-sized monolithic gas bubble look rather identically in the microscope white light. In Figure XXIV.2 (*c*) one can see the particles, which were spontaneously nucleated after a prolonged settling in filtered 1 M NaCl solution. The size of such particles is ~ 1 micron, which exceeds the pore diameter. Thus such particles were generated in the liquid after filtration. These particles are seen very faintly, distinct interphase boundary misses, contrary to the

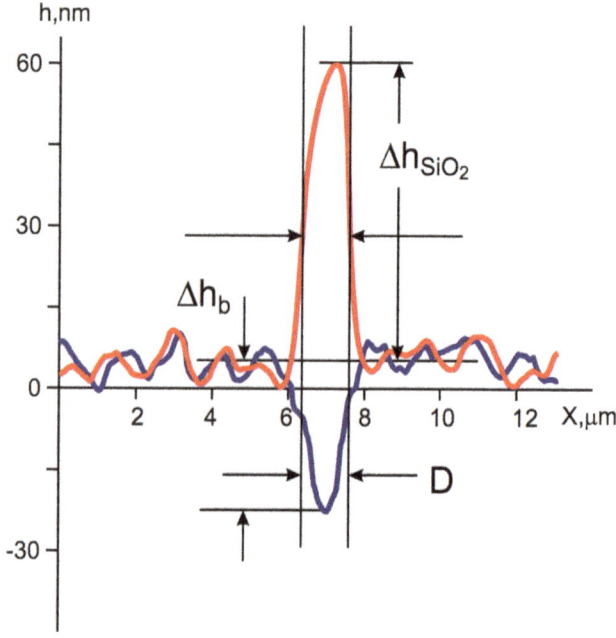

Figure XXIV.3. Optical density profiles for the colloid silica particle of radius $R = 0.8$ mm (red color) and for the spontaneously-nucleated particle in NaCl solution (0.8 M, blue color). The maximum and minimum values of optical density Δh are measured from the conventional zero level, which is not coincident with the OX-axis.

particles of silica and monolithic air bubbles. In the phase microscopy experiment we compared the spatial optical density profiles for one of the colloid silica particles at the micron scale ($R = 0.8$ mm), and for one of the long-living particles of low optical density, spontaneously nucleated in 0.8 M NaCl solution and having approximately the same size. In Figure XXIV.3 we present the 1D distributions of the optical density for them. In this Figure the optical density is plotted along the Y-axis. It is worth noticing that in our experiments the optical density is scaled in nanometers, i.e. in the units of wavelength of the radiation used. It is known that refractive index is dimensionless, so the data along the Y-axis should be multiplied by a certain dimension factor, which can be found from the calibration measurements. The optical density profile for the spontaneously-nucleated particle is concave, while for the silica particle it is convex, i.e. the refractive index for colloid silica is higher than that for water, whereas the refractive index for the spontaneously-nucleated particle is lower. Since we know the refractive index of fused quartz and assuming the shape of the low density particle is spherical, we can directly measure the absolute value of its refractive index.

The estimate of the refractive indices of the particles is expressed as

$$n = \gamma \left(\frac{\Delta h}{2R} \right) + n_0 \qquad (6)$$

where $n_0 = 1.332$ (for water), Δh is the maximum / minimum value of the optical density h, scaled in nm; for the silica particles $\Delta h_{SiO2} > 0$, while for the spontaneously-nucleated particle $\Delta h_b < 0$. Here γ is a corrective dimensionless factor, which is determined from the calibration; in our case $\gamma = 1.63$. It can be easily deduced that $n_{SiO2} = 1.46$, which agrees with the literature. At the same time, for the spontaneously-nucleated spherical particle $n_b = 1.26$. We see that such particles cannot be purely gaseous spheres, but they can therewith be the clusters composed of the long-living gas nanobubbles with the refractive index $n = 1$; the presence of liquid films between the gas cores causes a slight rise of refractive index, i.e. $n_b > n = 1$. It is clear that in the case of the colloidal silica particles the assumption of sphericity is quite reasonable, whereas for the particles, spontaneously arisen in salt solution, this supposition should be specially tested. Indeed, the violation of sphericity could result in reducing the value of n_b, e.g., when we deal with sliced micron-sized gas bubble, adhered to the cover plate. However, if it were the monolithic gas bubble of that size, having the refractive index $n = 1$, it should have a distinct interphase boundary, as is seen in Figure XXIV.2 (b). Thus the absence of such boundary indicates that these particles cannot be monolithic micron-sized particles.

A question arises of whether there exists some experimental evidence of spontaneous nucleation of gas nanobubbles in the bulk of salt solutions? To answer this question, we performed the light scattering experiments. We started with the dynamic light scattering (DLS).[46-48] The DLS technique is based upon the photonic correlation spectroscopy and is widely used for finding the sizes of nanoparticles in liquids. The idea of this technique consists in measuring the intensity I of light scattered by stochastically moving Brownian particles in liquid at a certain angle θ. The value I fluctuates in time, and we therefore can express it as a random process $I(t)$. The intensity is measured by a photomultiplier, which is connected with a quadratic detector; the latter gauges the time-correlation function

$$G_I(\tau) \equiv \langle I(t)I(t+\tau) \rangle$$

Based upon the Wiener – Khinchin theorem,[49] the function $G_I(\tau)$ can be expressed as the Fourier-transform of the spectral density of the scattered light, which can be presented by a Lorentzian of half-width Γ:

$$G_I(\tau) = I^2(1 + A|g_E(\tau)|^2)$$
$$= I^2(1 + A\exp(-2\Gamma|\tau|))$$

where I is the average intensity of the scattered light, A - optical apparatus factor,

$$g_E(\tau) = \frac{\langle E(t)E^*(t+\tau)\rangle}{I}$$

is the dimensionless correlation coefficient for the electric field of optical wave. Assuming that the intensity fluctuations are caused by the Brownian diffusive motion of scatterers across the laser beam, and the dispersion of displacement of a Brownian particle $\langle \Delta r^2 \rangle$ obeys the diffusion law $\langle \Delta r^2 \rangle = Dt$, where D is diffusivity, we obtain that the correlation time t_c of the function $G_I(\tau)$ can be expressed as:

$$1/\tau_c = \Gamma = Dq^2 \qquad (7)$$

where q is the scattered wave vector,

$$q = \frac{4\pi n}{\lambda}\sin\left(\frac{\theta}{2}\right) \qquad (8)$$

n is the refractive index of medium.

If the diffusivity D is known, we can calculate the mean radius of scattering particles r_p. For instance, for spherical particles we can use the Stokes – Einstein formula:

$$D = \frac{k_B T}{6\pi \eta r_p} \qquad (9)$$

where k_B is the Boltzmann constant, T is the temperature, η is the viscosity of liquid, where the particles of radius r_p are suspended. If we deal with the particles of various sizes, the time correlation function $G_I(\tau)$ is just the sum of several exponents.

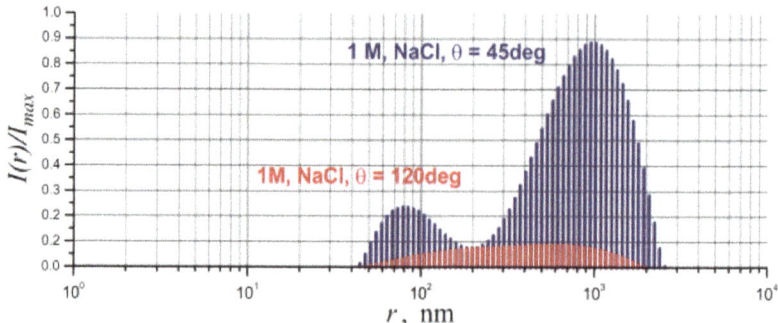

Figure XXIV.4. Distribution of the intensity of the light scattered at the angles $\theta = 45°$ and $120°$ over the sizes of particles inside the sample of 1 M NaCl solution; $\lambda = 532$ nm.

As earlier, we investigated aqueous solution of NaCl with different concentration within the scattering angular range $\theta = 30 - 130°$; this setup was described in detail.[36,50]

Figure XXIV.4 shows the scattering intensity versus the scatterer radius; the spectra are related to filtered 1 M NaCl solution and were taken at $\theta = 45°$ (blue color) and $\theta = 120°$ (red color). The scattering-intensity distribution at $\theta = 45°$ is apparently bimodal: the low-size maximum is related to 70 – 80 nm, while the large-size maximum is related to 1 micron. The pattern at $\theta = 120°$ looks rather smooth. The difference between the two distributions is explained by the fact that the scattering indicatrix for the scatterers in the micron-sized mode depends on the scattering angle much stronger than for the Rayleigh scatterers (their size is less than the wavelength, i.e. this is the low-size mode). We associate the short-size mode with the self aggregation of bubstons, while the large-size mode is associated with the aggregation of bubstons at non-filtered solid nanoparticles. Thus the presence of the large-size mode should depend on the quality of filtration. In Figure XXIV.5 we demonstrate the DLS size distribution after an additional filtration: we can see that the ratio between the intensities of these modes is changed apparently due to reducing the large-size mode intensity. It can be associated with decreasing the density of solid nanoparticles.

It turned out that the DLS size distribution is controlled by the presence of dissolved gas. We investigated the DLS spectra in 5 ml of initially degassed physiological solution (C = 0.16 M NaCl) after a day of settling under air-ingress conditions in the dust-free area. Initially zero DLS distribution was observed at this angle, but after settling we can see scatterer sizes from 50 to approximately 300 nm. According to our model, the low limit of the spectrum is related to the size of separate bubstons (the monomers), while higher sizes correspond to dimers, trimers and more coarse aggregates.

Thus we claim that the micron-sized and nanometer-sized particles are indeed present in the bulk of salt solution; we assume the low optical density particles observed in the phased microscopy experiment, and the micron-sized particles found in the DLS experiment are the same particles. The main result obtained in the DLS experiment is the presence of rather broad size distribution for the nanometer-sized particles. However we still cannot say anything about an internal structure of the micron-sized particles. To answer this question, we carried out an additional experiment on the laser

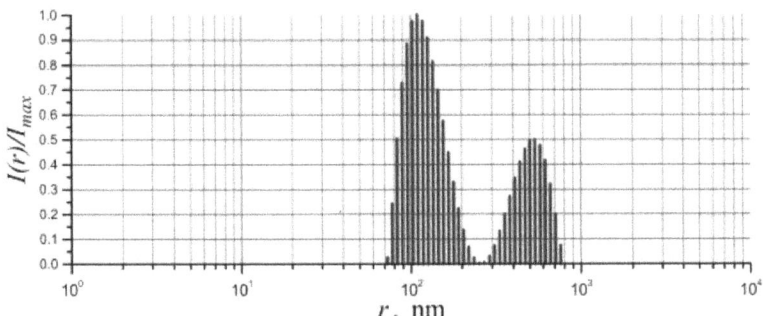

Figure XXIV.5. Bimodal distribution of the scattering intensity from 1 M NaCl solution after additional filtration, $\theta = 45°$.

polarimetric scatterometry. In this experiment we measured the angular dependencies of the scattering matrix elements.

As is known,[51-53] the scattering $F_{ij}(\theta)$ matrix is (4x4) matrix, transforming the Stokes vector of incident wave into the Stokes vector of scattered wave. The analysis of angular dependences of the scattering matrix elements not only gives the information about the effective size and geometric shape of scatterers, but also allows us to distinguish monolithic particles from the particles of cluster type. The experimental setup used by us was described in detail.[32-37] In this experiment we employed the laser radiation with λ = 532 nm. The setup was calibrated by aqueous monodisperse suspensions of colloidal silica and Polystyrene latex; the angular dependences of the scattering matrix elements obtained in these calibration experiments were in good agreement with the corresponding theoretic graphs,[32] which allows us to claim that our experimental results are quite reliable. In Figure XXIV.6 the angular dependences of the matrix elements for distilled water (black circles) and aqueous NaCl solution (0.8 M, blue circles) are plotted. The element $F_{11}(\theta)$ is the scattering indicatrix; other elements are normalized to

$$F_{11}(\theta): f_{ij}(\theta) = \frac{F_{ij}(\theta)}{F_{11}(\theta)}$$

As is known,[52] the scattering matrix for the particles possessing a symmetry plane (e.g., spherical particles) has a block-diagonal form. Additionally, as was shown by our numeric modeling,[54,55] the scattering matrix for clustered particles composed of spherical Rayleigh-type monomers has a block-diagonal structure as well. This is why it is not necessary to display here the graphs for all 16 elements, and we restrict ourselves to the elements $F_{11}(\theta), f_{12}(\theta), f_{22}(\theta), f_{33}(\theta), f_{34}(\theta), f_{44}(\theta)$. In the same diagrams we exhibit the result of numeric modeling for monolithic gas bubbles with the following parameters of the lognormal distribution: 1 – (R_{eff} = 100 nm, v_{eff} = 0.01), 2 – (R_{eff} = 0.5 μm, v_{eff} =

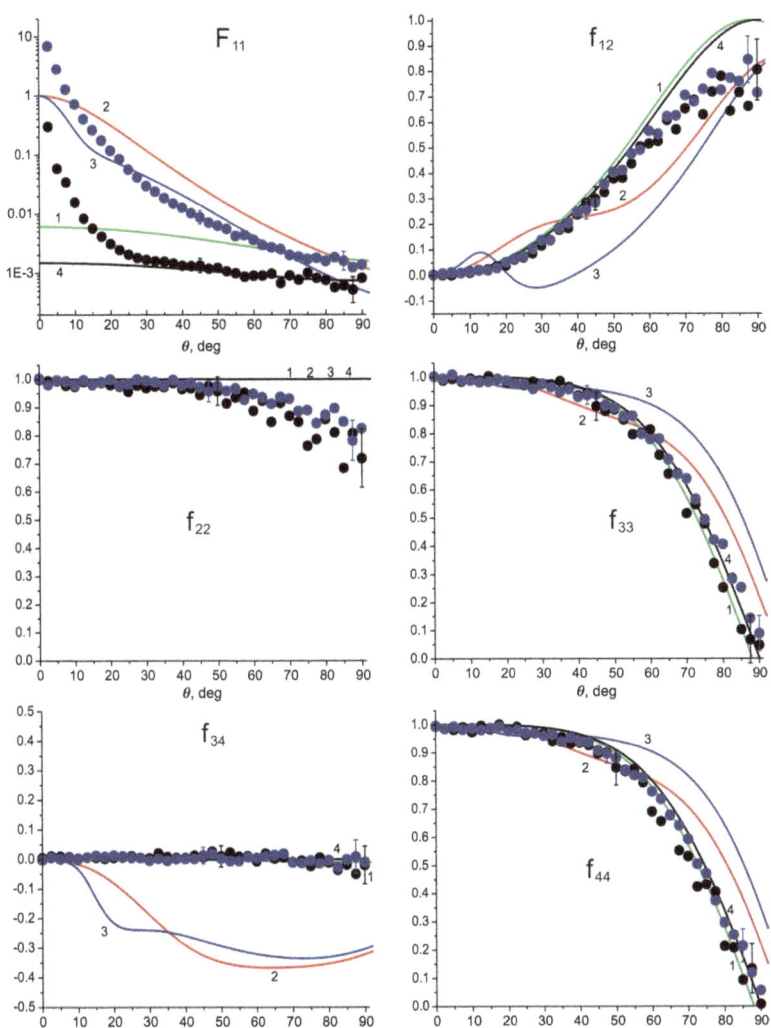

Figure XXIV.6. Experimental data: black circles (doubly distilled water); blue circles (NaCl solution, 0.8 M). Theoretic curves: gas spheres with the lognormal parameters: 1 – (R_{eff} = 100 nm, v_{eff} = 0.01), 2 - (R_{eff} = 500 nm, v_{eff} = 0.01), 3 - (R_{eff} = 1 μm, v_{eff} = 0.1); 4 – Rayleigh particles.

0.1), 3 – (R_{eff} = 1.0 μm, v_{eff} = 0.1), 4 – Rayleigh particles. Here R_{eff} and n_{eff} are the effective radius and the dimensionless effective width of the lognormal distribution respectively. The lognormal distribution parameters have been chosen in accordance with the data of DLS experiments. As follows from the dependences in Figure XXIV.6, the element $F_{11}(\theta)$ of the solution of NaCl behaves somewhat similarly to the curve calculated for the monolithic micron-sized bubbles. At the same time, this element definitely cannot be connected with the scattering by the Rayleigh particles or, for instance, the gas nanobubbles with the lognormal parameters R_{eff} = 100 nm and v_{eff} = 0.01. However, as is seen in other diagrams, the experimental dependences for the rest elements are best approximated by the Rayleigh particles or gas nanobubbles with the radii up to 100 nm. As is known, such a behavior is specific for clustered scatterers at the micron scale:[56-58] scatterer as a whole has a size exceeding the wavelength of the incident light, i.e. it is similar to a Mie scatterer (as is observed in the behavior of the element $F_{11}(\theta)$), whereas the size of constituent particles (monomers) is essentially smaller than the wavelength, i.e. these are the Rayleigh particles, or (in our case) the gas nanobubbles of radius ~ 50 nm, see below. In our model, such micron-sized scatterers are the bubston clusters. Thus the microscopy and DLS data provide an additional argument in favor of clustering the nanobubbles.

A question arises of what is the scenario of formation of such clustered particles? The data obtained with the phase microscope allows us to suppose that the clusters composed of bubstons should have rather dense packing, and their shape should be close to a spherical one. We modified the hierarchic model for the spherical monomers clustering.[59] The problem consists in the numeric modeling of clusters composed of spherical monomers having a certain distribution over their sizes. We approximated this distribution by the lognormal function. In our model we iteratively generated sequences of clusters, starting with N separate spherical particles. The values of radii of these spheres are the realizations of the stochastic process with the specified lognormal distribution. At each step of the iteration routine two clusters were randomly chosen; these clusters aggregate to one another forming thus a new cluster. The probability P of choosing the given cluster obeys the power law in the form $P = C \cdot V^{-\alpha}$, where V is the cluster volume, α is the parameter of the model, and C is the normalizing factor. In this hierarchic cluster-cluster aggregation model the average fractal dimension D of generated ensembles of the clusters depends monotonically upon the parameter . This parameter is just a new "degree of freedom" that allows us to bring the angular dependence of the scattering indicatrix (the element $F_{11}(\theta)$) into the maximum proximity with the experimental points.[55]

To simulate the structure of the clusters composed of the ion-stabilized bubbles, we tried to find a solution to the inverse scattering problem for stochastically generated 10^3 hierarchic-type clusters. For that we calculated numerically a set of the scattering matrices as the averages over computer-generated cluster samples, whose statistical parameters were preset over a uniform grid. A solution was found by minimizing the divergence between the measured angular profiles of the scattering matrix and the calculated ones for an entire cluster sample. In Figure XXIV.7 the results of simulation of scattering matrix by this routine are shown. Here R_{eff}, v_{eff} are the effective radius and effective width of the lognormal distribution over the bubston sizes D is the average fractal dimension of the bubston clusters, $\langle N \rangle$ is the average number of bubstons in the cluster, σ_N is the dispersion of number of the bubstons in the cluster. As follows from the graphs, the calculated theoretic dependences are very close to the experimental ones for almost all matrix elements. Thus the data obtained in the microscopy and scatterometry experiments allows us to conclude that the particles, spontaneously nucleated in the salt solutions, are actually hierarchic bubston clusters.

The angular profiles of the matrix elements of cluster ensemble as dependent upon the statistical parameters R_{eff}, v_{eff}, $\langle N \rangle$, σ_N and α were

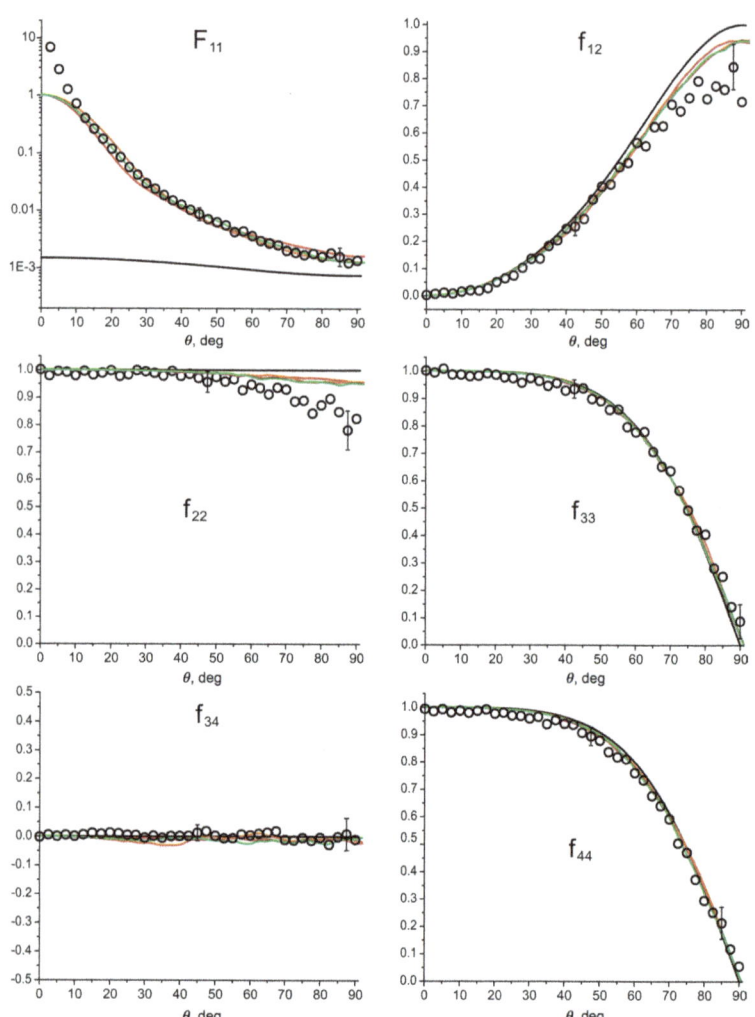

Figure XXIV.7. Angular dependences of the scattering matrix elements for the ensembles of the bubston clusters with the parameters $\langle N \rangle = 420$, $R_{eff} = 50$ nm, $v_{eff} = 0.02$; here $\alpha = -1.1, D = 2.35$ (red color), $\alpha = -1.4, D = 2.45$ (green color), $\alpha = -1.7, D = 2.56$ (orange color). The scattering by the Rayleigh particles is shown by black color. The experimental dependences for NaCl solution (0.8 M) are given by the open circles.

investigated by the numeric experiments. For $v_{eff} = 0.01 - 0.04$ and $\sigma_N = 60 - 90$ the results of calculations were steady, i.e. the changes of the element $F_{11}(q)$ were not essential. This is why in the computer simulation we put $v_{eff} = 0.02$ and $\sigma_N = 70$. For arbitrary values of R_{eff} in the range $33 < R_{eff} < 55$ nm it appears possible to find such values of $\langle N \rangle$ and α that the mean-square deviation of the theoretic diagrams for the element $F_{11}(\theta)$ from the corresponding experimental data does not exceed the experimental error: $\sigma_{F_{11}} \leq \sigma_{\exp} = 0.01$. In Figure XXIV.8 a stochastic computer model of hierarchic cluster with the parameters $\langle N \rangle = 420$, $R_{eff} = 50$ nm, $v_{eff} = 0.02$, $\alpha = -1.4$ ($D = 2.45$) is displayed. We suppose that the bubston clusters look like this.

As was obtained in the phase microscopy experiment, refractive index of clustered particles is $n = 1.26$. For the bubston cluster with the same type of aggregation we numerically simulated its refractive index as the function of the cluster radius r. At $r = 0.5$ micron, which corresponds to the effective radius of the micron-sized particle

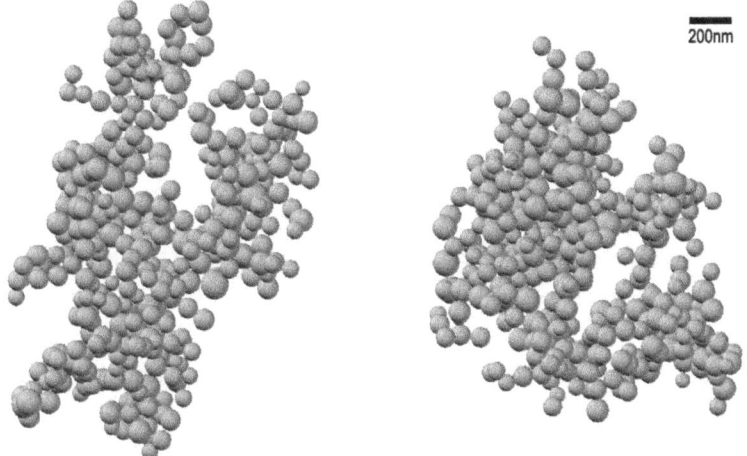

Figure XXIV.8. Mutually perpendicular projections of a stochastic realization of the hierarchic bubston cluster with the parameters $\langle N \rangle = 420$, $R_{eff} = 50$ nm, $v_{eff} = 0.02$, $\alpha = -1.4$ ($D = 2.45$).

observed with the phase microscope, the refractive index of this cluster is $n = 1.29$, which is very close to $n = 1.26$. Thus our assumption that the scattering particles in the bulk of liquid and the low-density particles observed in very thin liquid layers (phase microscopy) are the same particles is confirmed. We can claim that the existence of the bubstons and bubston clusters in aqueous salt solutions was confirmed in a direct experiment.

As was shown in the beginning, the bubstons and the bubston clusters make an additive contribution to the molecular scattering. To estimate this contribution, we measured the scattering coefficient $(R_{90})_{exp}$ at $\lambda = 532$ nm for doubly distilled water with resistivity of 4.5 M$\Omega\times$cm. It occurred that $(R_{90})_{exp} = 1.43\times 10^{-6}$ cm^{-1}. At the same time, the theoretical value of the scattering coefficient at this wavelength $(R_{90})_{theor} = 0.95\times 10^{-6}$ cm^{-1}, i.e. $(R_{90})_{exp} \approx 1.5(R_{90})_{theor}$. In the model of additive contribution from the bubston clusters we arrive at $(R_{90})_{clust} \sim 0.5(R_{90})_{theor}$. By using the known program codes,[52] we can calculate the mean scattering cross section $\langle C_{sca} \rangle$ at $\lambda = 532$ nm for an ensemble of clusters, which is the solution to the inverse scattering problem. The bulk density of the bubston clusters in doubly distilled water can be estimated according to the formula

$$n_0 = 0.5 \frac{16}{3}\pi \frac{(R_{90})_{theor}|_{\lambda=532\,nm}}{\langle C_{sca} \rangle} \sim 10^3 \text{ cm}^{-3}$$

A question arises of whether it is possible to remove completely the bubston clusters from a liquid? As follows from our model, a substantial reduction of the ionic content results in decreasing the bulk density of the bubston clusters. However, the experimental techniques based upon the light scattering are not sufficiently effective at the ultra-low ionic concentrations, because the scattering by the bubston clusters is very weak. At the same time, the low-threshold acoustic cavitation (see the Section I) can be stimulated in very dilute salt solutions. Furthermore, the low-threshold optical (laser-induced) breakdown is also observed

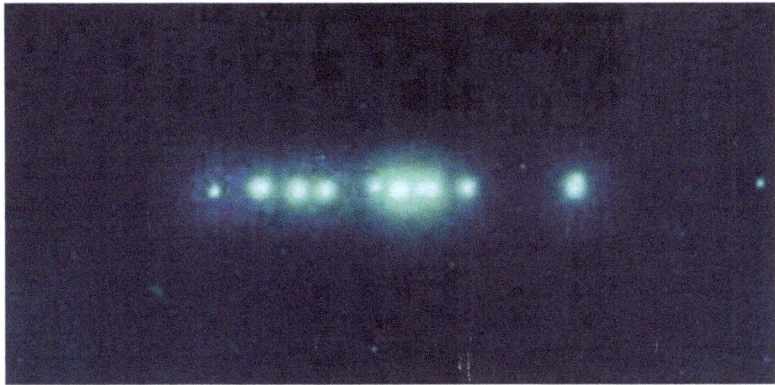

Figure XXIV.9. Photo of the breakdown flash in the sample of doubly distilled water.

in such liquids. According to our approach, the bubston clusters are the centers of such breakdown and acoustic cavitation, which makes possible to study the bubston clusters at ultra-low concentrations of dissolved ions.[45]

The pattern of optical breakdown in doubly distilled water with a resistivity of 4.5 MΩ×cm, saturated with dissolved air and induced by laser pulsed radiation with the pulsewidth of 21 ps and $\lambda = 1.064$ microns (the radiation at this wavelength is invisible) is given in Figure XXIV.9. It is very important that the radiation should be focused, because the breakdown can be induced only provided that the radiation intensity exceeds a certain threshold value.

If the caustic length amounts to 1 – 2 cm, i.e., the focusing is not tight, (in the photo the caustic is ~ 1 cm), the pattern of breakdown flash consists of multiple separate sparks distributed along the caustic. It may be assumed that each spark is related to the breakdown of a single bubston cluster. This is the so-termed multiple optical breakdown.[38] Furthermore, the optical breakdown does not necessarily occur in each laser pulse, i.e. it arises with certain probability W. The multiple and sporadic character unequivocally indicates that the breakdown is initiated by the centers randomly moving inside a focal volume V during a laser shot. This volume is found from the condition that the radiation intensity in this volume exceeds the breakdown threshold (see Ref. 38, where the threshold value was estimated). For a fixed pulsewidth this volume can be increased or reduced by changing the energy in the pulse. For the particular case shown in Figure XXIV.9, the breakdown threshold is 10^{10} W/cm^2, the plasma sparks shielded the incident laser light only slightly, and about 90% of radiation goes behind the caustic. Thus the volume V should include the caustic.

We assume that free electrons always exist inside individual bubstons in the cluster, and these electrons are driven by the electric field of optical wave. These electrons experience elastic collisions with the liquid walls of a bubston. Note that we neglect the collisions of the electron with gas molecules inside the bubston as the mean free path of electrons in the atmosphere is ~ 1 micron,[60] whereas the bubston radius amounts to a few tens of nanometers. In the process of elastic collisions the electrons are effectively heated due to the antibremsstrahlung effect, their kinetic energy increases, and as soon as this energy becomes comparable to the potential of the surface ionization of water molecules (according to Refs. 61, 62, this value does not exceed 6.5 eV), and the avalanche-like ionization of the cluster inner shells starts. The shells become unstable and subject to tearing due to the Coulomb repulsion forces. This results in the collapse (the coalescence) of the cluster, which is accompanied by the formation of monolithic micron-sized vapor-gas bubble. The coalescence time is estimated as $\tau = \dfrac{\Delta l}{v}$, where Δl is the thickness of the liquid shell, v is the speed of sound in water, $t < 1$ ps. This effect was termed stimulated optical coalescence, predicted,[38] and later observed in a direct experiment.[63] If we deal with the pulses with nanosecond or picoseconds time-scale, the stimulated coalescence comes to the end prior to the laser pulse completion, and the electrons keep oscillating in the optical wave field, but now the electrons collide with heavy particles – the fragments of the liquid shells. These collisions are no longer elastic, and in the process of these collisions the electrons lose their kinetic energy and emit the bremsstrahlung photons. The observer sees these photons (Figure XXIV.9). It is very important that the bubstons are stabilized due to the adsorption of anions. Indeed, only in this case free electrons can appear inside a bubston. This is a result of asymmetric vibrations of bubstons in a cluster, when the bubston spherical shape is violated (the vibrations of bubston in the cluster are discussed in Ref. 31), which leads to generation of local low frequency electric fields inside the bubstons. These fields stimulate the emission of electrons from the negatively charged cluster shells. Furthermore, in the case of negatively charged shells the electrons do not adhere to these shells in the process of collisions, but repel from them, which eventually results in the heating of electrons. Thus, more preferable adsorption of anions is indirectly confirmed in the optical breakdown experiments.

As was noted above, under certain conditions the optical breakdown occurs in a sporadic mode, i.e. takes place only when one or two or three (and so on) bubston clusters appear inside the focal volume V during the laser shot. Assuming that the average bulk density of the bubston clusters is equal to the n_0, the theoretical probability W_{theor} of the breakdown within the volume V can be found from the Poisson distribution:

$$W_{theor} = \sum_{n=1}^{\infty} \frac{\exp(-n_0 V)}{n!} (n_0 V)^n$$
$$= 1 - \exp(-n_0 V) \qquad (10)$$

see Ref. 45 for more detail. For the focusing conditions, typical for our experiments, the volume V can be estimated as the double volume of a cone with height $h \approx 0.5$ cm and radius $R \approx 100$ microns, i.e. $V \sim 10^{-4}$ cm^3.

This approach has been used for measuring the bulk density of the bubston clusters for doubly distilled water, i.e. for the liquid with very low value of n_0. Bearing in mind the estimate $n_0 \sim 10^3$ cm^{-3} (see above), we arrive at $n_0 V \sim 0.1 \ll 1$. Thus, expanding Eqn. (10) in a Taylor series, we obtain $W_{theor} \approx n_0 V$. At the same time, since the optical breakdown occurs in the sporadic regime, i.e. the breakdown is not induced by each laser shot, we can measure the breakdown probability $W_{exp} = \dfrac{N}{N_0}$ in a direct experiment; here N_0 is the total number of laser shots (the liquid sample is shot by laser pulses with a certain pulse repetition frequency within certain time T_0) and N is the total number of the breakdown events that have occurred within the time T_0. Assuming $W_{theor} \approx W_{exp}$, we can measure n_0 in the liquids with low ionic concentration. It is clear that this approach does not work at the high ionic concentration.

In the optical breakdown experiments we examined the dependence of W_{exp} upon the content of dissolved gas when the temperature and pressure above the liquid surface were fixed. For this experiment we first evacuated a sample of doubly distilled water, and then saturated this sample with various gases at atmospheric pressure and room temperature, i.e. the value n_0 was measured as a function of solubility of gases explored in this experiment. If the ideal gas model were applicable, we would have to get a monotonic dependence; the more gas is dissolved the higher value of W_{exp} should be measured. The corresponding graph is shown in Figure XXIV.10.

As is seen, this dependence is non-monotonic: for two gases with similar solubilities the bubstons are generated for such gas molecules (the nucleation centers), which have higher electronic polarizability. Indeed, ethane has the highest solubility, while the highest W_{exp} is realized for NO, which is polar molecule, the polarizability for nitrogen is higher than that for hydrogen, the same is true for the pair "argon and oxygen". We thus can conclude that the bubstons are nucleated more effectively for the gas molecules, which are capable

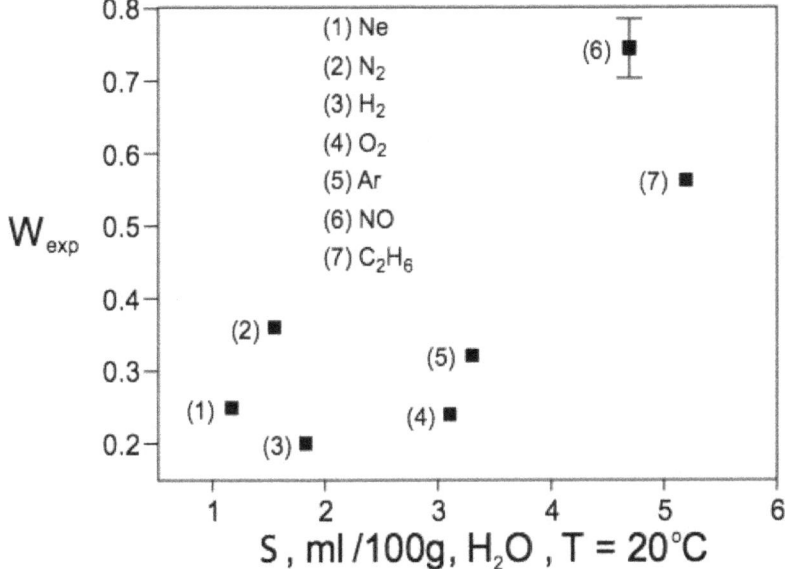

Figure XXIV.10. Probability W_{exp}, measured after saturation of the water sample by gases of various solubility.

of better interacting with water molecules. Note that we did not obtain the optical breakdown, when we saturated the initially evacuated water with helium. This implies that the bubstons and bubston clusters cannot be generated if the molecular gases in liquid are replaced with helium. Indeed, helium is chemically inert atomic gas, its atoms are so small that can freely diffuse between water molecules without distortion of the hydrogen bond network. It is important that helium atoms possess minimum electronic polarizability among other elements of the periodic table. We have developed and patented the technique of helium washing of deionized water, which allows us to replace completely the dissolved molecular gas with helium. Mass spectrometry data shows that after the helium processing the content of dissolved nitrogen and oxygen is reduced by more than hundred times. This is why, for helium water we have pH = 7, and the redox potential (accurate to our measurements) is zero. For comparison, for the initial water (prior to the helium processing) we had pH = 5.6 due to the presence of dissolved CO_2, and the redox potential was 400 - 500 mV due to the presence of dissolved oxygen.

Thus the oxidation processes are essentially suppressed in the helium-washed water. We split a steel plate into two approximately equal parts and placed these parts into the processed and non-processed water. The steel plate was covered by a corrosion film in the ordinary water, whereas in the processed water no corrosion was observed. Thus, pharmaceutical products, for example, antioxidants, can be stored in the helium water for very long period of time.

Summary

The experiments, carried out by using four independent techniques (phase microscopy, DLS, polarimetric scatterometry and optical breakdown) revealed the existence of the micron-sized particles in aqueous solutions of salt; the refractive index of such particles is less than that for the ambient liquid, but slightly higher than that for gas medium. Such particles can be associated with the bubston clusters. The bubstons are generated by gas molecules, which are capable of interacting with water molecules. The atoms of helium cannot be centers of the bubston nucleation.

Acknowledgment

This study was supported in part by the Russian Foundation for Basic Researches, Grant No. 13-02-00731 A, and Presidium of Russian academy of sciences Program No. 28, "The origin of life and the formation of the biosphere" (sub-program I, the subject "Physics, Chemistry and Biology of Water").

References

1. Fabelinskii I. *Molecular Scattering of Light*; Plenum: NY, 1981.
2. Epstein P.; Plesset M. *J. Chem. Phys.* **1950**, *18*, 1505-1509.
3. Ljunggren S.; Eriksson J. *Col. and Surf. A* **1997**, *129–130*, 151–155.
4. Crum L. *Nature (London)* **1979**, *278*, 148-149.
5. Crum L. in *Cavitation and Inhomogeneities in Underwater Acoustics*; ed. by Lauterborn W. Springer-Verlag: NY, 1980, pp. 3-12.
6. Crum L. *Appl. Sci. Res.* **1982**, *38*, 101-115.
7. Crum L. in *Mechanics and Physics of Bubbles in Liquids*; ed. by van Wijngaarden L. Martinus Nijhoff Publishers: The Hague, 1982, pp. 101 – 115.
8. Sirotyuk M. in *High Intensity Ultrasonic Fields*; ed. by Rozenberg L. Plenum Press: NY, 1971, pp. 319-337.
9. Carambassis A.; Jonker L.; Attard P.; Rutland M. *Phys. Rev. Lett.* **1998**, *80*, 5357-5360.
10. Lou S.; Ouyang Z.; Zhang Y.; Li X.; Hu J.; Li M.; Yang F. *J. Vac. Sci. Technol. B* **2000**, *18*, 2573-2575.
11. Ishida N.; Inoue T.; Myiahara M.; Higashitani K. *Langmuir* **2000**, *16*, 6377-6380.
12. Tyrell J.; Attard P. *Phys. Rev. Lett.* **2001**, *17*, 176104, 1-4.
13. Yang J.; Duan J.; Fornasiero D.; Ralston J. *J. Phys. Chem. B* **2003**, *107*, 6139-6147.
14. Simonsen A.; Hansen P.; Klösgen D. *J. Col. Int. Sci.* **2004**, *273*, 291-299.
15. Agrawal A.; et al. *Nanolett.* **2005**, *5*, 1751-1756.
16. Zhang X.; Khan A.; Ducker W. *Phys. Rev. Lett.* **2007**, *98*, 136101, 1-4.
17. Craig V. *Soft Matter* **2011**, *7*, 40-48.
18. Ball P. *ChemPhysChem.* **2012**, *13*, 2173–2177.
19. Wang S.; Liu M.; Dong Y. *J. Phys.: Condens. Matter* **2013**, *25*, 184007, 1-5.
20. Yang C.; Lu Y.; Hwang I. *J. Phys.: Condens. Matter* **2013**, *25*, 184010D, 1-10.
21. Weijs J.; Lohse D. *Phys. Rev. Lett.* **2013**, *110*, 054501, 1-5.
22. Jungwirth P.; Tobias D. *Chem. Rev.* **2006**, *106*, 1259-1281.

23. *International Critical Tables*; ed. by Washbur E. McGraw-Hill: NY, 1928, Vol. 4.
24. Wagner C. *Phys. Z.* **1924**, *25*, 474-477.
25. Onsager L.; Samaras N. *J. Chem. Phys.* **1934**, *2*, 528-536.
26. Bunkin N.; Bunkin F. *JETP* **1992**, *74*, 271-278.
27. Bunkin N.; Bunkin F. *JETP* **2003**, *96*, 730-746.
28. Bunkin N.; Bunkin F. *Z. Phys. Chem.* **2001**, *215*, 111-132.
29. Bunkin N.; Shkirin A. *J. Chem. Phys.* **2012**, *137*, 054707, 1-10.
30. Bunkin N.; Bunkin F. *Phys. Wave Phen.* **2013**, *21*, 81-109.
31. Bunkin N.; Lobeyev A.; Lyakhov G.; Ninham B. *Phys. Rev. E* **1999**, *60*, 1681-1690.
32. Bunkin N.; Suyazov N.; Shkirin A.; Ignatiev P.; Indukaev K. *J. Chem. Phys.* **2009**, *130*, 134308, 1-12
33. Bunkin N.; Suyazov N.; Shkirin A.; Ignatiev P.; Indukaev K. *JETP* **2009**, *108*, 800-816.
34. Bunkin N.; Shkirin A.; Kozlov V.; Starosvetskiy A. *Proc. SPIE* **2010**, *7376*, 73761D, 1-11.
35. Bunkin N.; Ninham B.; Shkirin A.; Ignatiev P.; Kozlov V. Starosvetskij A. *J. Biophotonics* **2011**, *4*, 150-164.
36. Bunkin N.; Shkirin A.; Ignatiev P.; Starosvetskij A.; Chaikov L. Burkhanov I. *J. Chem. Phys.* **2012**, *137*, 054706, 1-10
37. Bunkin N.; Shkirin A.; Kozlov V. *J. Chem. Eng. Data* **2012**, *57*, 2823-2831.
38. Bunkin N.; Bunkin F. *Las. Phys.* **1993**, *3*, 63-78.
39. Bunkin N.; Lobeyev A. *Quantum Electron.* **1994**, *24*, 297-301.
40. Vinogradova O.; Bunkin N.; Churaev N.; Kiseleva O.; Lobeyev A.; Ninham B. *J. Col. Int. Sci.* **1995**, *173*, 443-447.
41. Bunkin N.; Lyakhov G. *Phys. Wave Phen.* **2005**, *13*, 61-80.
42. Bunkin N.; Bakum S. *Quantum Electron.* **2006**, *36*, 117-124.
43. Bunkin N.; Kochergin A.; Lobeyev A.; Ninham B.; Vinogradova O. *Col. Surf. A* **1996**, *110*, 207-212.
44. Bunkin N.; Kiseleva O.; Lobeyev A.; Movchan T.; Ninham B.;

Vinogradova O. *Langmuir* **1997**, *13*, 3024-3028.
45. Bunkin N.; Ninham B.; Babenko V.; Suyazov N.; Sychev A. *J. Phys. Chem. B* **2010**, *114*, 7743-7752.
46. Chu B. *Laser Light Scattering*; Academic Press: NY, 1974.
47. Berne B.; Pecora R. *Dynamic Light Scattering with Applications to Chemistry, Biology and Physics*; Willey-Interscience: NY, 1976.
48. Schmitz K. *An Introduction to Dynamic Light Scattering by Macromolecules*; Academic Press: NY, 1990.
49. Bendat J.; Piersol A. *Random Data: Analysis & Measurement Procedures*; Wiley-Interscience: NY, 2000.
50. Kovalenko K.; Krivokhizha S.; Masalov A.; Chaikov L. *Bul. Lebedev Phys. Inst.* **2009**, *36*, 95-103.
51. Gerrard A.; Burch J. *Introduction to Matrix Methods in Optics*; Wiley: London, 1975.
52. Mishchenko M.; Travis L.; Lacis A. *Scattering, Absorption, and Emission of Light by Small Particles*; Cambridge University Press: Cambridge, 2002.
53. Doicu V.; Wriedt T.; Eremin Y. *Light Scattering by Systems of Particles*; Springer-Verlag: Berlin, Heidelberg, 2006.
54. Bunkin N.; Yurchenko S.; Suyazov N.; Starosvetskiy A.; Shkirin A.; Kozlov V. in *Classification and Application of Fractals*; edited by Hagen W., Nova Science Publishers: NY, 2011, pp. 3-52.
55. Bunkin N.; Shkirin A.; Suyazov N.; Starosvetskij A. *J. Quant. Spectr. Rad. Trans.* **2013**, *123*, 23-29.
56. Menguc M.; Manickavasagam S. *Int. J. Eng. Sci.* **1998**, *36*, 1569-1593.
57. Kimura H. *J. Quant. Spectr. Rad. Trans.* **2001**, *70*, 581-594.
58. Kolokolova L.; Kimura H.; Ziegler K.; Mann I. *J. Quant. Spectr. Rad. Trans.* **2006**, *100*, 199–206.
59. Jullien R., Cont. Phys. **1987**, *28*, 477-493.
60. Raizer Yu. *Gas Discharge Physics*; Springer: Berlin, New York, 1997.
61. Boyle J.; Glormley J.; Hockanadel C. *J. Phys. Chem.* **1969**, *73*, 2886-2890.
62. Grand D.; Bernas A.; Amouyal E. *Chem. Phys.* **1979**, *44*, 73-79.
63. Babenko V.; Bunkin N.; Suyazov N.; Sychev A. *Quant. Electron.* **2007**, *37*, 804-812.

www.ingramcontent.com/pod-product-compliance
Ingram Content Group UK Ltd.
Pitfield, Milton Keynes, MK11 3LW, UK
UKHW021253180426
11947UKWH00010B/753